X-RAY MICROSCOPY
IN CLINICAL AND
EXPERIMENTAL MEDICINE

Publication Number 851

AMERICAN LECTURE SERIES®

A Monograph in

The BANNERSTONE DiVISION *of*
AMERICAN LECTURES IN LIVING CHEMISTRY

Edited by

I. NEWTON KUGELMASS, M.D., Ph.D., Sc.D.

Consultant to the Departments of Health and Hospitals
New York City

X-RAY MICROSCOPY

IN CLINICAL AND

EXPERIMENTAL MEDICINE

—————————— *By* ——————————

T. A. HALL

Senior Technical Officer
Cavendish Laboratory
University of Cambridge
Cambridge, England

H. O. E. ROCKERT

Associate Professor of Histology

University of Göteborg
Faculty of Medicine
Göteborg, Sweden

R. L. deC. H. SAUNDERS

Professor of Anatomy
Dalhousie University
Faculty of Medicine
Halifax, Nova Scotia, Canada

C H A R L E S C T H O M A S · P U B L I S H E R
Springfield · Illinois · U.S.A.

Published and Distributed Throughout the World by

CHARLES C THOMAS · PUBLISHER

Bannerstone House

301–327 East Lawrence Avenue, Springfield, Illinois, U.S.A.

© *1972, by* CHARLES C THOMAS · PUBLISHER

ISBN 0–398–02458–8

Library of Congress Catalog Card Number: 72–187659

Printed in the United States of America

K-8

FOREWORD

Our Living Chemistry Series was conceived by Editor and Publisher to advance the newer knowledge of chemical medicine in the cause of clinical practice. The interdependence of chemistry and medicine is so great that physicians are turning to chemistry and chemists to medicine in order to understand the underlying basis of life processes in health and disease. Once chemical truths, proofs and convictions become sound foundations for clinical phenomenon, key hybrid investigators clarify the bewildering panorama of biochemical progress for application in everyday practice, stimulation of experimental research and extension of postgraduate instruction. Each of our monographs thus unravels the chemical mechanisms and clinical management of many diseases that have remained relatively static in the minds of medical men for three thousand years. Our new Series is charged with the *nisus elan* of chemical wisdom, supreme in choice of international authors, optimal in standards of chemical scholarship, provocative in imagination for experimental research, comprehensive in discussions of scientific medicine and authoritative in chemical perspective of human disorders.

Dr. Hall of Cambridge, England, Dr. Röckert of Göteborg, Sweden, and Dr. Saunders of Halifax, Novia Scotia, learned enough from their application of x-ray microscopy to medical research to share this newer knowledge with medical scientists in the light of actual biological research, experimental procedures, clinical possibilities and established limits of performance. Enlarged images obtained directly with x-rays reveal the internal structure of optically opaque objects, living or dead. The images are totally different in character to the light microscope and are caused by a pure absorption phenomenon, hence the information is more specific. It involves five techniques because no one system of x-ray microscopy offers all possible capabilities, for example, contact microradiography, projection microradiography, x-ray absorption microanalysis, x-ray microanalysis. Each procedure is applicable to objects which cannot be viewed microscopically by means of any other radiation. The preparation of even smaller surface and thickness of tissues enables the pathologist to proceed from a microradiograph of the living subject to close-contact microradiography of blood vessels, tissues and cells from the skin, muscle, bones, teeth, liver, kidneys, glands and other organs.

We have reason to believe that the research physicists' neglect of modern techniques condemns clinicians to amateurism while their overemphasis on instrumentation makes their work a treadmill blindfolded. Actually, the major tool of research is the human brain, our pacemaker. Problems are solved, hypotheses formulated and worthy ideas are born largely in the subconscious, and the richer the material on which it can play, the more likely is the emergence of original and inspired thought. The application of newer laboratory techniques, however powerful; the interpretation of observations in the light of a preconceived theory, however respectable, is likely to lead to results which are not more than ordinary. But the perception of distant analogies, the bodily transfer from one field to a startlingly different one of answers to problems already solved, the unused bit of knowledge which at the critical moment sparks off an illuminating train of thought, are the real modes of great advances revealed in this unique scientific contribution. The doctor's brain is a world of a number of explored continents and great stretches of unknown territory available for the clinical application of modern physics.

> The moral of x-rays is this—
> that a right way of looking at things
> will see through almost anything.

I. NEWTON KUGELMASS, M.D., Ph.D., Sc.D., *Editor*

PREFACE

SEVERAL METHODS of x-ray microscopy have now graduated from the physi- cist's laboratory. The main purpose of this book is to describe how these methods have been used in medicine and to indicate the range of problems to which they may be applied reliably. Since progress in x-ray microscopy is rapid, we shall also try to assess the improvements which may come in the near future.

Why use x-rays for microscopy? Every system of microscopy depends on radiation characterized by some wavelength, which limits the spatial resolu- tion attainable. The limit has been reached for microscopy with visible light. Since x-ray wavelengths are smaller by a factor of hundreds or thousands, it was hoped at one time that resolution could be proportionately improved, without the bad effects of vacuum and electron-damage inherent in electron microscopy. But the potential superiority in resolution has not been realized. Under the most favorable conditions, x-ray microscopy now achieves a spatial resolution of approximately 1,000 Å (10^{-5} cm)—not radically superior to optical microscopy—and no great improvement is expected in the im- mediate future.

It is not the short wavelength, but other properties of x-rays which have been exploited most successfully for microscopy. The most obvious prop- erty, the one which led to Röntgen's discovery of x-rays, is their great pene- tration. Penetration and image-contrast are readily controlled through an operator's choice of wavelength. Based on this fact, there are x-ray methods which provide high-resolution images of optically opaque, thin specimens like sections of bone and of thick specimens as well, including living tissues and whole organisms. Another useful feature of x-rays, the simple equation describing their absorption, has led to reliable measurements of density in microscopic entities such as single cells and chromosomes. Finally, since characteristic x-ray wavelengths are associated with each chemical element, virtually independently of the state of chemical combination, there are methods which reliably localize elements and measure their concentra- tions *in situ* in microscopic areas. These capabilities, confirmed by experi- ence, make x-ray microscopy worthwhile even though it is slower and more involved than optical microscopy.

No one system of x-ray microscopy offers all of the capabilities just men- tioned. Consequently, four quite distinct methods are now being used in medicine: contact microradiography, projection microradiography, x-ray fluorescence microanalysis and electron-probe microanalysis. Since these methods do not all produce magnified images, only a generalized definition of microscopy can encompass them. We may define x-ray microscopy as the use of x-rays to produce a magnified image or to provide data about individual, microscopically identified parts of an object. Of the four methods, projection microradiography directly produces a magnified image. Contact microradiography produces a one-to-one, high-resolution x-ray image which may then be enlarged. The electron probe can produce an image by a scanning technique, or else may be used to analyze, *in situ,* individual components identified through non–x-ray microscopy. Conventional fluorescence microanalysis cannot produce an image but again provides a chemical analysis of microscopic components which are identified by other methods.

Formally, the powerful technique of x-ray diffraction analysis is a method of microscopy, since the end result is an image of structure on a very fine scale indeed. Of course, this image is not produced directly but is obtained through computation and is materialized only in a diagram. Furthermore, the method reveals only structures which repeat periodically in space. The technique of x-ray diffraction is a separate discipline which is not discussed in this book.

Before discussing the medical applications of x-ray microscopy, in our first chapter we review the underlying physical principles. Chapters 2, 3 and 4 are devoted to medical studies. In the final chapter, we consider the physics of the x-ray methods more closely, in order to assess the place of x-ray microscopy in medicine.

In trying to present critically what x-ray microscopy may offer to medicine, we have chosen to draw heavily on our own experience. We have not tried to describe or cite all of the work in the field; that would require a larger book. The chief medical topics of our exposition, namely microangiography, bone and other calcified structures and specific chemical elements in soft tissues, illustrate the strong points of x-ray microscopy. While the possibilities for the study of calcified material may be fairly well represented, undoubtedly we are only at the beginning of applications to soft tissues, especially studies of thick specimens and of particular chemical elements. We hope that this exposition will suggest and assist applications to many fields of medicine.

CONTENTS

X-RAY MICROSCOPY
IN CLINICAL AND
EXPERIMENTAL MEDICINE

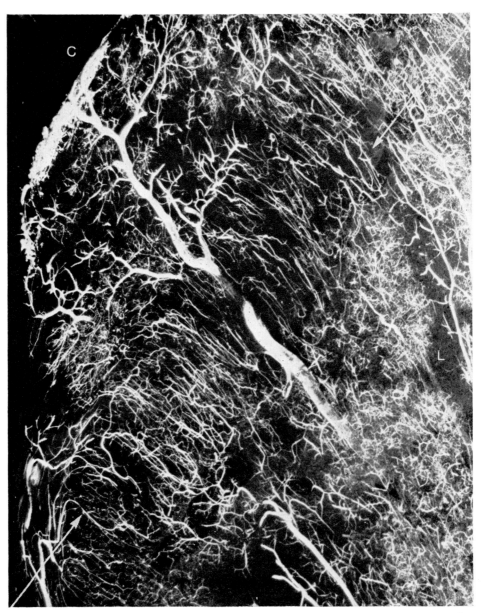

Frontispiece. Microangiogram of a coronal section of the human brain, showing the open capillary mesh of the internal capsule (between arrows) and the slender striate arteries which supply it. The denser capillary nets of the caudate (C) and the lentiform (L) nuclei border it on either side. CMR. Mag. x7.

CHAPTER 1

PHYSICAL PRINCIPLES

1.A. METHODS OF MICROSCOPY AND THEIR APPLICABILITY WITH X-RAYS

THE VARIOUS METHODS of microscopy are all based on refraction, reflection from focussing surfaces, diffraction, the formation of shadows, or techniques of scanning with some type of readout and remote display.

Refraction is the change of direction of rays of radiation which occurs with the change of speed as radiation passes through an inhomogeneous medium or crosses the boundary between different media. The phenomenon is exploited by lenses, which bend the rays in such a way that radiation leaving any one point in an object is brought back together (focussed) in a corresponding image point. The prime example of the refracting microscope is the ordinary optical microscope, and conventional electron microscopes are also based upon refraction. However, as the basis of a method of x-ray microscopy, refraction is entirely ruled out because of the extremely low refractive index of x-rays.

Reflecting microscopes generally use curved mirrors instead of lenses as focussing elements. Optical reflecting microscopes are commonplace; they are especially useful for near-ultraviolet or ultraviolet wave lengths which are not well transmitted by glass lenses. The possibilities for x-ray reflecting microscopes are dominated by the unfortunate fact that x-rays can be efficiently reflected from a surface only when the angle of incidence is very low (i.e. nearly grazing incidence). Consequently the ray diagrams of x-ray reflecting systems are unconventional; large aberrations tend to occur; and the reflecting surfaces must conform very accurately to peculiar shapes. X-ray reflecting microscopes have been built (cf. Cosslett and Nixon, 1960), but at present they seem to offer little advantage over easier alternatives. The reflection method has not come to be widely applied and is not discussed further in this book.

Diffraction microscopy is based upon the principle of wave-front reconstruction which provides a way of reconstructing the image of an object from a record of its diffraction pattern. Wave-front reconstruction has recently become a subject of great interest in the form of the laser-produced hologram, which is usually a one-to-one rather than a magnified image. But it has been shown that magnified images can be produced and that it may be possible to obtain images with extremely good spatial resolution, of the order of 1 Å, if x-rays can be used to form the diffraction pattern (Stroke, 1966). Naturally this possibility is being studied intensively, but the work is still in an early experimental stage. As shown in a recent theoretical study (Rogers and Palmer, 1969) there are formidable difficulties blocking the realization of this approach as well as the related method of focussing with "zone plates."

The more widely practiced methods of x-ray microscopy have been based on the formation of shadows, or on scanning techniques, or on a separate tactic, the x-ray analysis of areas identified by conventional microscopy. These methods have been extensively applied in the forms of contact microradiography, projection microradiography, absorption microanalysis, fluorescence microanalysis and electron-probe microanalysis.

1.B. PRINCIPLES OF THE MAIN X-RAY MICROSCOPIC METHODS

In conventional *contact microradiography* (Fig. 1), x-rays from a small source pass through the specimen to form an image in a fine-grained

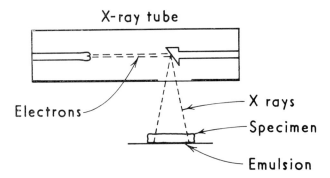

Figure 1. Contact microradiography.

photographic emulsion, which must be in direct contact or at least extremely close. The procedure is analogous to medical radiography but on a finer scale. The image in the developed emulsion is an unmagnified shadow, revealing the differing absorptions along the various ray paths through the specimen.* The resolution of such images is much finer than that of the

---

* When shadows are formed by visible light, generally only outlines are revealed because the radiation is not transmitted through the substance of the object to any significant degree. In the case of x-ray shadows, the penetration may be selected to give a wide variety of transmissions along ray paths through different parts of the object, so that internal structure can be seen.

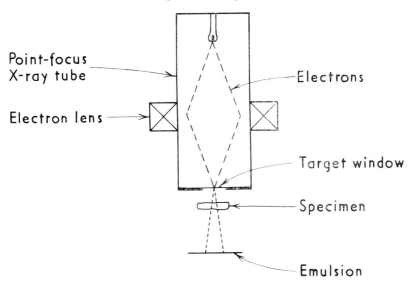

Figure 2. Projection microradiography.

human eye, and the image can be used—observed, analyzed or enlarged—only with the help of an optical microscope.

In conventional *projection microradiography* (Fig. 2), the x-rays must come from a very small source. This is usually produced by focussing a beam of electrons into a very small spot in a thin metallic target, which is so thin that the x-rays generated within it can emerge from its other side.

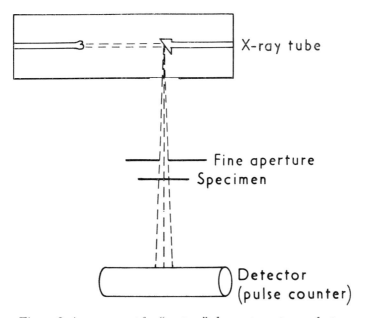

Figure 3. Arrangement for "contact" absorption microanalysis.

The specimen can then readily be placed quite close to the source. The image is again a shadow, which may be recorded in photographic emulsion, viewed on a fluorescent screen, or reproduced remotely by electronic means from an x-ray-sensitive photoelectric screen (e.g. a television-type "pickup tube"). But, as the distance from source to specimen is much less than the distance from specimen to image, the image is substantially magnified.

In *absorption microanalysis*, apertures restrict the analyzed region to an area ranging from a diameter of several tens of microns down to a minimum

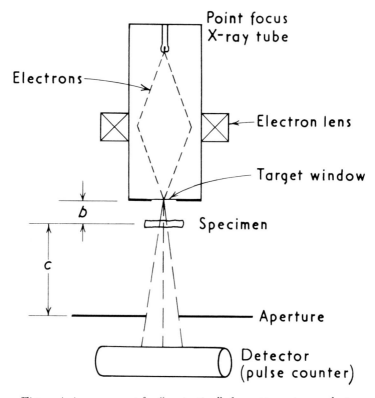

Figure 4. Arrangement for "projection" absorption microanalysis.

of about one micron, and the photographic emulsion is replaced by a detector which counts the number of incident radiation quanta. The contact geometry (Fig. 3) and the projection geometry (Fig. 4) can both be used. By means of the counter, one may measure the x-ray transmission of the analyzed area more directly, quickly and accurately than by microdensitometry of a photographic emulsion, and theory or calibration then enables one to deduce the local total mass per unit area (see Secs. 3.A and 5.C) or even the local mass per unit area of individual chemical elements (see Sec. 5.C.3). However, no x-ray image is formed, and the analyzed area must be identified histologically by an auxiliary microscopical means, such as an optical microscope built into the apparatus.*

Conventional *x-ray fluorescence microanalysis* (Fig. 5) uses a "point source" of x-rays, as in projection microradiography, but for the purpose of chemical elemental analysis; again no x-ray image is formed.† The x-rays incident on a specimen always excite fluorescent x-rays characteristic of the constituent elements, and the spectroscopic analysis of these characteristic

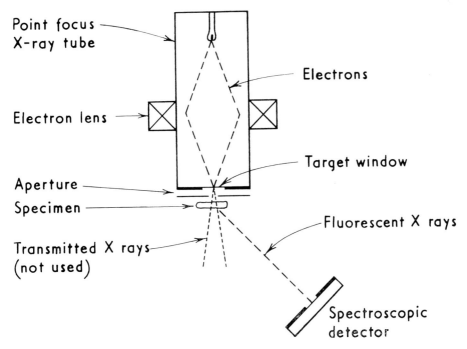

Figure 5. X-ray fluorescence microanalysis.

rays provides a measure of the amounts of one or more chemical elements within the entire irradiated area. In the disposition of components, the method differs from projection absorption microanalysis chiefly in that the detector must be offset to avoid the transmitted radiation while it samples the much less intense characteristic x-rays. The region which is to be ir-

* In this chapter and in Chapter 5, we use the term "absorption microanalysis" in a restricted sense, to refer to procedures where one does no more than to measure directly the absorption in a selected microarea. In technical discussion such procedures must be distinguished from microradiographic methods where absorption is determined for selected areas by measurements on microradiographic images. In the general literature (including some passages in this book) the distinction is often unimportant and both procedures may be called "absorption microanaly-sis."

† However, with an instrument suitably designed around an x-ray point source, one may view a field by projection microradiography, and then insert apertures to analyze a selected area by absorption or fluorescence without displacing the specimen.

radiated and analyzed, usually selected by means of an optical microscope, may be limited in practice to a minimum diameter in the range 10μm to 100μm by the apertures placed between source and specimen.

In *electron-probe x-ray microanalysis* (Fig. 6), one analyzes the characteristic x-rays excited in the specimen by the direct impact of a beam of electrons, the "probe," which is usually focussed into a spot approximately one micron in diameter. As in fluorescence microanalysis, the probe may be fixed on a spot selected by optical microscopy, and the x-rays generated in this region may be analyzed to give a measure of the local concentrations of one or more chemical elements. Alternatively, the probe may be scanned over the surface of the specimen and any one of various signals may be used to control the brightness and thereby form a magnified image on a syn-

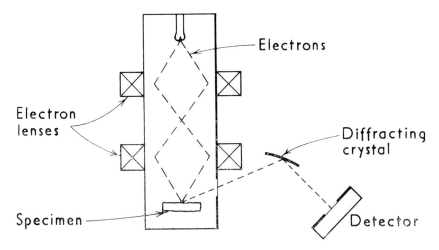

Figure 6. Electron-probe x-ray microanalysis.

chronously scanning cathode-ray tube. For example, when the signal is the instantaneous output from an x-ray detector set to respond to x-rays from one particular chemical element, the cathode-ray tube produces a magnified image of the distribution of that element. This is a system of x-ray emission microscopy: although the specimen is not irradiated by x-rays, the image is based on emitted x-rays. The scanning technique is explained more fully in Section 5.E.1. It is quite versatile and there are many other modes of operation, described and exemplified in Section 5.E.

The reader may wonder about certain differences among Figures 2, 4, 5, and 6. A single electron lens suffices for fluorescence microanalysis where electron beams less than 10μm in diameter are not needed; projection microradiography can also be practiced with a single lens although two lenses are needed for the highest performance; two lenses are essential for electron-probe work, where large currents must be focussed into spots smaller than

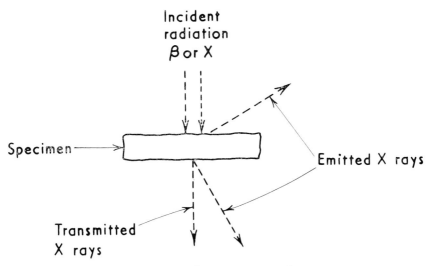

Figure 7. Radiations used in the various types of x-ray microscopy.

1μm. As to the spectroscopic analysis of emitted x-rays, a diffractor may be incorporated for fluorescence microanalysis just as in Figure 6, and *vice versa* the diffractor may sometimes be omitted in the probe system, but the depicted arrangements are more common for the reasons discussed in Sections 5.D, 5.E and A.7. Finally it should be noted that the specimen must be in a vacuum for electron-probe study, but not for the other methods.

Figure 7 and Table I recapitulate the simple relationship between the physical principles of the various x-ray methods. In every case the specimen is irradiated, and one observes either the transmission of the incident radiation or the x-rays which are generated within the specimen. When the recording system merely measures a single intensity, the result can only be a single datum (one mass or elemental amount). When the recording system preserves the record of the different signals from the different regions of the

TABLE I

FUNDAMENTAL FEATURES OF THE METHODS OF X-RAY MICROSCOPY

Method	Incident Radiation	Exposed Field	Detected X-rays	Result
Microradiography (Contact or Projection)	X	macro	transmitted	Image. Measurement of local mass and elemental analysis possible.
Absorption Microanalysis	X	micro	transmitted	Local analysis, mass and elemental.
Fluorescence Microanalysis	X	micro	emitted	Local elemental analysis .†
Electron-probe (Static)	electron	micro	emitted	Local elemental analysis .†
Electron-probe (Scanning)	electron	*	emitted	Image, elemental analysis.†

* Microspot scanned over macrofield.

† The technique may be extended to provide analysis of mass, as described in Section 5.D.3 (fluorescence microanalysis) and Section 5.E.3 (electron-probe).

specimen—either recording from all regions simultaneously, as with photographic film, or recording from one spot after another, as with the scanning technique—x-ray images may be obtained.

1.C. PHYSICAL CAPABILITIES AND LIMITS OF THE MAIN X-RAY MICROSCOPIC METHODS

Before considering quantitatively the capabilities of the different methods of x-ray microscopy in biological work, let us note the salient features here. In microradiography, one may use more or less penetrating x-rays at will, so that a favorable degree of contrast can be obtained for specimens varying over wide ranges of thickness and density. Because the transmission of x-rays along any path through a specimen is related in a simple way to the local mass per unit area, calibration enables one to measure the masses of histological and cytological structures. This has been done most often with the contact method. The projection method has the special advantage that even for thick specimens, all parts are "in focus," with little blurring. Hence one may obtain revealing stereoscopic images of thick but open objects, such as small insects or networks of blood vessels.

The analysis of chemical elements by x-rays is favored by two facts: the x-ray emission and absorption spectra are relatively simple, and are almost entirely independent of states of chemical binding. Elemental microanalysis is practiced in microradiography (see Sec. 5A3) and is the purpose of the fluorescence and electron-probe methods. Fluorescence microanalysis provides an integral chemical analysis over an area of the order of $10\mu m$ in diameter, with a minimum of specimen preparation, no need to place the specimen in a vacuum, and no danger of specimen damage inflicted by the incident radiation. Electron-probe analysis has the highest sensitivity among the x-ray analytical methods.

The physical properties and capabilities of the different x-ray methods are characterized in Table II.

Needless to say, the entries of Table II are generalizations which can serve only as a rough guide. For a given method and specimen it is usually not possible to achieve all of the listed sensitivities simultaneously, and even the individual limits can only be achieved with favorable specimens. In general, we have tried to indicate what can be done under the most favorable conditions encountered in actual practice.

The data and the calculations underlying the entries of Table II are presented in Chapter 5, along with more detailed exposition and critical evaluation of the x-ray methods, as well as consideration of practical aspects such as cost, convenience and availability. At this point we shall proceed directly to consider medical applications.

TABLE II

CAPABILITIES OF THE METHODS OF X-RAY MICROSCOPY

Method	X-ray Image?	Spatial Resolution, μm	Depth of Focus	Minimum Measurable Mass, gm	Minimum Amount of Element, gm	Minimum Weight— Fraction of Element, ppm	Specimen Must Be in Vacuum?	Destructive?	Can Study Living Material?
Contact Radiography	Yes	¼ (5.A.1)	50 μm (5.A.1)	2×10^{-13} (5.A.2)	10^{-12} (5.A.3)	10,000 (5.A.3)	No	No	Yes
Projection Radiography	Yes	0.1 (5.B.1)	cms (5.B.1)	10^{-14} (5.B.2)	? (5.B.2)	? (5.B.2)	No	No	Yes
Absorption Analysis {Contact	No	1 (5.C.1)	—	7×10^{-14} (5.C.1)	10^{-12} (5.C.3)	10,000 (5.C.3)	No	No	Yes
Projection	No	0.1 (5.C.2)	—	3×10^{-15} (5.C.2)	10^{-12} (5.C.3)	10,000 (5.C.3)	No	No	Yes
Fluorescence Analysis	No	10 (5.D.4)	—	? (5.D.3)	10^{-10} (5.D.1)	10 (5.D.2)	No	No	Yes
Electron Probe	Yes	1* (5.E.5)	10 μm (5.E.6)	2×10^{-16} (5.E.3)	10^{-16} (5.E.2)	100 (5.E.4)	Yes	Maybe	No

The numbers in parentheses denote the sections of this book where the entries are discussed.
* Spatial resolutions as good as 0.1μm have been achieved in special applications, but the available resolution is approximately 1 μm in the overwhelming majority of cases.

CHAPTER 2

MORPHOLOGICAL MICRORADIOGRAPHY

INTEREST IN X-RAY MICROSCOPY arises from the fact that x-ray wavelengths are shorter than those of light. Consequently their use may afford higher resolution than is obtainable with the light microscope, and their penetrating effect, which is likewise due to their short wavelength, permits study of the internal structure of optically opaque objects. Because x-rays are absorbed and emitted in a manner specific to the nature and amount of the elements present, x-ray microscopy also permits the localization and chemical analysis of particular elements on the microscale.

X-ray microscopy is finding many morphological and quantitative applications in general biology, medicine and dentistry. In the past decade, no less than five international conferences have been devoted to the subject (see references).

Contact microradiography and projection x-ray microscopy, otherwise referred to as the contact and projection methods of x-ray microscopy, are the two methods most widely used today. These methods may be applied to the study of tissues and cells as in historadiography, or to the study of the microcirculation as in microangiography. This chapter will be devoted to these applications, but will commence with a short historical survey to show how these techniques evolved and mention will be made of some of the fundamental problems of this branch of microscopy.

2.A. HISTORY

The discovery of x-rays by Röntgen (1895) was immediately followed by attempts to apply them to microscopy. It was soon found however that x-rays cannot be readily focussed by any sort of lens to form an image, and since no refractive effect could be detected, the idea of making some form of x-ray microscope was early abandoned. Nevertheless attempts were made to study the microscopic structure of biological specimens by enlarging radiographs taken with the specimen almost in contact with the photographic plate (Meek, 1896; Burch, 1896; Ranwez, 1896).

Early in the century Pierre Goby (1913) developed apparatus for taking

12

miniature x-ray pictures which he termed microradiographs. His camera consisted of two sliding tubes with a device to observe and regulate a spot on a fluorescent screen formed by a pencil of x-rays. He showed that small objects placed in contact with a fine grain photographic plate, when exposed to a small x-ray source, gave sharp images, which could be considerably enlarged ($\times$12 to $\times$17). He studied the microscopic structure of various diatoms, foraminifera, plants, and insects, and pointed out that the technique permitted a nondestructive dissection of optically opaque objects. Later Combeé (1955) named the technique contact microradiography.

With the discovery that soft x-rays or x-rays of long wavelength provided better microradiographic contrast of biological specimens, it became necessary to use a vacuum camera to prevent the absorption of such rays by air. Dauvillier (1927, 1930) and Lamarque (1936) were the first to construct special low voltage x-ray tubes to produce soft radiation, and perform histological studies *in vacuo*. Soft x-rays in turn required the use of thinner samples. This, when coupled with the use of fine grain Lippmann emulsion for recording, produced better image detail and permitted higher magnification. Dauvillier obtained good resolution of the cells in histological sections of elder tree pith at a magnification of $\times$600 diameters. Lamarque produced remarkably detailed pictures of the cells in histological sections of human skin, and named the technique historadiography. He, and his coworker Turchini, recognized that the variation of absorption in different areas depended on the atomic weight of the cellular components, and foresaw that historadiography would enlarge the field of histology and histochemistry. At this time Applebaum (1932) started to apply soft x-ray technique to studies of normal and pathological calcification in human teeth.

Microradiography then began to be applied to vascular studies. Pioneer work was performed by Bohatirchuk (1942) and Barclay (1947, 1951) who used hard x-radiation to study the small blood vessels in various organs following the injection of radiopaque materials. Barclay's demonstration of the microvascular patterns of the stomach and kidney attracted much attention, and stimulated Tirman (1951) and others to perform similar vascular studies on various animal organs. In Belgium, Collette (1953) used this approach to study lymphatic vessels and introduced the technique of microlymphangiography.

Attempts to correlate the microradiographic appearance of histologically fixed blood vessels with their functional state in the living animal were unsound, and led Bellman (1953) to devise microradiographic technique for the study of small blood vessels in the living animal. He introduced the term microangiography for all types of microradiography in which the microscopic blood vessels are filled with contrast medium, and demonstrated beautifully the microcirculatory transit of contrast medium through living organs under experimental treatment.

Engström (1950 et seq.) and others of the Swedish school greatly advanced microradiography with the demonstration that it was a valuable tool for quantitative cytochemical analysis. By absorption measurements with soft x-rays they showed that both the total mass and the amount of certain elements can be determined in very small structures. For example, gastric mucosa cells were weighed by estimating the attenuation of x-rays by means of densitometry in the microradiograph (Engström and Glick, 1950).

The first mirror or reflection microscope was built by Kirkpatrick and Baez (1948) following the demonstration by Ehrenberg (1947) that x-ray image formation by concave reflectors was possible. Unfortunately this instrument, which gives high resolution over a small field, suffers from spherical aberration and astigmatism. Soft x-rays are used in a helium atmosphere to minimize absorption in the image throw. It is an exacting technique as yet of limited application to biological studies.

Cosslett and Nixon (1952) built the first x-ray projection microscope, circumventing the difficulty of focussing x-rays by using electromagnetic lenses to focus electrons into a microbeam on to a metal foil target and so produce a point source of x-rays. This point source of x-rays is used to project an enlarged image of a nearby object on to a distant fluorescent screen or photographic plate. Low and high voltage models of this microscope have been built (Saunders, 1962; Saunders and Ely, 1966) which have proved very valuable for general biological, medical and dental research. Projection and contact microradiographs taken with these instruments are reproduced in this book.

The scope of x-ray microscopy may be gathered from Ely's *Micro X-Radiography and Analysis Bibliography (1913–1962)*, Cosslett and Nixon's (1960) *Monograph on X-Ray Microscopy*, and the published proceedings of the five international conferences that cover the period 1957 to 1969 (see references)[*].

2.B. MICROANGIOGRAPHY

2.B.1. Introduction

Microangiography or x-ray microscopy of microscopic blood vessels filled with contrast medium may be performed by either contact microradiography or projection x-ray microscopy. The contrast medium or radiopaque solution used for this purpose must necessarily be of small particle size to permit its flow through the smallest blood vessels and capillaries. Owing to the small caliber of the vessels involved it should be highly radiopaque to

---

[*] The *sixth* in this series of conferences was held in Osaka, Japan in September, 1971, and the Proceedings will be published soon.

give good contrast between the vessels and surrounding tissue, and be read-ily miscible with blood for work on the living animal.

2.B.2. Choice of Contrast Medium

Contrast media suitable for microangiography were reviewed by Harri-son (1951) and Bellman (1953). Some of the earlier injection masses, such as the bismuth and silver compounds, are unsuitable because of their vary-ing radiopacity and tendency to flocculate. Solutions of electrolytes are use-less owing to their tendency to diffuse through the vessel wall and produce a blurred x-ray image. Most of the radiopaque solutions used for micro-angiography are therefore either of a particulate or colloidal nature. The choice depends on whether a partial or total injection of the microvascular bed is desired, and whether the study will be performed on dead or living tissue.

Micropaque, which is a colloidal suspension of barium with a particle size of $0.1\mu m$ to $0.5\mu m$, is an excellent contrast medium for microangiography in general. Injected intravascularly into fresh animal or human material, such as surgical or autopsy specimens, it gives detailed microangiograms of the microcirculatory bed of all organs and tissues. In practice a 20% to 25% solution of Micropaque made up in either 5% to 10% formalin, or saline warmed to slightly above body temperature ($40°C$), is used for work on the dead or living animal respectively. If the specimen is to be sectioned histologically it is advisable to make up the Micropaque in a warm solution of gelatin (2% to 3%), since the gelatin sets on cooling and formalin fixa-tion, so holding the injectant within the vessels. The white color of a Micro-paque solution makes it easy to follow the vascular distribution during in-jection, with either a hand lens or dissecting microscope. Micropaque made up in warm saline can be used for microcirculatory studies in the living ani-mal, although eventually it causes intravascular clotting.

Chromopaques, which are highly colored contrast media, are very useful in microangiography since they can be used to differentiate arteries from veins, or the several parts of a complicated vascular system. If two are used at different dilutions, and are injected into the arterial and venous sides of an organ, the difference in their x-ray opacity permits the systematic analy-sis of a complex vascular pattern. For example injection of the coronary arteries with a 40% solution of yellow Chromopaque, and the cardiac veins with a 60% solution of blue Chromopaque, permits the microangiographic differentiation of these vessels down to the arterioles and venules (Fig. 8). Moreover the surface distribution of the vessels can be recorded by color photography for comparison with the microangiogram.

The particle size of the Chromopaques is about $1\mu m$, so that they can readily penetrate all parts of the capillary bed if desired. However their vis-

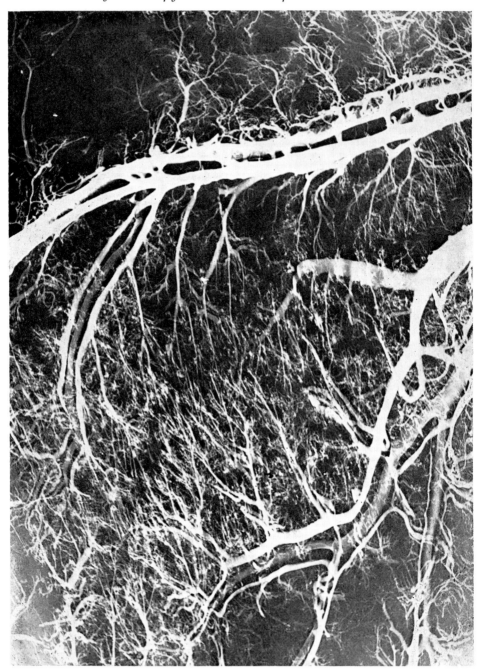

Figure 8. Contact microradiograph of the anterior ventricular wall of a lamb heart, taken after injection with 40% yellow and 60% blue Chromopaque to differentiate the coronary arteries and cardiac veins. Triads, consisting of coronary arteries flanked by two veins, can be easily distinguished. CMR 30 kV 25 mA. Ilford H.R. Plate. Mag. × 5.

cosity is held at a level (approx. 550 centipoises) that normally prevents their passage beyond the arterioles, unless diluted with an equal volume of water to reduce this viscosity to near that of blood, and so permit passage into the capillaries. Their high radiopacity factor, which is slightly in excess of 100% w/v barium sulfate, makes them highly suitable for microangiography. Anatomical, embryological and pathological material injected with Chromopaques can be x-rayed and then cleared by the Spalteholz (1914) method for study under the light microscope. Both Micropaque and Chromopaque are obtainable from Damancy & Co. Ltd., Ware, Herts., England, or the Picker X-Ray Co. in America and Canada. Chromopaque is available in red, yellow, white and blue solutions.

Microangiography on the living animal requires a contrast medium of high atomic number, and hence high radiopacity. This is because the column (vol) of contrast medium circulating through a small caliber vessel such as a capillary at a given moment is small. Good contrast between the small vessels and surrounding tissue is obviously essential for visualization of the microcirculatory bed. Moreover the small caliber, responsiveness, and permeability of the microcirculatory vessels calls for a contrast medium of small particle size that is readily miscible with blood, and capable of traversing the capillary bed without producing vascular irritation or excessive transmural diffusion.

Water soluble iodinated radiopaques such as used in standard radiography usually give disappointing results, unless Bellman's blood displacement technique is employed (Bellman and Adams-Ray, 1956). In this technique microangiography is performed immediately after the injection of contrast medium into the regional vessels, but this only demonstrates the state of the blood vessels at a given moment and not the actual microcirculation. Umbradil® (35%; Astra Chemical Co.) and Hypaque® (50%; Winthrop Laboratories) were successfully used by Bellman and his coworkers to study vascular reactions in the living rabbit ear after vascular occlusion and cold injury (Bellman and Velander, 1959; Bellman and Strombeck, 1960; Bellman *et al.*, 1959). Iodinated radiopaques are unsuitable for serial microangiography because they have a vasodilating effect and diffuse through the vessel walls so blurring the vascular image.

Thorotrast® (Hyden Chemical Co., New York) is the best contrast medium for microangiography on the living animal, because it is vascularly nonirritant and has the highest radiopacity factor among commercially available radiopaques. It is a colloidal suspension of thorium dioxide (24% to 26%) (atomic weight 232) and hence possesses alpha activity. Other important features are its small particle size (0.1μm), which permits penetration into all parts of the microcirculation, and its long persistence in the blood stream. The last mentioned makes it particularly useful for serial recording of vessel responses to experimental treatment. Transmural diffusion is slow, but Thorotrast eventually diffuses through capillary vessels to produce a perivascular halo. When surgical procedures arc envisaged Thorotrast should be colored with Berlin Blue, otherwise it may escape unnoticed from severed vessels and ruin the microangiogram by opacifying the tissue field.

2.B.3. Injection Methods

Injection of formalin-fixed specimens for microangiography is seldom satisfactory, because the fixative causes tissue shrinkage and so obstructs the fine vessels. The best results are obtained by injecting fresh autopsy or surgical specimens, and carrying out microangiography as soon as possible. Such material can be reexamined after fixation, but it must be remembered that fixatives produce vessel changes and alter x-ray transmission; longer exposure times will therefore be required. Organs such as the spleen and brain are difficult to handle in the fresh state, hence it is usually necessary to perform microangiography after the specimen has been injected and either fixed or frozen. It should be understood that no injection method can guarantee complete filling of the vascular bed within an organ, and hence it is unwise to use microangiograms for quantitative studies.

To reduce resistance a short piece of polyethylene tubing should be used in the injection system. One end of the tube should be fitted with a needle and the other connected to a T junction leading to a manometer and hand syringe or pulsatile pump. The needle should not be inserted into the selected vessel until the injection system has been flushed with contrast medium, to test the resistance of the system, ensure free flow, and remove air. On intubating the selected vessel, the needle should be tied in with cotton thread, because surgical ligatures slip and leak. The microangiography tray should carry polyethylene tubing of various calibers and matching needles. Polyethylene tubing can be reduced in caliber by drawing it to a fine point over a spirit lamp or cigarette lighter; this procedure is useful when attempting to cannulate very small vessels. Useful sizes of polyethylene tubing are Intramedic P.E. # 10, 20, 50, 60, 100; these accept needle gauges # 20, 23, 27, 30. Both are obtainable from Clay Adams Inc., 299 Webro Road, Parsippany, New Jersey 07054.

Manual injection by syringe tends to rupture small vessels, so producing radiopaque "hemorrhages" within the tissue that spoil the microangiogram. Injection is therefore best performed by gravity feed or pulsatile pump, since these allow pressure and flow regulation. An excellent pulsatile pump for microangiography is that manufactured by the Holter Co., Bridgeport, Pennsylvania (Model RD # 044). Injection pressures should be maintained within physiological ranges for the animal under study; useful information on the arterial blood pressure of man and animals other than man is given in Spector's (1956) *Handbook of Biological Data*. A lower pressure range should be used for human fetuses (e.g. 25/55 mmHg at 26 weeks); otherwise capillary rupture will occur.

Certain organs, such as the human brain, may require prolonged perfusion for some hours with a dilute solution of Micropaque (7% to 10%) to secure satisfactory injection of the small vessels of the microcirculation. Some of the variables of microangiographic injection have been reviewed by Rubin *et al.* (1964) and Kormano (1967) who found that the addition of a few drops of detergent to the injectant improved injection and reduced the tendency of Micropaque to form aggregates.

Micropaque (20% to 25% in saline or in 5% to 10% formalin) injected intra-arterially until whitening of the venous return is noted will usually produce satisfactory filling of the vascular bed of an organ. In practice the progress of the injection should be closely watched with a binocular loupe or dissecting microscope to assess capillary filling and avoid capillary rupture. If organs, such as the brain or teeth, are injected *in situ,* it is advisable to watch the filling of regional capillaries, such as those of the eye, lip or gum, since these provide a useful index of good injection. A certain radiopacity differential may be noted between the arteries and veins in the microangiogram due to the Micropaque undergoing further dilution as blood accumulates on the venous side. The duration of injection naturally depends on the size of the animal or organ, but the only satisfactory guide is direct observation of the capillary bed and draining veins, on the exposed cutaneous, mucosal or other surfaces.

To distinguish arteries from veins microradiographically, these vessels should be injected with two Chromopaques differing in color, dilution and radiopacity. Here vessel differentiation depends on opacity (absorption) differences, so it is advisable to start experimentation with two markedly different dilutions (e.g. 50% and 75%) until the desired effect is obtained. The opacity differential of the prepared dilutions may be assessed by drawing them up into fine polyethylene tubes and taking a test microradiograph. Specimen preparation may then be carried out with the selected dilutions, one dilution being injected intra-arterially, and the other by retrograde venous injection. Red or white Chromopaque may be used on the arterial side, and yellow or blue on the venous side, to provide a good contrast for color photography. Body warm (approx. 40°C) solutions of Chromopaque made up in saline are useful for work on the living animal. If specimens are to be sectioned histologically, such solutions should contain 3% gelatin to hold the Chromopaque within the vessels. Partial or complete injection of the vascular bed is obtainable with Chromopaque. Its viscosity normally prevents passage beyond the arterioles, but dilution with an equal volume of water reduces the viscosity to nearly that of blood, so permitting capillary penetration.

Microangiography *in vivo* presents special injection problems owing to the labile nature of the small vessels of the microcirculation and the rapid transit of contrast medium through these vessels. Intravenous injection involves minimal operative interference, but a large volume of contrast medium has to be administered and is effective only for a short time. While performing microangiographic experiments on the rabbit ear, Bellman (1953) found that repeated intravenous injections of contrast medium (35% Umbradil) were not well tolerated, and unless a relatively long injection time was used difficulty was experienced in hitting the arterial phase of the microangiogram with a single exposure. Direct intra-arterial injection into the posterior auricular artery of the rabbit ear seldom gave a uniform or predictable result, owing to variation of injection pressure. The difficulty of mixing contrast medium with flowing blood caused Bellman to develop a special blood displacement technic for work on the rabbit ear.

In the blood displacement technic, the posterior auricular artery is surgically exposed at the base of the rabbit ear, and intubated with fine polyethylene tubing (Intramedic # 190; Clay Adams Inc., New York) drawn to a fine point (0.5 to 0.1 mm). Microangiography is then performed while contrast medium (1.5 to 3 ml, 50% Hypaque; Winthrop Laboratories) is injected with a hand syringe at a sufficiently high pressure to displace the blood circulating within the ear vessels. Satisfactory injection is indicated by blanching of the distal part of the ear; the injectant is washed away with warm saline between x-ray exposures. While the blood displacement technique does not record the actual microcirculation, it does provide a detailed microangiogram of the state of the small blood vessels at a given moment. Bellman and his coworkers used this technique with great success to study vascular reaction to experimental cold injury and vessel occlusion, as well as vascular transformations within skin grafts (Bellman and Adams-Ray, 1956; Bellman and Velander, 1959; Bellman and Strombeck, 1960; Bellman *et al.*, 1959).

Because intra-arterial injection of local vessels produces a blood displacement effect, contrast medium should be injected at some distance from the microcirculatory field under study. The site of injection becomes a matter of importance when working with surgically created skin flaps, muscle bridges, isolated loops of gut and exteriorized organs. An intra-aortic cannula positioned to produce a "spill over" of contrast medium into one of the larger aortic branches may be useful in such instances. In general it is best to use a contrast medium such as Thorotrast, which has a long persistence in the blood stream, and carry out either an intravenous injection close to the heart or an intracardiac injection. Difficulty may be experienced in catching the arterial phase, but serial exposures with a rapid plate changer can overcome the problem.

In contact microradiography it is essential to maintain close contact between the living specimen and plate emulsion, as well as protect the emulsion from body fluids. Serial microangiography by the projection method presents fewer problems since the plate does not lie in contact with the specimen, and the x-ray beam can be switched on synchronously with the injection pump to record the vascular transit of the radiopaque.

The volume of contrast medium injected for microangiography *in vivo* should not be so large as to cause circulatory disturbances. Little is known on this subject, but in animal experiments it seems advisable to use a known fraction of the estimated plasma volume. For example in microangiography of the living rabbit ear a single intravenous injection of Thorotrast (38 ml/kg; approx 0.1 plasma vol) suffices to permit serial study of microcirculatory reaction to vasomotor drugs and freezing injury for an hour or more. This is because Thorotrast has a long persistence in the blood stream and of course recirculates for some time. The contrast medium should be held at body temperature prior to injection to avoid unwanted vascular reactions such as vessel spasm. Contrast media vary in toxicity and, as already mentioned, iodinated radiopaques tend to be unsuitable for serial microangiography in the living animal both because of their low radiopacity and vasodilator effect. Soila (1963) has outlined the problems connected with microangiography in the living human subject.

2.B.4. Recording the Microangiogram

2.B.4.a. Contact Microangiography

Microangiograms of the injected specimen should be taken as soon after injection as possible. Otherwise the injected specimen should be wrapped in saline-soaked gauze and kept in a refrigerator until microradiography can be conveniently carried out. It will be assumed that the worker has access to an x-ray diffraction tube, such as used by crystallographers, with which to carry out studies by contact microradiography.

Working under red safelight conditions, the injected specimen is positioned on the emulsion side of a photographic plate with a piece of thin plastic film interposed to protect the emulsion from specimen fluids. Kodak Contrast Lantern Slides (3¼ × 4 in) covered with a thin plastic film such as Bexphane P (1/1000 in thick), which has a low x-ray absorption factor, give excellent microangiograms.* The plate and specimen must then be positioned at a suitable tube-specimen distance (10 to 14 in) below the window of the x-ray tube.

A Philips Norelco x-ray diffraction tube, fitted with a beryllium window and copper target and possessing a focal spot of 1 mm square or less, is highly suitable for contact microangiography. Since the x-ray beam comes off the target at a sharp angle (approx. 7°) the specimen and plate must be

---

* Bexphane P is obtainable from B. X. Plastics, Ltd., Manningtree, England.

tilted to lie at right angles to the beam. A small fluorescent screen will be found useful for determining the throw of the x-ray beam prior to positioning the plate and specimen.

Operators unaccustomed to working with x-ray diffraction tubes must bear in mind that they deliver a dangerously intense beam, capable of rapidly causing a serious x-ray burn. Some workers screen the beam by uniting the x-ray tube and plate holder with a wide-bore metal collimation tube. This tends to be inconvenient for experimental work on animals. It is preferable to use an open system, in which the x-ray tube is screened by a lead-lined cubicle fitted with a lead glass observation window. The x-ray tube should be bolted to a wall to reduce vibration, and an adjustable shelf constructed beneath it to permit alteration of the target-plate distance. The shelf should be wide enough to accommodate small animals and accessory equipment, and bear a cross mark indicating the center of the x-ray beam; the last can be determined with the aid of a fluorescent screen. Such a mark saves time when positioning a plate.

A Philips diffraction tube operated at 30 to 40 kV, 20 to 30 mA, gives a satisfactory microangiogram within a matter of minutes. Exposure factors will depend on the tube voltage, specimen thickness, and whether the specimen is in a fresh or fixed state. After the exposure has been completed, the x-ray tube should be switched off, and the specimen carefully removed to avoid scratching the plate. The plate is then developed in Kodak Dektol (1 to 2 min), followed by an Acetic Acid stop bath, and Kodak Rapid Fixer. All processing solutions should be filtered beforehand, otherwise crystals or dirt may settle on the emulsion and spoil the image. Foreign bodies become very obvious when the microangiogram is dried and enlarged photographically.

The developed plate should be examined in the wet state under a dissecting microscope, to check both the exposure and microangiographic field. The plate should then be thoroughly washed, preferably in a distilled water bath fitted with an aquarium pump or circulator, and dried in a dust free chamber. The contact microradiograph or microangiogram is then ready for study; a Zeiss dissecting microscope fitted with a zoom lens is very useful for this purpose. The plate image can be enlarged by microphotography, using a fine grain film such as Adox K.B. 17 (35 mm, 40 ASA).

Microangiograms recorded in this way provide an overall view of the vascular field, and indicate whether stereographs should be recorded before sectioning the specimen. If desired the injected specimen may be cut into thick sections (1 to 10 mm) with an electric bacon slicer. Such sections will be found to provide clearer microangiograms by helping to overcome the problem of vascular superimposition.

Stereoscopic microradiographs of the intact or sectioned organ can be readily recorded with a tilting plate holder. This is constructed to give a right and left tilt of 4° to both the plate and specimen. Such a plate holder can be easily made from plastic sheet such as Perspex (Fig. 9). To support the specimen, the U-shaped plate bay is covered with thin plastic foil (1/1000 in Bexphane P or Mylar), which also permits plate exchange without disturbing the specimen. This foil is fixed to three sides of the plate bay with cellulose tape and is replaced when either soiled or scratched. The plate bay should be of the same depth as the plate to ensure good contact between plate and specimen.

After the specimen has been positioned on the plastic foil, a plate is inserted into the bay beneath it. The plate holder is then tilted 4° to the

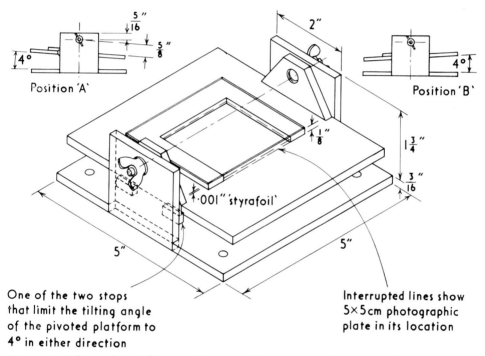

Position 'A'

Position 'B'

·001" 'styrafoil'

One of the two stops that limit the tilting angle of the pivoted platform to 4° in either direction

Interrupted lines show 5×5 cm photographic plate in its location

Figure 9. Tilting plate holder used for recording stereoscopic microradiographs. Constructed from clear Perspex to give a right and left tilt of 4°.

right (R) and the first plate exposed. The exposed plate is then removed, replaced with another, and the holder tilted 4° to the left (L) before making the second exposure. The R and L tilts, or stereograms, on development can be viewed either as plates or prints under a Zeiss stereoscopic viewer.

Better microangiographic definition can be obtained from the injected specimen by preparing and utilizing thin frozen or paraffin sections. Microtome sections (5μm to 200μm) recorded on fine grain emulsion such as Kodak High Resolution (H.R.) and Kodak Spectroscopic (649–0) plates give excellent vascular detail and permit considerable photographic enlargement of the microangiogram. Such plates, owing to their finer grain, require considerably longer x-ray exposure times ($\times 3$ to $\times 8$) than lantern slides. Microangiograms recorded on fine grain emulsion can be readily copied with a Zeiss Photomicroscope.

Paraffin sections must be deparaffinized before microangiography owing to the high x-ray absorption of paraffin. Bellman (1953) found it unnecessary to remove the embedding material provided beeswax was used, and the microangiogram was recorded with a slightly higher kilovoltage. Frozen or paraffin sections prepared for microangiography must be mounted on thin plastic foil (Bexphane P 1/1000 in) to protect the emulsion. Such foils are electrostatic but become more manageable if breathed on gently before placing them on the plate. When recording thin sections (5μm to 200μm) it is necessary to operate the x-ray tube at a low kilovoltage and so reduce the penetration. The Philips miniature contact microradiography unit (CMR 5) is very useful for this type of work, since it can be operated in the 1 to 5 kV range; it yields very soft x-rays which necessitates the use of a vacuum camera (Fig. 10).

The recording of a contact microradiograph of some part of a living animal does not differ markedly from the microangiographic procedure outlined above for the study of injected specimens. Structures such as the rabbit ear, experimental tissue flaps, or exteriorized organs, can be taped or pinned to a cork mat in such a way as to permit a plate being inserted beneath the area of interest. A simple plate tunnel constructed of plastic sheet (Perspex) is a useful accessory for sliding plates under a vascular field and can be automated if desired; it should be designed to ensure close contact between the plate and tissue.

Some difficulty may be experienced in positioning plates within the body of a living animal, especially in the case of a small animal such as the rat. It will then be necessary to use a small plate (2×2 in), or even subdivide the plate with a diamond pencil. The plate emulsion must be protected from body fluids; this may be done by wrapping the plate in either thin plastic or aluminum foil (12μm)—the latter will also protect the plate from light. Small lantern slides (2×2 in) such as the Kodak Standard and Ilford High Resolution plates, and also Agfa Printon Rapid Film, are useful for such studies. Long exposure times are unsuitable for microangiography on the living animal because of vessel movement. Hence it generally becomes necessary to compromise between exposure time and vessel definition and record the microangiogram on fast or large grain emulsion at the cost of some vessel definition.

2.B.4.b. *Projection Microangiography*

To perform projection microangiography it is necessary to have access to an x-ray projection microscope such as that designed by Cosslett and Nixon

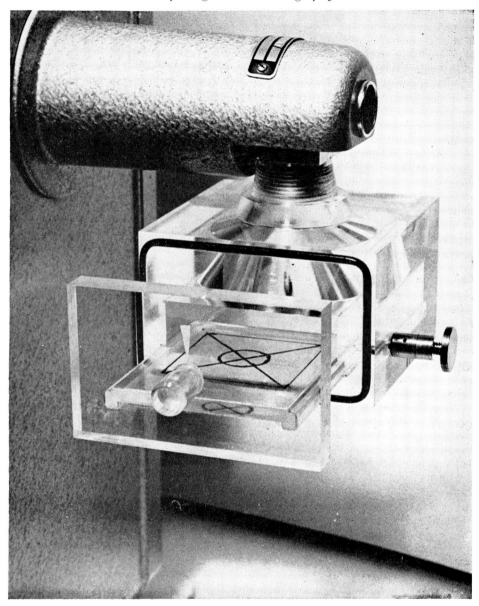

Figure 10. Vacuum camera for Philips contact microradiography unit (CMR5) constructed from clear Perspex. Note the sliding plate holder, specimen centering device, and air admittance valve. For particulars see text.

(1952, 1960). The point projection method not only provides primary magnification but also sharper definition of the small vessels that comprise the microcirculation. Owing to the great depth of field of this microscope, all blood vessels remain in focus. Consequently their volume pattern can be studied and recorded stereographically. Contact and projection microangiography are complementary techniques. Where possible both methods should be used, for the former lends itself to wide field survey and the latter to study of microcirculatory detail.

Specimen preparation for projection microangiography is the same as that already described for contact microangiography. The actual recording of a projection microangiogram is not difficult, differing from the contact method only in the positioning of the specimen in relation to the plate and x-ray source. The x-ray microscope must be properly focussed prior to its

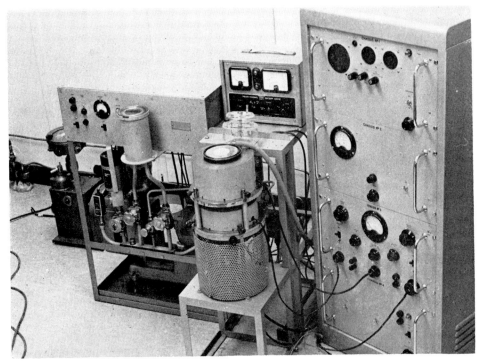

Figure 11. Cosslett-Nixon x-ray projection microscope (Mark II, Model XM30) showing the lens column, vacuum and electrical controls. Note the circular fluorescent screen overlying the target assembly and below it successively the objective and condenser lenses. Voltage range 5 to 30 kV.

use, and it is essential that the operator understands the principles of x-ray point projection. To this end it is necessary to describe briefly the x-ray microscope and the method of focussing the instrument.

The x-ray projection microscope consists of an electron gun surmounted by two cylindrical magnetic lenses, which together form the microscope column (see Fig. 11). The center of the upper or objective lens is machined to hold a funnel-shaped polepiece, within which lies the target holder and a thin metal foil target. The magnetic lenses are used to focus a minute electron beam on to this thin target, so producing an intense x-ray point source which may range from 0.1µm to 1µm in diameter. The x-ray beam emitted by this point source is used to project an enlarged image of a nearby object on to a small fluorescent viewing screen or photographic plate, as desired. The resolution is determined by the source size, hence at present the instrument has about the same resolution as the optical microscope.

The degree of x-ray magnification observed on the fluorescent viewing screen or recorded on the photographic plate is determined by the ratio of the target-specimen and target-plate distances. Both the primary magnification and field size can therefore be altered by simply varying these two distances. A magnification range of ×100 to ×1000 is obtainable in a camera length of 1 to 10 cm. Owing to heat production it may be impractical to bring biological objects too close to the target, but if the nature of the specimen permits it may be positioned 1 mm or less from the point source to secure maximal magnification. Various target metals, such as 4µm to 12µm copper, silver, gold or aluminum may be used to secure different types of x-radiation. A copper foil target operated with a 10 to 25 kV accelerating voltage gives excellent microangiograms.

The x-ray microscope is focussed in two stages. First, the target holder is removed and replaced by a small fluorescent screen. Working with a low beam current, less than 5µA at 10 kV, the electron beam is focussed to a point on this screen by adjustment of the lens controls. The target holder is then replaced and an x-ray beam produced by working with a higher beam current of about 10µA to 20µA at 10 to 30 kV as required. A coarse grid, such as an Athene 200 E.M. grid, is then placed on the target and brought to a sharp focus by further adjustment of the lens controls, while viewing the image on a fluorescent screen with a ×10 ocular. Final focussing is achieved by using a fine test grid, such as 1500 silver mesh with 3µm bars and 14µm spaces, while adjusting the lens controls to secure a sharp projected image on the screen. These procedures should be carried out under red safelight conditions with the eye well adjusted to darkness. A test plate taken of the grid will quickly inform as to whether or not the focus is satisfactory. Once the instrument has been focussed, usually few other adjustments are necessary other than switching the x-ray beam on or off, so that attention can be directed to the experiment in hand.

Projection microangiograms are recorded by placing the injected specimen a short distance (b = 1 to 15 mm) above the microscope target, while the plate is positioned at a suitable distance (a = 5 to 25 mm) above the target. These distances will determine the primary magnification (M), for M = a/b, and also the size of the image field. Greater distances can be used but reduce the intensity of the beam, so increasing exposure time. An open rectangular frame, made of brass and stepped to allow positioning of the plate at various distances above the target, serves as a useful camera. Once the specimen has been positioned over the target, a plate is inserted into the camera under red safelight conditions. An exposure is then made by switching on the x-ray beam of the microscope. After the plate has been processed photographically, the image can be examined with a hand lens or preferably under a dissecting microscope. Secondary or photographic enlargement may be carried out on fine grain film by microphotography.

When formalin-fixed specimens are being recorded, it is advisable to cover the specimen with a thin plastic foil (1/1000 in Bexphane P) to prevent formalin fumes from crazing the gelatin emulsion of the overlying plate. This practice also reduces specimen dehydration, which can cause specimen movement during a lengthy exposure and so blur the vascular image. The target-specimen (TS) and target-plate (TP) distances should be entered in a record book since these are required for calculating the primary or x-ray magnification. This factor, multiplied by any secondary or photographic magnification of the plate, will give the total magnification of the printed out microangiogram. A piece of test grid such as 1500 silver mesh grid (1500 lines per in), if placed beside the specimen is a useful way of checking microscope focus and calibrating magnification. The bars and spaces of commercial 1500 grid are respectively approximately 3μm and 14μm wide.

Because the x-ray microscope has an extremely small focal spot or point source, it also gives excellent contact microradiographs. Contact plates are recorded by first positioning the specimen over the microscope target and then placing the plate directly on top of the specimen, with a thin plastic foil (1/1000 in Bexphane P) interposed to protect the emulsion. Microangiograms showing extremely sharp capillary detail can be obtained by this method.

Projection microangiograms of injected specimens are usually recorded with a copper target operated at 20 to 30 kV, using standard lantern slide plates (Kodak Contrast 2 $\times$ 2 in, or 3¼ $\times$ 4 in). Such plates can be used for most projection studies because the limitation introduced by the grain size of the recording emulsion is here overcome by placing the specimen close to the x-ray source instead of the plate. Processing should be performed with a developer such as Kodak Dektol or D19, followed by a dilute Acetic Acid stop bath and Kodak Rapid Fixer.

Thick sections (1 to 15 mm) of injected specimens, prepared with an electric bacon slicer usually yield good projection microangiograms. This is because the volume pattern of the microcirculatory vessels is usually complete in thick sections. The volume pattern can be recorded stereographically by simply shifting the specimen laterally (1 or 2 mm) over the microscope target between successive exposures. The resultant pair of plates may be viewed with a simple stereoscope, or printed out and viewed at a higher magnification with a Zeiss aerial stereoscope.

Highly detailed projection microangiograms may be obtained by using thin frozen or paraffin sections (5μm to 200μm) mounted on plastic foil

(1/1000 in Bexphane P), or stretched across a brass ring, over the microscope target. Study of thin histological sections necessitates operating the x ray microscope at a lower kilovoltage (7 to 10 kV). The combination of thin sections and soft x-radiation requires the use of a vacuum camera, such as shown in Figure 12. The vacuum camera is screwed on to the target

Figure 12. Interior view of vacuum camera used on the projection x-ray microscope. The stepped, funnel-shaped specimen holder is threaded onto the target assembly. The surrounding platform supports a clear Perspex plate chamber, which contains a moveable plate holder to secure different magnifications.

holder of the microscope, and the beam refocussed with the aid of a small fluorescent screen placed within the funnel-shaped well of the camera. Histological sections, mounted in the manner described above, are then lowered into the camera well and positioned over the microscope target. Finally, the upper part of the vacuum camera is loaded with a plate, placed over the camera well and briefly pumped down (1 min) with a vacuum pump. The x-ray microscope is then switched on and the plate exposed. Microangiograms of this type are best recorded on a fine grain emulsion, such as Ilford High Resolution (H.R.; 2×2 in) or Kodak Spectroscopic (Type 649-0; 1×3 in) plates, processed in Ilford Contrast Developer.

Projection microangiography can be performed on the living animal while operating the x-ray microscope with a copper target (5μm to 10μm thick) at an accelerating voltage of 10 to 25 kV or higher. Tissue flaps can usually be positioned directly over the microscope target, thus facilitating area selection and experimental manipulation. Before attempting projection studies on the living animal, it will be found helpful to study the regional vascular pattern on a freshly killed animal. This will assist in interpreting the fleeting and occasionally bizarre appearances seen on recording the living microcirculation.

Microcirculatory studies of this type are best performed on thin tissue flaps such as the shaven rabbit ear. The anaesthetized rabbit is placed beside the x-ray microscope with one ear positioned about 15 to 20 mm above the target. The animal should be wrapped to reduce shock and the ear taped in position to minimize movement produced by respiration. A U-shaped brass frame about 25 mm high is then placed over the ear; this serves as a plate holder and avoids pressure on the base of the ear. Contrast is produced by intravascular injection of a radiopaque solution, such as Thorotrast, and exposures made immediately after injection. Intravenous injection may be made into the marginal vein of the opposite ear or into an artery close to the heart, thereby avoiding vascular spasm and shock in the area under observation. Intra-arterial injection may be performed with an electrically operated pump of the Holter type, while microangiograms are recorded serially.

Microangiograms of the normal ear have been studied in detail by this method and used as a control for experiments in which the skin has been subjected to cold and other forms of injury. Plates have been recorded of the living rabbit ear showing the microcirculatory pattern just before freezing and fifty minutes later.

Microlymphangiography or x-ray visualization of the lymphatic vessels may be performed on the living animal by the projection method. The ear of the living rabbit, after being positioned over the microscope target, is injected subcutaneously with a small volume of Thorotrast (.02 to .05 ml) which serves as a depot of contrast medium. If this depot is then massaged, the contrast medium enters the lymph vessels ruptured by the needle, and its transit from the tip to the base of the ear can be recorded by taking serial exposures. The peripheral lymphatic capillary network, collecting vessels and their nonreturn valves have been so recorded. Lymphatic and blood vessels have also been imaged simultaneously, by combining microlymphangiographic and microangiographic injection technics (Saunders, 1961). Contact and projection x-ray microscopy thus provide another method of studying the anatomy and physiology of the lymphatics. Workers interested in microlymphangiography should consult the work of Bellman and of Oden (Bellman and Oden, 1957, 1958–1959; Oden, 1960, 1961; Oden *et al.*, 1958).

2.B.5. Applications

Microangiography, until the past decade, had been performed solely by the contact method. The projection method has a number of advantages, among these being that it provides primary magnification, a marked depth of field, and better resolution or definition of the smallest blood vessels (Saunders and coworkers, 1957 et seq., especially Saunders 1957a,b, 1960a,b, 1964, 1969; Saunders *et al.*, 1957a,b).

The literature on contact microangiography is too large to be reviewed here, but some especially interesting studies have been performed on the kidney (Barclay, 1948, 1951; Bellman and Engfeldt, 1955a; Ljungqvist and Lagergren, 1962; Ljungqvist, 1963), stomach (Barclay, 1949, 1951; Key, 1950; Doran, 1951); heart (Bellman and Frank, 1958; Kinley and Saunders, 1968), ovary (Bellman *et al.*, 1953; Bellman and Engfeldt, 1955b), liver (Daniel and Pritchard, 1951; Sousa and Mirabeau Cruz, 1957; Bettencourt and Mirabeau Cruz, 1963; Machado, 1965), eye (Francois *et al.*, 1955; Pattee *et al.*, 1957), and bone (Göthmann, 1960a,b; Juster *et al.*, 1963).

Results obtained by contact and projection microangiography while studying various organs and tissues with various types of equipment will now be described and illustrated.

A microradiographic study of the intramuscular vascular patterns of human muscles was undertaken (Saunders *et al.*, 1957a,b), since little was known concerning the internal blood supply of muscle. It became evident that the internal patterns were dependent on the external pattern of blood supply, and consequently both were studied. Human bodies were injected intravascularly, within several hours of death, employing a 25% solution of Micropaque made up in formalin. Standard radiography was performed on muscle groups, both *in situ* and following their excision, to determine the gross features of their blood supply. Contact microradiography was then performed on the isolated muscles to record their external and internal blood supply in detail.

Contact microangiograms of isolated human muscles showed that the trunk and limb muscles were supplied by segmental arteries, derived directly from adjacent main arterial axes or from their named and unnamed muscular branches. It soon became apparent that such muscles are not dependent solely on principal arteries as previously reported (Campbell and Pennefather, 1919; Brash, 1955). Muscles with a free belly showed a grouped blood supply (Class 1), those with a fixed belly a segmental or dispersed blood supply (Class 2), and those with free and fixed parts a mixed type of blood supply combining features of the two preceding classes (Class 3). These microangiographic findings clearly revealed the principle underlying the pattern of the external blood supply of human muscles and suggested the surgical importance of such patterns when attempting to preserve the blood supply of a single muscle or group of muscles.

The internal blood supply of human muscle as revealed by contact microradiography was particularly interesting. The large size of certain muscles made it necessary to take multiple microradiographs and piece them together, to demonstrate an intramuscular pattern in its entirety. Large lantern slides (Kodak Contrasty $3\frac{1}{4} \times 4$ in) proved useful for this purpose. The microradiographs were recorded with a Philips Norelco X-ray Diffraction tube operated at 30 kV 20 mA, using a 14-in target-plate distance. The tube employed possessed a copper target, 1.0×1.2 mm focal spot, and a beryllium window.

Such microradiographs showed that the various arteries of supply usually divide on entering a human muscle. The resultant branches then course for a considerable distance (2 to 3 cm), while maintaining a uniform caliber, and eventually enter into the formation of a coarse intramuscular network which consists of anastomosing arterioarterial arcades. This coarse and hitherto undescribed vascular network was named the macromesh. The macromesh was generally rectangular in outline within the large flat trunk muscles, but markedly elongated within the long bellied limb muscles. For example, measurements performed on microradiographs of the trapezius and latissimus dorsi muscles showed that the meshes of this coarse rectangular net ranged between 1 to 2 cm and 1 to 5 cm in size.

The macromesh was found to enclose another finer vascular net formed by the smallest blood vessels and accordingly named the micromesh. Microradiographs demonstrated the relationship between the micromesh and capillary bed and showed that the precapillary arterioles arose almost at right angles from the micromesh to terminate in a capillary leash about the longitudinally disposed muscle bundles. A good example of the micromesh lying within the macromesh is illustrated by a microradiograph of the human diaphragm (Fig. 13). The small veins which accompany the arteries of both the macromesh and micromesh should be noted.

As a result of these findings in human muscle, microangiographic studies were then performed on skeletal muscle in the living rabbit. Muscle flaps

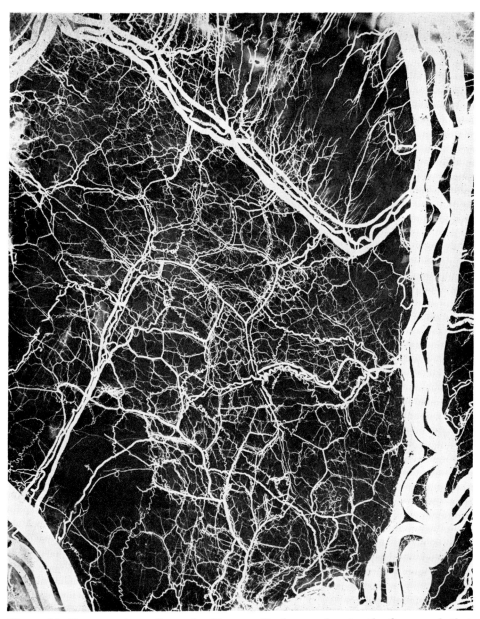

Figure 13. Contact microradiograph of human diaphragm showing the fine vessels that form the micromesh. This fine vascular net is seen to consist of tortuous arterioles and nontortuous venules. Small terminal arterioles can be seen arising from this net. The larger vessels at the margin are part of the coarse net or macromesh. The large tortuous artery on the right measured 1 mm on the original plate.

and bridges were created surgically so that small plates could be positioned beneath them. Contact microradiographs were then recorded serially following the intracardiac injection of Thorotrast. Such studies strikingly demonstrated the flow of contrast medium through the arteriolar arcades of the micromesh and the manner in which it passed from precapillary arterioles into the capillary bed within the mesh space. Contrast medium was also observed to leave the micromesh by passing directly via an arteriovenous bridge (preferential channel) into a nearby vein; peripheral streaming of contrast medium was detected in such veins. Two arteriovenous routes of transfer from the micromesh were thus demonstrated, namely via the capillary bed and via arteriovenous bridges.

It consequently became evident that the blood supply of skeletal muscle consisted of an anastomotic or high level distributor circulation, as well as a low level or nutritive circulation, subserved respectively by the macromesh and micromesh patterns. These intramuscular patterns are admirably suited to meet the changing blood supply demanded by muscle contraction and relaxation, whether on a widespread or local scale. That they have been overlooked is not surprising, since little attention has been paid to the peripheral behavior of contrast media in either animals or man. Microangiography now supplies the necessary technique.

Microangiography has proved an excellent technique for studying the blood supply of the jaw and teeth in monkey and man. Study of these vessels has hitherto suffered from severe technical difficulties imposed by the hard surrounding tissue and limited depth of focus of the optical microscope. In the case of developing or adult teeth which are *in situ,* their periodontal and dental pulp vessels may be injected with contrast medium either post mortem or *in vivo,* via the heart, aorta, common or external carotid arteries.

Freshly extracted human teeth can be injected with contrast medium by micropipette or by suction injection using a modification of Kramer's (1951) India ink technique. Micropipette injection may be performed through a window in the root, but this is difficult owing to the small caliber (0.1 to 0.2 mm) of the root vessels and their tough perivascular connective tissue. Suction injection is performed by first drilling a hole in the tooth crown and severing several of the pulpal capillary loops with fine scissors. The root of the tooth is then immersed in a contrast medium such as Thorotrast or 25% Micropaque. Suction is then applied to the tooth crown and the contrast medium drawn up the root vessels into the dental pulp vessels. A rotary vacuum pump capable of giving a negative pressure of about 150 mm Hg is used for the purpose.

Microradiographic studies of the dental pulp vessels in the developing and adult human tooth (Saunders 1957, a,b) revealed the vascular anatomy of the tooth in a manner unobtainable by other methods. For example, microradiographs of human fetal jaws injected with 25% Micropaque demonstrated a peridental vascular plexus about the dental sac of the developing tooth, which was closely applied to the outer layer of the enamel organ. Another network of vessels, namely the intradental vascular plexus, was demonstrated within the dental papilla of the developing tooth or precursor of the dental pulp. This intradental plexus was seen to consist of arterioles and venules, as well as a capillary network that lay just beneath the cusps of the developing crown (Fig. 14). These intradental and peridental plexuses are concerned with the developmental production of dentine and enamel.

Contact microradiographs of freshly extracted adult human teeth injected with Thorotrast by suction injection provided detailed information on the dental pulp vessels. Thick longitudinal sections prepared from injected and decalcified teeth showed that the dental pulp contained a peripheral or subdentinal vascular plexus. This plexus consisted of capillary loops as well as an anastomotic network that lay near the pulp surface, and extended from the upper or cornual part of the pulp chamber down into the root canal. Individual capillary loops lay within dentinal bays, hence the plexus presented an irregular subdentinal contour. These capillaries ranged between 7μm and 10μm in caliber, but their caliber increased in size on approaching a draining venule.

Microradiographs of injected human teeth showed that the apical arteries divided on entering the tooth root to give rise to a bundle of central or principal arteries which ascended the root canal to the pulp chamber. The smaller caliber and straight course of these arteries helped distinguish them from the broader and beaded silhouette of the pulpal veins. Transverse sections of single rooted teeth (e.g. canine, incisor) showed that the larger pulpal vessels were centrally located and surrounded by a zone of smaller paraxial vessels. The paraxial arteries gave off fine radially disposed branches which terminated in the subdentinal capillary plexus. This plexus drained into wide subjacent venules, which in turn passed centrally to join large tributaries of the major pulpal veins. The major pulpal veins in their descent to the root tip became markedly reduced in caliber; this suggests a sluggish venous return which would favor transvascular diffusion of calcium, fluorine and other ions.

Two main types of pulpal vascular pattern could be demonstrated microradiographically in the adult human dentition. Single rooted teeth showed large centrally located vessels surrounded by a paraxial zone of smaller vessels with the subdentinal capillary plexus at the periphery. Multirooted teeth such as molars exhibited a similar general pattern, except that anastomoses extended across the base of the pulp chamber to link the vessels ascending the various roots.

Better definition of the dental pulp vessels was obtained by projection

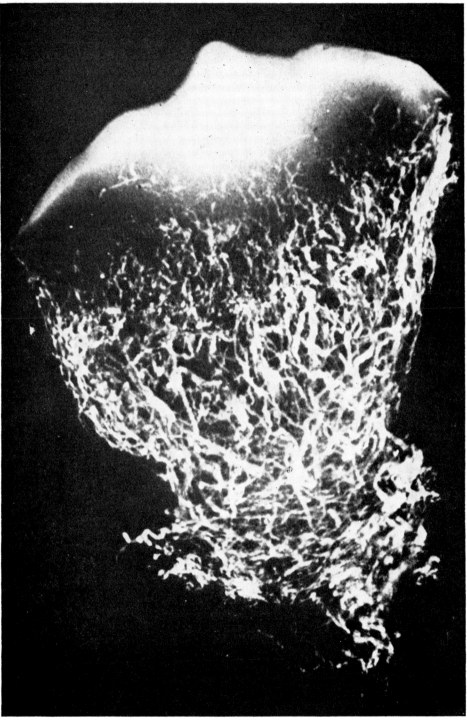

Figure 14. X-ray micrograph of injected and isolated human foetal lower lateral incisor showing the intradental vascular plexus beneath the developing crown. This plexus lies within the dental papilla or future dental pulp; here, part of it is seen through the uncalcified organic matrix of the crown. Human fetus, six months, C.R. 9.5 ins. Mag. × 41.

x-ray microscopy (Saunders 1957, a,b), hence further studies of the perio-dontal and dental pulp vessels in man and the rhesus monkey were per-formed by the projection method (Saunders, 1966; Saunders and Röckert, 1967; Saunders, 1967a,b).

Projection microangiograms of the sectioned jaw of the young rhesus monkey showed that the inferior dental artery gave off germinal or pulpal arteries. These germinal arteries entered the wide open root canals of the developing teeth to terminate in the intradental plexus of the developing pulp. Crypt arteries could also be seen passing to the peridental plexus which lines the bony crypt and surrounds the developing tooth. Micro-angiography revealed an interesting "sharing" of vessel leashes by the two dentitions. For example, as an apical artery ascends toward an erupted deciduous tooth, another artery from the same stem branches off to supply the dental sac of the adjacent permanent successor. Presumably growth of the permanent tooth germ gradually produces a vascular "steal" at the cost of the blood supply of the erupted deciduous tooth.

Projection x-ray studies of erupted monkey teeth of either dentition clearly demonstrated the blood supply of the root tips and the distribution of the apical, periodontal, interdental and neural branches. In anterior teeth the pulpal vessels formed a vascular bundle within the root canal and "fanned out" on entering the pulp chamber to terminate in the subdentinal capillaries. In posterior teeth (e.g. deciduous molars) the pulpal vessels ascending within each root canal give off a "fanlike" leash toward each pulpal horn and its subdentinal capillaries. They also form a "vascular bridge" across the lower part of the pulp chamber. High power projection studies revealed that the peripheral or subdentinal capillary bed lay in close contact with the deep face of the dentine and its dentinal tubules.

Projection x-ray microscopy of injected and isolated human fetal teeth showed the intradental vascular plexus lying within the uncalcified organic matrix of the developing crown. Large subcuspal vessels appeared to be selected early from this plexus to become the future principal vessels of the root and pulp. Microangiograms of developing human molar teeth were particularly interesting, for they demonstrated that the capillary net of the intradental plexus actually precedes the appearance of the enamel and the mineralization of the crown.

Freshly extracted adult human teeth when injected and examined by projection x-ray microscopy demonstrated that the dental pulp has a richer blood supply than is at first apparent on examining the pulp. In a single rooted tooth for example (Fig. 15) the larger centrally located pulpal vessels are surrounded by smaller paraxial vessels which radiate and subdivide to form the rich peripheral or subdentinal capillary plexus. This plexus extends over the whole coronal part of the pulp and down into the root canal. Its richness is an index of pulpal metabolic activity. Further work is required

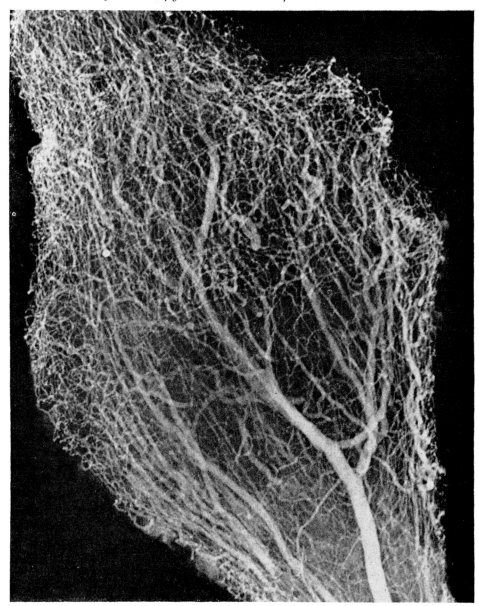

Figure 15. Projection X-ray micrograph of an adult human bicuspid tooth showing the small pulpal arteries and larger pulpal veins within the pulp. Note the rich peripheral or subdentinal capillary plexus both in the coronal part (above) and the radicular part (below). Female aged 35 years. Mag. × 36.

regarding the "take up" of nutrients and minerals from the pulpal vascular system, and their diffusion outward in both the growing and mature tooth.

Microradiographic studies of the small blood vessels within the lung were first reported by de Sousa, Mirabeau Cruz and Morais (1958). Working at the Instituto do Rocha Cabral in Lisbon, Portugal, they demonstrated the

intrapulmonary vessels of the dog lung and diseased human lung and established the value of contact microradiography as a means of studying lung
pathology. Their microangiograms showing the vascular features of tuber-
culous disease are particularly interesting.

Projection x-ray studies of the pulmonary microcirculation in the healthy
human lung were first reported by Saunders and Carvalho (1963). Adult
human lungs removed at autopsy were injected via the pulmonary arteries
or veins or both. Colored solutions of differing radiopacity were used to
differentiate arteries from veins within the lung. Injection was performed
and controlled by observing under a dissecting microscope the flow of the
injectant within the pleural capillaries. Intubation of the pulmonary segmental arteries and veins with fine polyethylene tubing was also used to
secure selective injection of various lung areas. Bronchial artery intubation
and injection was performed in a similar manner.

Contact microradiographs provided an overall view of the terminal distribution of the segmental branches of the pulmonary artery. Among other
things they showed that the penultimate branches of the pulmonary artery
follow the respiratory bronchiole, and then undergo a ramification which
imitates that of the respiratory bronchiole and the alveolar ducts. A lobular
distribution could consequently be recognized in the microangiogram. Contact microradiographs also demonstrated the mural blood supply of the
bronchi and bronchioles in a striking manner. Figure 16, for example, shows
that the lobular branches of the pulmonary artery not only supply the
peribronchial lung tissue but also the intramural vascular plexuses within
the small bronchi.

Projection microangiograms of the human lung showed the fine terminal
branches of the pulmonary artery and also clearly demonstrated how they
contribute to the formation of the pulmonary capillary network within the
walls of the air sacs or alveoli. Such microangiograms (Fig. 17) showed
that the lobular branches of the pulmonary artery give off small ductular
branches about the alveolar ducts, which then subdivide into a leash of
saccular branches or precapillary arterioles that supply the air sacs. As can
be seen, the pulmonary capillaries arise abruptly from the side of the saccular branches, as well as from their terminals, and so form a close meshed
network about the air sacs. The pulmonary capillary network is common to
two or more adjacent air sacs, and the interalveolar septa when viewed
end-on impart a honeycomb appearance to the air sacs. The pulmonary
capillaries drain into small veins or venules which unite in a distinctive
V-shaped manner at the angles of the polygonal air sacs.

High power projection micrographs of lung were obtained by placing
samples immediately above the target of the x-ray microscope. These
showed that the mesh spaces of the pulmonary capillary net are often
smaller than the diameter of the boundary capillaries, so forming a close
mesh on the wall of the air sac. Interalveolar septa when viewed end-on

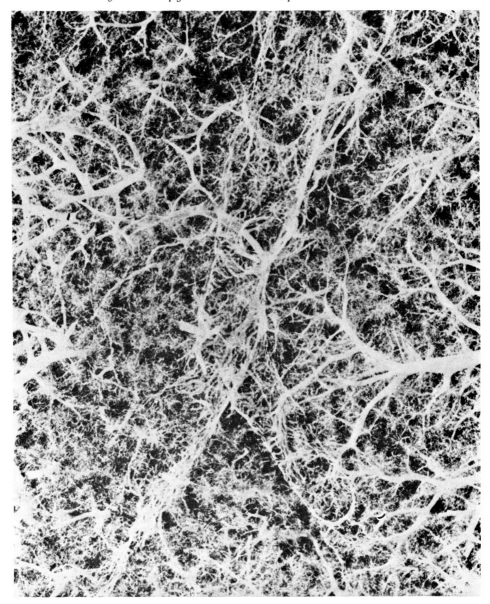

Figure 16. Contact microradiograph of human lung showing lobular branches of the pulmonary artery supplying peribronchial lung tissue and also the intramural vascular plexuses of several small bronchi. Mag. × 13.

clearly showed that the capillary vessels actually bulge into the interior of the air sac. Counts of eight to fifteen capillary meshes have been made in a single alveolar wall.

Another interesting feature of the human lung revealed by x-ray micro-scopy was the abutment pattern produced by the contact of the polygonal

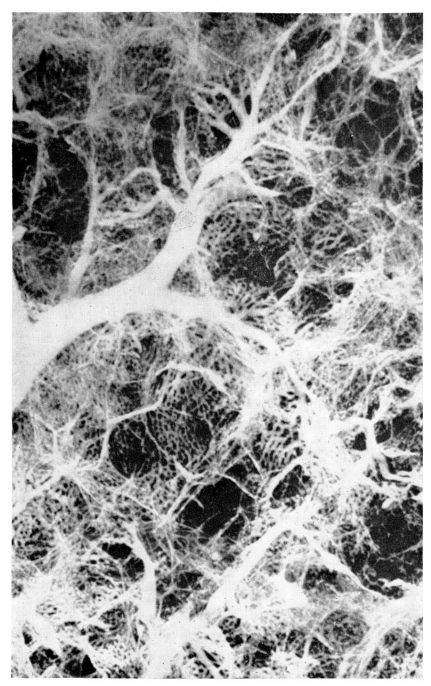

Figure 17. Projection microangiogram of human lung, showing the terminal distribution of the pulmonary artery branches, the capillary bed in the walls of the air sacs or alveoli, and draining venules uniting at the angles of the air sacs. Mag. × 90.

air sacs with the wall of either a bronchus or vein, with the consequent sharing of pulmonary arterioles, venules and capillaries at the abutment site. This structural feature appears to have bearing on routes of infection within the lung, and the genesis of pulmonary hemorrhage.

The methods of studying cutaneous blood vessels are to inject them with opaque substances, to stain them selectively, or to observe them directly from the living surface of the skin. To study the vessels which have been injected or stained, one has to section the skin or clear it, and neither method is completely satisfactory. Direct observation with the light microscope allows a very limited view of these vessels. In contrast the technique of x-ray projection microscopy offers unparalleled opportunities for studying the volume and pattern of vessels in whole skin, without the necessity of sectioning it. The technique thus provides a new approach to the study of normal, pathologic and experimental skin conditions that is not generally known (Saunders and Montagna, 1964).

Human skin may be prepared for microangiography by injecting the main limb arteries of human fetuses or adult subjects with a 10% to 25% solution of Micropaque. Injection should be performed as soon as possible after the amputation of a limb or death of the subject. Micropaque, owing to its fine particle size readily enters and traverses the cutaneous capillaries, hence injection should be continued until whitening of the subcutaneous veins is noted. Following injection the skin should be carefully dissected off in the subcutaneous plane, and examined in the fresh and unfixed state by x-ray microscopy. Samples taken from the palm of the hand or sole of the foot give particularly interesting results owing to the strongly developed friction ridges in those regions. Frozen sections are useful to record cross-sectional appearances. Contact microradiography provides good survey pictures of injected skin, especially if recorded on fine grain plates, but for detailed studies of the skin vasculature it is better to use the projection method.

All parts of the cutaneous vascular system can be demonstrated within a given skin area provided it has been well injected with contrast medium. Microangiograms of injected fetal skin for example clearly show the various vascular components such as the subpapillary arteriolar and venous plexuses, and cutaneous capillary beds (Saunders, 1961).

Microangiograms of the whole thickness of the adult human skin, excised from areas that display longitudinally disposed skin ridges such as the ball of the foot, show that there is also a longitudinally disposed blood vessel pattern underlying and repeating the skin pattern. Capillary loops possessing thick and thin limbs could be seen to project from this longitudinal vessel pattern. Microangiograms of skin derived from the finger or toe tip showed that the subpapillary vascular pattern repeated that of the overlying friction ridges of the skin to a surprising degree, so that the capillary pattern repeated that of the finger or toe print. Basic finger print patterns such as the plain or tented arch, loop and whorl therefore appear as vascular prints. Figure 18 provides a good example of a looped skin pattern reproduced by its underlying capillaries. As might be expected the underlying vessel pattern bifurcates wherever the skin ridge divides or "forks."

The microcirculatory patterns within the skin of the dead and living rabbit ear have been recorded by projection x-ray microscopy following injection of the central artery with contrast medium (Saunders, 1964). Such studies showed the small terminal arteries, capillaries and draining venules with great detail and clarity. They also revealed patterns of microvascular organization such as the coarse distributor networks formed by interconnecting arteries and the finer nutritive networks of capillaries enclosed by them. A microangiogram of the central area of a dead rabbit ear, such as Figure 19, demonstrates these features, and also shows the clarity with which small vascular shunts can be recorded. Here in the center of the field can be seen two S-shaped arteriovenous anastomoses which connect two small arteries directly to an H-shaped subcutaneous vein.

Arteriovenous anastomoses have also been recorded in the living rabbit ear over the x-ray microscope. Arteriovenous anastomoses commonly arise from the secondary and tertiary branches of the central artery of the rabbit ear and can be rendered more obvious by freezing the skin of the ear beforehand. Figure 20 shows contrast medium passing through an open S-shaped arteriovenous anastomosis in the living rabbit ear. The wider venous end of the shunt can be readily seen and usually assists in the identification of such anastomoses. Measurements performed on microangiograms show that such shunts have a caliber of 0.05 to 0.1 mm and length of 0.5 to 1 mm. A cluster of arteriovenous anastomoses may sometimes be seen to arise from a small artery and empty into an adjacent vein (Saunders and James, 1960).

Projection x-ray studies of living rabbit skin, previously injected with warm saline and sodium nitrite to produce vasodilatation, demonstrate that thoroughfare channels (Zweifach, 1959) and capillaries proper can both be demonstrated by microangiography (Fig. 21). Comparison of the microangiographic appearance of the cutaneous capillary bed in the living and dead rabbit is instructive (Fig. 22), hence after completing an *in vivo* study it is advisable to record microangiograms of the same area after death. An unexpected finding was that large and small cutaneous nerves could be located by the pattern of the intraneural and perineural vascular plexuses. The intraneural vascular plexus of the great auricular nerve for example forms a conspicuous bundle in the rabbit ear which can be identified in microangiograms of both the dead and living rabbit ear.

Many gaps exist in our knowledge of the microcirculation. With improved experimental technique, the capillary patterns within various tissues and their regulating mechanisms may be profitably explored by microangiography.

Figure 18. Projection microangiogram of skin of human toe showing that the subpapillary blood vessel pattern repeats the loop pattern of the overlying skin friction ridges (e.g. toe or finger prints). Mag. × 14.

Egaz Moniz (1931, 1934) the Portuguese Nobel Laureate used the now classic technique of cerebral angiography to demonstrate the radiological anatomy of the large cerebral vessels. Recently it has been shown that x-ray microscopy provides a new approach to the study of the small blood vessels within the human brain (Saunders, 1959, 1960c; Saunders, Feindel and Carvalho, 1965; Saunders, Bell and Carvalho, 1969).

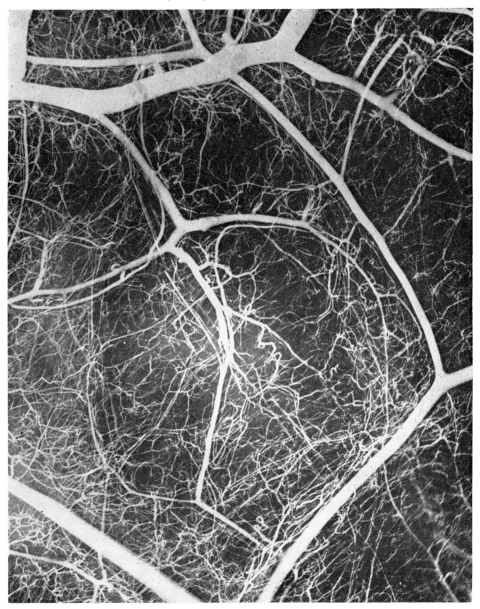

Figure 19. Projection microangiogram of the central area of the rabbit ear showing the coarse and fine microcirculatory distributor networks (macromesh and micromesh) as well as two S-shaped arteriovenous anastomoses. Mag. × 21.

During these studies it was found that the blood vessels of the human brain could be injected satisfactorily within the intact skull by using the vessels of the iris of the eye as an index of filling (see Fig. 23). This is because the origin of the anterior and middle cerebral artery lies close to that of the ophthalmic artery. On performing injection of the internal carotid artery with 25% Micropaque it was therefore customary to watch the filling

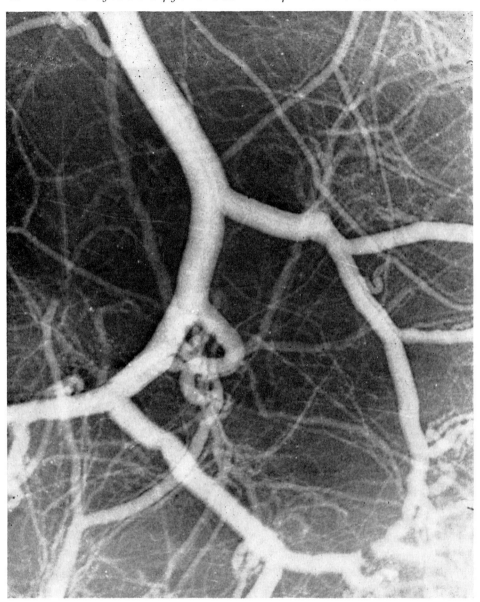

Figure 20. Projection microangiogram of living rabbit ear showing an open S-shaped arterio-venous anastomosis. Note the wider venous end of the shunt. Mag. × 14.

of the small iridial vessels of the eye for signs of overdistension or rupture. This technique is especially useful when injecting the cerebral vessels of the human fetus or newborn infant.

Injected human fetal brains (4th to 7th month) when examined by x-ray microscopy showed that the pial arteries and veins form an anastomotic network which encloses a complex and lace-like capillary bed. No arterio-

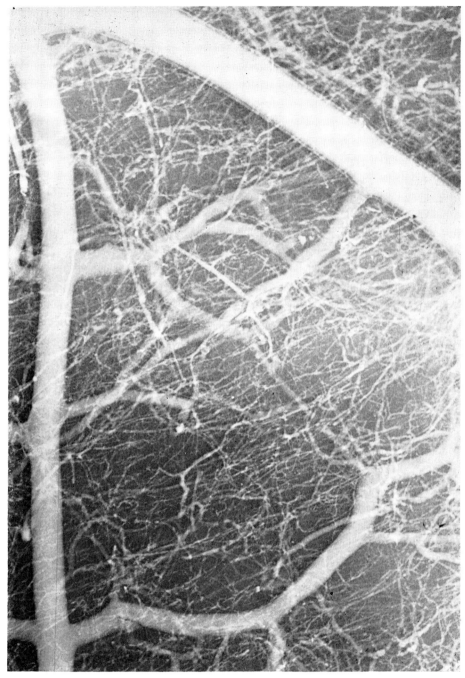

Figure 21. Projection microangiogram of living rabbit ear showing the thoroughfare channels and capillaries proper. Mag. × 50.

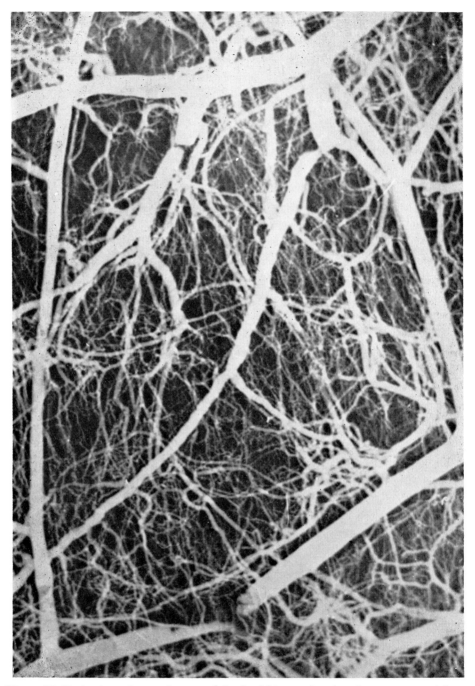

Figure 22. Projection microangiogram of dead rabbit ear showing the coarse arterial net (macromesh), and finer net (micromesh) formed by the arterioles, venules and capillary bed. Mag. × 50.

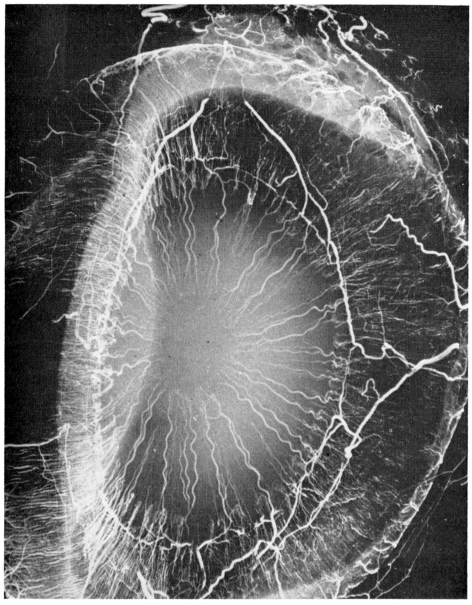

Figure 23. Contact microradiograph showing the appearance of the iridial vessels of the human eye that is used as an index of satisfactory injection of the cerebral vessels. The vessels of the iris, ciliary body and choroid can be identified. Female infant aged 9 months. Mag. × 9.

venous anastomoses were observed, but arteriovenous bridges or preferential channels of short transit distance were common. Numerous small club-like projections could be seen to arise from the pial network. Rotation of the specimen over the x-ray microscope however showed that these projections were in fact the commencement of short cortical and transcerebral arteries that penetrate the underlying grey and white matter of the brain.

Projection microangiograms of the injected human fetal brain provided a striking demonstration of the mode of origin of the transcerebral arteries from the pial net on the brain surface, and of their course through the grey and white matter of the cerebral hemisphere. The transcerebral arteries were seen to terminate in a rich periventricular plexus that lay around and adjacent to the lateral ventricle of the brain (Fig. 24). Sectional studies made in various planes further showed that these transcerebral arteries formed a "vascular corona radiata" that coincided directionally with the fiber pattern of the main projection systems.

Microangiographic studies performed on injected adult human brains showed that the intracerebral vessels displayed a similar vascular pattern. Here the transcerebral arteries were seen to arise from both the large cortical arteries and pial net, and traverse the subjacent grey and white matter in a graceful sweep before ending in the periventricular capillary bed about the lateral ventricle. Microangiograms showed that the transcerebral arteries in the adult brain measured about 2 to 3 cm in length.

Projection x-ray microscopy demonstrated the blood supply of the spinal cord and spinal nerves in a manner not obtainable by other technics. It made possible the study of both the peripheral system of vessels surrounding the cord, and the brush-like central arteries that penetrate it to supply the grey matter. The radicular arteries could be traced along the spinal nerve roots and the capillary bed within the spinal root ganglia clearly demonstrated. Figure 25, for example, shows the peripheral or pial anastomotic net which surrounds the spinal cord, and the anterior spinal artery giving off central arteries to the right and left halves of the cord. Each central artery can be seen to break up into ascending, transverse and descending branches to supply different levels of the grey columns.

Following these studies a detailed investigation of the arterial and venous components of the human cerebral microcirculation was performed by contact and projection x-ray microscopy (Saunders, Feindel and Carvalho, 1965). The large cerebral vessels of adult human brains (age range 25 to 89 years) were injected with Micropaque, or else with selected Chromopaques to differentiate the intracerebral arteries from the veins. General injection of the cerebral arteries was performed via the carotid and basilar arteries, and regional injection of the three cerebral arteries by individual cannulation. Localized injections were achieved by introducing a micropipette into a short segment of a cortical artery isolated between ligatures. Venous injections were performed by injecting the cerebral veins on the brain surface until the injectant appeared in the Galencial system, or vice versa.

Thick sections (1 to 10 mm) prepared with an electric bacon slicer, as well as thin microtome sections (200μm to 400μm) were then made from

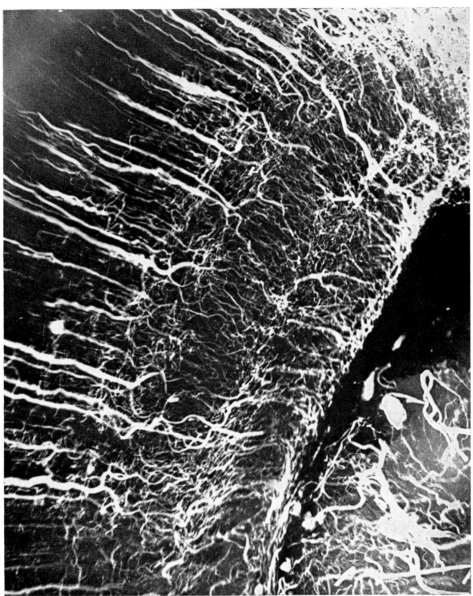

Figure 24. Projection micrograph of human fetal brain showing the long transcerebral arteries coursing through the white matter to terminate in the rich periventricular plexus that surrounds the lateral ventricle. The cavity of the lateral ventricle is seen as a dark area. Mag. × 26.

formalin-fixed material for microangiographic study. Survey pictures were recorded by the contact method and vascular detail of selected areas studied by projection x-ray microscopy. Selective injection has made it possible to distinguish between intracerebral arteries and veins and record their char-

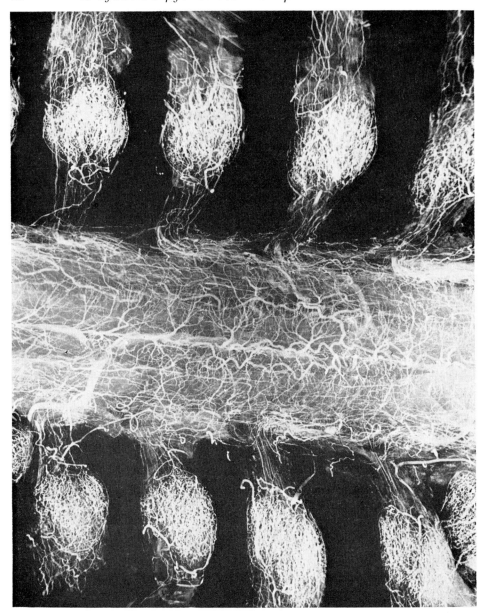

Figure 25. Projection micrograph of human spinal cord showing the peripheral system of blood vessels that surround the cord. Note the central arteries that pass to right and left of the anterior spinal artery to supply the central grey matter. The radicular vessels in the spinal nerve roots are well shown, as well as the rich capillary bed within the spinal root ganglia. Human fetus. CR 19 cm. Mag. × 9.

acteristic features. Correct designation was deemed important, owing to the confusion introduced by Pfeifer's (1928, 1930) work on the cat, in which he mistook arteries for veins.

The cerebral arteries of the human brain subdivide into large named and unnamed cortical branches which are distributed over the pial surface of the brain. From these branches arise numerous short cortical arteries which immediately penetrate the underlying brain substance, to form a striking palisade of short parallel arteries which supply the upper, middle and deeper layers of the grey matter or cortex (Fig. 26). These slender arteries terminate in a grapnel-like leash of precapillary arterioles which spread upwards and outwards as they join the capillary bed within the cortex.

Among the short cortical arteries can be seen the longer transcerebral arteries (Fig. 26), which arise from the pial surface and sweep with graceful curves through both the grey and white matter of the cerebral hemisphere to terminate in the periventricular plexus about the lateral ventricle. In their course the transcerebral arteries give off collateral branches, which supply the elongated capillary bed that surrounds the nerve cell fibers within the white matter. Such is the density of the short cortical and transcerebral vessels that they together impart the appearance of a cascade, as shown in Figure 26.

Microangiography demonstrated the rich blood supply of the grey matter of the human cortex, and showed the cortical capillary bed to be a continuous three-dimensional net. The meshes of this cortical net appeared as dark radiolucent spaces or neuronal areas occupied by the nerve cells (Fig. 27). The capillary bed of the grey matter was not sharply demarcated from that of the underlying white matter owing to the existence of fine anastomoses that connected the cortical and medullary vessels. The greater density of the cortical capillary bed was however immediately obvious, and assuming that capillary density is a crude index of metabolic rate, it is evident from microangiograms that the cerebral cortex has a much higher metabolic rate than the white matter.

Localized injection of a short segment (1 cm) of a cortical artery isolated between ligatures demonstrated a vertical vascular parcellation, that appeared on the microangiogram as a long slender wedge of blood vessels extending from the brain surface down to the lateral ventricle. Obstruction of this vertical vascular pattern by an intracerebral tumor probably accounts for the shunt of red arterial blood into adjacent cerebral veins, so that the appearance of red cerebral veins on the brain surface provides a telltale of a subjacent disease process (Feindel and Perot, 1965).

Microangiograms recorded after injection of cerebral veins showed that the short cortical arteries within the grey matter were accompanied by short cortical veins. These short cortical veins could be identified in the microangiogram by their "ragged root" appearance, produced by the abrupt right angled junction of their tributaries. More interesting were the transcerebral veins, which could be seen to connect the cerebral veins on the brain surface with deep seated paraventricular veins which lay alongside the lateral ventricle. These transcerebral veins coursed in a fan-like manner through the grey and white matter, and drained the medullary capillary bed. The

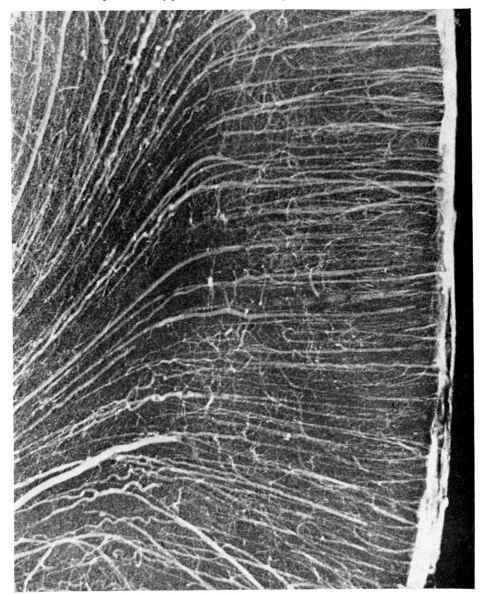

Figure 26. Contact microangiogram of human brain showing the palisade of cortical arteries within the grey matter of the motor cortex. These vessels break up to supply the cortical capillary bed that surrounds the nerve cells. The longer transcerebral vessels can be seen traversing the subjacent white matter in a striking cascade (Carvalho). Mag. × 20.

medullary capillary bed within the white matter showed an elongated mesh that reflected the regional fiber pattern. The longitudinally arranged para-ventricular veins received the transcerebral veins either singly or in leashes.

Thus although the microvascular bed of the human cortex and its sub-

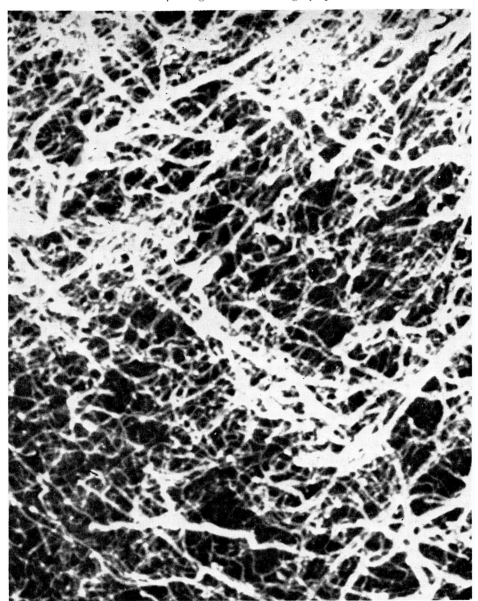

Figure 27. Projection microangiogram of human cerebral cortex showing the vascular detail obtainable with the new high voltage x-ray microscope XMPJ. Note the dark neuronal areas within the capillary net. Human brain J1; 74yr. Mag. × 81.

jacent white matter appears incredibly complex on the microangiogram, it does lend itself to anatomical analysis by selective injection of isolated arterial and venous segments. Such preparations indicate that the basic pattern in man consists essentially of columns or palisades of arterial and venous components disposed vertically, which run from the cortical surface toward the periventricular plexus. Microangiographic studies are now being correlated with radioisotope and fluorescein studies of the cerebral microcirculation being performed on the surgically exposed human brain (Feindel *et al.*, 1967).

2.C. HISTORADIOGRAPHY

2.C.1. Introduction

Historadiography or x-ray microscopy of biological tissues and cells may be performed by either the contact or projection method. The study of histological sections by contact microradiography dates from the work of Lamarque and Turchini (1936, 1937), and the method has since been applied to a wide range of biological problems. It is only recently however that exploratory studies of tissue with the x-ray projection microscope have been reported (Mosley, Scott and Wyckoff, 1957; Saunders and Van der Zwan, 1960). Methods of preparing specimens for historadiography by the projection method have been described by Le Poole and Ong (1957), Ong (1959) and Jongebloed (1966). Utilization of ultrasoft x-ray techniques as described by Henke (1967), and the combining of histochemical and x-ray microscopy techniques in what may be termed x-ray histochemistry (Bell, 1969), should yield much information on tissue structure and composition not otherwise obtainable.

The recording of a historadiograph by the contact method is essentially a simple procedure. Here the tissue section is placed in close contact with a fine grain or high resolution photographic plate, which is then exposed under vacuum conditions to a source of soft x-rays. After the plate has been developed the image of the contact microradiograph may be examined with an optical microscope, and if desired greatly enlarged by microphotography. Since historadiography is a nondestructive technique the tissue section can be mounted on the glass plate alongside the microradiograph for comparison of optical and x-ray appearances.

To record a historadiograph by the projection method, the tissue section is placed close to the target of the x-ray microscope, so that the point source can be used to project an enlarged image of the section on to a distant photographic plate. Since soft x-rays are used both the specimen and plate are enclosed within a vacuum camera mounted over the microscope. The enlarged image on the developed plate may be examined with a hand lens or optical microscope, and if desired further enlargement may be obtained photographically.

Contact microradiographs of tissue sections are readily prepared and may prove adequate for the study in hand provided the resolution required falls within the limits set by the grain of the recording emulsion. Projection microscopy on the other hand provides better resolution and primary magnification of the object, and so lends itself to study of selected tissue areas. A point source of 0.1μm to 0.2μm is obtainable with the x-ray microscope, hence details of this order can be resolved.

In historadiography it is important to use an x-ray tube with a small focal spot to reduce image unsharpness, and also work at low voltages to secure soft radiation and so better tissue contrast. It is essential to employ soft x-rays because the biologically interesting elements have low atomic numbers, and their characteristic absorption edges fall within the soft x-ray region. Suitable x-ray apparatus is now commercially available for historadiography, and calculations by workers such as Engström (1957) and Lindström (1955) have established the optimal tube voltages and x-ray wavelengths for tissue studies.

2.C.2. History

The term historadiography was introduced by Lamarque and Turchini (1936, 1937), who studied tissue sections with a special demountable x-ray tube designed to operate at 500 to 5000 volts and deliver soft x-rays of 2–12 Å wavelength. Both the tissue section and recording plate were enclosed within a vacuum chamber attached to the x-ray tube. Owing to the use of a fine grain emulsion of the Lippmann type, they were able to study contact microradiographs of tissue under the optical microscope and enlarge them photographically. Lamarque recognized that the differential absorption visible in the tissue image was related to the atomic number of its cellular constituents, and that the technique might be applied to histochemical analysis. This was the most significant application of microradiography to histology prior to the work of Engström and the Swedish school.

Engström (1946 et seq.) and his coworkers demonstrated that contact microradiography could be applied to many morphological and quantitative problems in histology and cytology. It was shown, for example, that cellular structures could be weighed by comparing the x-ray absorption of cells with that of a reference system of nitrocellulose foils exposed at the same time. An ordinary or polychromatic beam of soft x-rays was used, and its attenuation estimated by photometric measurement performed on the microradiograph.

Advantage was also taken of the fact that the x-ray absorption curve of each element exhibits at certain critical wavelengths a number of discontinuities or absorption edges, where a sharp change in the amount of absorption occurs. By performing microradiography with monochromatic x-rays, and using selected wavelengths on either side of such an edge, Engström showed that it was possible to estimate the difference in absorption at the two wavelengths and hence the amount of an element present in a tissue (cf. Sec. 5.A.3). X-ray absorption methods were soon applied to the determination of the dry weight (mass) of various cellular structures and also cytochemical elementary analysis (Brattgard and Hydén, 1952; Lindström, 1955). About this time also microradiography began to be applied to histological and experimental studies of mineralized tissue such as tooth and bone, notably by Applebaum (1932–1948), Engfeldt *et al.* (1952), Bohatir-chuk (1954), Jowsey (1955), Röckert (1955–1958, Göthmann (1960a,b), Sissons *et al.* (1960a,b) and Scott *et al.* (1960).

Historadiography of soft tissue with emphasis on morphology received less attention. This was possibly due to the lack of suitable equipment. Some however recognized its potential and introduced new techniques for the study of healthy and diseased tissues (Fitzgerald 1957; Leborgne *et al.* 1957; Greulich 1960). Interest in x-ray microscopy of soft tissue has however increased due to the recognition that physical developments in the ultrasoft x-ray region must revolutionize histological research (Henke, 1967) and that x-ray methods can provide information of a type not obtainable by other means. Special techniques which utilize recent developments in x-ray technology and histochemistry will have to be devised for historadiography.

2.C.3. Relevant Principles of X-ray Absorption

While penetration of the specimen by the x-ray beam is important in x-ray microscopy, it must also be attended by adequate specimen absorption to provide a satisfactory historadiograph. When the amount of specimen absorption is inadequate, means must be found to increase it in order to secure a better image or picture. Hence it is necessary to understand the principles governing x-ray absorption, as well as to appreciate that x-ray wavelength, specimen density and atomic number, are interrelated factors that greatly influence absorption and thus contrast within the historadio-graphic image. A mathematical formulation and a quantitative discussion of the interrelationships are given in Sections 5A and 5B. At this point a simplified discussion will be presented from the standpoint of qualitative microradiography.

Historadiography is usually performed with the continuous or general radiation of a fine focus x-ray tube, which of course consists of many wavelengths. Mineralized tissues are studied with hard penetrating x-rays of short wavelength, whereas nonmineralized tissues are examined with soft and readily absorbed x-rays of long wavelength. When selecting apparatus for historadiography it is therefore necessary to know the x-ray spectrum obtainable with a given x-ray tube. Fortunately the critical minimum wavelength at which x-radiation commences is dependent on the maximum voltage (kVp) applied to the x-ray tube, and can be readily calculated from the simple relation

$$\lambda(\AA) = \frac{12.4}{kVp}$$

It will be evident that the minimum x-ray wavelength shifts to lower values with increasing voltage.

While x-ray absorption coefficients are usually discussed in relation to quantitative work, they also have decided bearing on qualitative studies. It should be noted that each absorption coefficient in the published tabulations refers to radiation of one particular wavelength. But although historadiography is usually performed with continuous radiation, such absorption data serve as a guide to the optimal x-ray wavelength for maximal absorption by a given element. Monochromatic microradiography may be applied with advantage to certain qualitative studies (Clark, 1947).

The intensity of an x-ray beam is reduced on placing a specimen in the beam, and falls off exponentially with the thickness of the specimen. This is expressed by the equation $I_f/I_o = e^{-\mu x}$, where I_o represents the original beam intensity, I_f the final intensity of the beam after traversing x cm of the specimen, and e the Napierian base of logarithms numerically equal to 2.718. The symbol μ, known as the linear absorption coefficient, is the fraction of the energy absorbed per unit path length in the absorber when an x-ray beam is incident on a thin layer of absorbing material (cf. Appendix, Section A.4).

The linear absorption coefficient μ can be used to calculate changes in intensity due to different thicknesses of a given substance at a given wavelength, but it is specific for that particular absorber (Robertson, 1956). It has a limited value because it varies with the physical state of the substance. Nevertheless it can be used to calculate "half layer value," or that thickness of an absorber which reduces beam intensity to half its initial value, so indicating the penetrating power of the beam. The tables of linear coefficients compiled by Zuppinger (1935) for standard radiography are not applicable to historadiography, owing to the longer wavelengths employed.

A more useful constant is the mass absorption coefficient μ/ρ, since it is concerned with the absorption per unit mass of substance in a beam of unit cross-section (see Appendix, Sec. A.4), and is the same whether the substance (element) is in a solid, liquid or gaseous state. The absorption equation now appears as

$$I_f/I_o = e^{-(\mu/\rho)\rho x}$$

where ρ is the density, x is the specimen thickness, and μ/ρ the mass absorption coefficient. Values have been tabulated for a number of elements at various wavelengths: 0.01 to 10 Å (Victoreen, 1950); 0.710 to 8.321 Å (Dyson, 1956); 8.34 to 44 Å (Henke, White and Lundberg, 1957); 2 to 200 Å (Henke, Elgin, Lent and Ledingham, 1967). When the absorber is a chemical compound its mass absorption coefficient may be calculated from its constituent elements by reference to such tables (see Victoreen, 1950; Robertson, 1956), and by reference to the Appendix, Section A.4, Equations A-4 to A-13.

Study of the absorption tables will reveal that the mass absorption co-efficient of an element is specific for a given wavelength and increases with wavelength. It also increases with atomic number. In general the mass absorption coefficient varies approximately with the cube of the wavelength (λ^3) and with the cube of the atomic number (Z^3), as expressed by the equation $\mu/\rho = K\lambda^3 Z^3$. This equation refers to absorption alone; to represent the total attenuation coefficient for a beam of radiation one must add a term to represent scattering, but this term is generally negligible in the conditions of x-ray microscopy.

A helpful exercise is to plot the mass absorption coefficient of an element (e.g. Cu, Sn) against a number of wavelengths (cf. Fig. 44, Sec. 3.A, or Fig. 113, Sec. A.4). It will be noted that the coefficient increases rapidly with wavelength until at a certain wavelength, critical for each element, there is a sudden decrease in the amount of absorption. The critical wavelengths are referred to as absorption edges, and the abrupt changes of coefficient as absorption jumps or discontinuities. Each edge corresponds to an energy level within the atom. Among the edges, the K edge occurs at the shortest critical wavelength, while L and M are encountered at longer wavelengths. In practice, microradiographs can be taken at wavelengths on the high-absorption side of an absorption edge, thereby ensuring maximal absorption by a given element within the specimen.

Absorption tables will therefore guide the microscopist to the most appropriate range of x-ray wavelengths with which to study a specimen and secure maximal absorption by tissue components, with consequent improvement of contrast and definition in the tissue image, as for example in the historadiographic study of a tissue containing a mineral inclusion or treated by a metallic stain histochemically.

The highly physical approach of some authors to x-ray absorption measurements should however be treated with some reservation when applied to biological problems (Brattgård, Hallén and Hydén, 1953). This is because the application of physical formulae to histological problems not infrequently necessitates assumptions due to the lack of precise basic information regarding specimen composition, section thickness and individual cell volume.

The work of Lindström (1955), though essentially a quantitative approach to total dry weight determinations and elementary analysis on a cellular level, provides instructive microradiographs that illustrate the effect of registering histological sections at different voltages. It shows for example that elastic lamellae within an artery wall may be barely visible in micro-radiographs taken at high voltages and give high contrast when registered at low voltages. This work includes useful tables on the elementary composition of some organic compounds, as well as the mass absorption coefficients of certain elements of low atomic number encountered in histological specimens.

The absorption of a histological section is determined by its thickness, density, elementary composition, and the wavelength of the x-ray beam used to study it. To secure good resolution and image contrast it is necessary to use thin sections and soft x-rays. Engström (1957) constructed a useful graph for tissue microradiography in which voltages suitable for thin sections were plotted against section thickness given in microns. He observed that histological sections of ordinary thickness (1μm to 100μm) of dehydrated tissue required voltages in the 500 to 2000 volt range. This graph (Fig. 28) is a basic guide for historadiography.

While some of the best results have been obtained in this low voltage range, good results can nevertheless be obtained from thin sections of soft

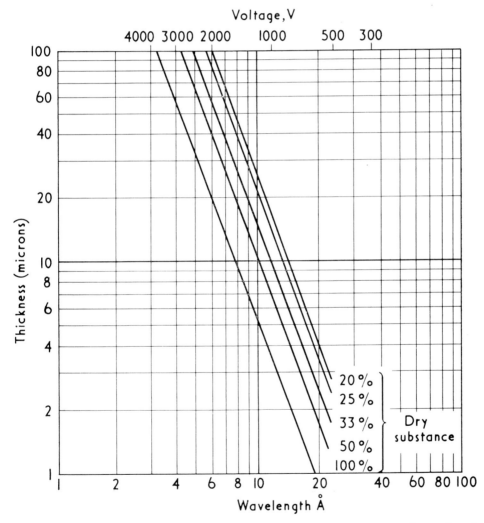

Figure 28. Diagram from Engström showing suitable voltages for microradiography of thin sections of soft tissues.

tissues while working at slightly higher voltages, as for example in the 1 to 5 kV range provided by the Philips contact microradiography unit (C.M.R.5) or the 5 to 15 kV range of the Cosslett-Nixon x-ray projection microscope (XM 30). Still higher voltages are usually needed to study sections of tooth or bone, owing to their greater density.

Thick sections (100μm to 500μm) of certain tissues, such as lung, can prove useful in historadiography, especially when it is wished to preserve internal structure. Structures can be traced over greater distances than is possible in ordinary histological preparations, and recorded stereographically. The marked depth of field of the x-ray projection microscope lends itself to the study of histological structures three-dimensionally.

The composition of some tissues is such that they yield good historadiographs without much preparation. This is because of marked intrinsic variation in either density or chemical content. Naturally occurring differences in density are encountered in tissues containing cellulose, chitinous or calcareous structures. Some tissues exhibit less obvious variations of this type, and may even appear unsuitable for historadiography. This however may be due simply to the thickness of the sample or its water content producing marked absorption with resultant masking of microscopic detail. The use of thin dehydrated sections coupled with a suitable x-ray spectrum may reveal an unsuspected absorption differential that changes the value of the historadiograph.

The elementary composition of a tissue is important in historadiography because of its effect on absorption. As is well known, different parts of the animal or human body contain different proportions of various elements, some of which lie fairly high on the atomic scale although most are of low atomic number. The protein component of soft tissues for example consists mainly of carbon, hydrogen, oxygen, nitrogen and sulfur. These elements being of low atomic number have a low mass absorption coefficient. Consequently such tissues yield a poor historadiograph unless soft x-radiation or special staining techniques are used to increase tissue absorption. Absorption tables, such as those of Cosslett and Nixon (1960), illustrate how the mass absorption coefficient of such elements increases with x-ray wavelength. Absorption of course also increases with atomic number, as exemplified by the marked historadiographic contrast obtained with mineralized tissues and tissues treated with metallic stains.

Historadiographs may be markedly improved by taking advantage of the rapid increase in absorption that attends the use of long x-ray wavelengths. In projection x-ray microscopy this can be achieved by selecting a suitable foil as a target metal and adjusting the operating voltage to give the desired range of wavelengths. A wide range of elements is now available in the form of thin metal foils, 4μm to 10μm thick, which can be used as targets in the x-ray microscope.

Much exploratory work on histological material remains to be done along these lines. Monochromatic microradiography promises greater sensitivity of contrast and definition, but the production of a sufficiently intense monochromatic beam presents difficulties, especially in the long wavelength region. Henke (1967) has recently described a vacuum spectrograph which permits contact microradiography with ultrasoft x-rays (10 to 150 Å). When monochromatic radiation is used the contrast is simply and quantitatively related to the mass and chemistry of the sample. Such microradiographs, owing to the mechanism underlying the x-ray contrast, reveal histological information not readily obtainable by optical microscopy. As already mentioned, chemical information may be obtained by recording microradiographs of the same specimen at different wavelengths on either side of a critical absorption edge (cf. Secs. 3.A.3 and 5.A.3).

2.C.4. Preparation of Material

Conventional histological techniques may be applied to x-ray microscopy, but are not always satisfactory. This is because such techniques often alter tissues physically and chemically and do not take into account that biological tissues transmit and absorb x-rays in a different way from light. Since x-ray microscopy is a recent development, few techniques of specimen preparation have as yet been evolved specifically for historadiography.

Hitherto most historadiographic studies of soft tissue have been performed on formalin-fixed material. Formalin fixation however introduces cellular changes and increases the overall x-ray absorption of the tissue. One of the best techniques of preparing formalin-fixed tissue for high resolution contact microradiography was developed by Greulich (1960). It gives good definition but unfortunately is a lengthy procedure that subjects the tissue to numerous solutions and hence chemical losses.

In Greulich's technique, thin paraffin sections (2μm to 8μm thick) are floated out on warm distilled water (37° to 40°C), and picked up on high resolution or spectroscopic plates (Kodak HR or 649GH, 649-0) under red safelight conditions. The plate emulsion has previously been protected by dipping the plate briefly (30 sec) in a solution of 1% nitrocellulose (Parlodion) in ether-ethanol (1:1), and drying it overnight (12 hours) in an upright position. The paraffin section on the plate is then dried in the dark for some time (4 hr).

The paraffin is then removed from the section by passing the plate through benzol (2 changes; 5 min each), next a series of descending alcohols (absolute alcohol, 95%, 80%, 70%; each 1 change of 1 min), and so to distilled water (1 change, 10 min). This water stage is important since paraffin wax contains a benzol-insoluble fraction, which if not removed, superimposes an x-ray image of the crystalline water-soluble residue upon the tissue image.

Finally the plate is transferred through ascending alcohols (70%, 80%,

95%; each 1 change, 30 sec) to absolute alcohol and benzol, whereupon the plate with its superimposed section is allowed to dry in the dark.

After exposure to x-rays, the plate is immersed in absolute acetone (2 min) to dissolve the celloidin film protecting the plate emulsion and so float the tissue section free. Care must be taken not to let the emulsion dry since acetone evaporation may cool and distort it. A sable brush or fingertip moistened with acetone may be used to dislodge the section, which must be removed completely before the plate is developed (Kodak D 19b, full strength, 5 min at 20°C). After development the plate is fixed and washed in the usual manner.

As regards fixatives, insufficient work has been performed with various types of fixative to assess their relative advantages and disadvantages in historadiography. Tissues fixed in Zenker's solution yield more microradiographic contrast than formalin-fixed tissues owing to the mercuric chloride content of this solution. This heavy metal salt is radiopaque, hence this fixative is precluded if the end point of decalcification is being tested by microradiography.

The effect of using a fixative of high atomic number is illustrated by the study by Recourt (1957a) of osmium tetroxide penetration into skeletal muscle. More historadiographic studies of this type are necessary before the value of protein-precipitant metals such as osmium, mercury and chromium can be truly assessed.

Freeze drying, which is a form of fixation, is the best method of preparing soft tissues for historadiography. Other than the loss of water, it produces few or no cellular changes or chemical losses, and hence results in a historadiographic image of sharp definition and good contrast. In practice tissues are quenched in isopentane cooled to −190°C in liquid air, and freeze dried in a tissue drier (e.g. Edwards model TD2) at −40°C for ninety-six hours, or transferred to absolute alcohol at −50°C and freeze substituted for a similar period. The techniques of freeze drying and freeze substitution are described in detail by Culling (1963).

Embedding agents such as paraffin wax must be removed from the section before historadiography, because they are usually x-ray absorbent and blur the tissue image. Flotation of paraffin sections on a water bath results in cellular changes and chemical losses (Hancox, 1957). Hence structural and density differences can be readily detected in comparison historadiographs made from tissue sections floated out over saline or water, and sections that have been dry mounted. The last more nearly approach an unmodified sample and so tend to give a truer tissue image. In this technique, unstained paraffin sections prepared from freeze-dried material are first deparaffinized, and then spread out on a thin plastic (Melinex) or formvar film. Formvar films are prepared by dipping a coarse grid, or brass rings, in a solution of 0.25% formvar in ethylene dichloride; such films have a low absorption factor and do not markedly impair the tissue image. Sections so mounted can be placed directly on the plate emulsion for contact micro-radiography, or introduced into the beam of the x-ray projection microscope.

As a rule tissue sections must be deparaffinized before historadiography. Bellman (1953) however carried out microradiography on beeswax-embedded tissues without removing the beeswax. Sections cut at $200\mu m$ from injected kidney, intestine and ovary, when mounted on thin mica sheets and recorded at a higher kilovoltage than is customarily used for historadiography, gave striking pictures of microvascular anatomy. The method however is better suited to microangiography, and thin plastic foil serves as a better mount than mica.

Sometimes it is best to dispense altogether with paraffin embedding and use fresh sections cut on a freezing microtome. Such sections should be mounted on a thin plastic foil such as Melinex. Marked x-ray absorption due to the water content of the tissue is to be expected, but the method may be useful for certain tissue components in the fresh state. If the water content of the tissues has an unfavorable effect on contrast, one may air-dry the tissue prior to microradiography by placing the section overnight in a desiccator.

Bell (1969) has devised a simple and convenient method of preparing histological sections for x-ray microscopy. The tissue section, after being placed on a slide and stained or otherwise treated, is dehydrated, cleared, and mounted with DPX synthetic mounting medium under a coverslip as for optical microscopy. Control microphotographs are taken at this point if desired. A day or two before historadiography the slide is soaked in xylene long enough to remove the coverslip and excess DPX. The slide is then immersed in a mixture of approximately equal parts of thick DPX and xylene for 10 to 60 min, after which the slide is drained and propped upright to dry for about 30 to 60 min. The thin DPX film containing the tissue section may then be trimmed and peeled off the slide with a razor blade. The filmed section is labelled and stored in a folded lintless paper between dry slides; if necessary the film can be compressed and flattened with a light object.

The optimal concentration of the DPX-xylene mixture and the time of soaking both vary with different types of tissue and section thickness, and hence must be determined by trial and error. The drying time for the film varies too, and it is possible for the film to become too dry to permit easy peeling. Sections thicker than $50\mu m$ may require a second coat, in which case the second immersion in the mixture should be shorter than the first and performed before the first coat is quite dry, to prevent the tissue cracking. If the section was initially very firmly attached to the slide, it may be necessary after filming, drying and trimming, to soak it in 95% alcohol for a few minutes, after which the peeling is done with the slide immersed in a shallow dish of water.

Such mounts are easily handled with forceps, and can be placed directly on the plate emulsion for contact studies, or inserted into the beam of the x-ray microscope for projection studies. This is a practical method for handling soft tissue sections, such as brain, for historadiography. The thin DPX film has a low x-ray absorption factor and hence does not interfere appreciably with cell detail.

DPX tissue mounts can be x-rayed repeatedly, and then remounted on a glass slide for comparison of historadiographic and histological appearances. Remounting is performed by first dissolving the DPX mount with xylene or benzol, and then gently pressing the wet section down on to a glass slide with lens paper, before sealing it with a xylene-soluble mountant and coverslip.

Historadiography may be performed with stained or unstained sections. Unstained sections of soft tissues, such as skin, kidney, and salivary gland, when prepared by the freeze drying technique, reveal a surprising absorption differential on microradiography. This may be due simply to uneven distribution of dry weight or localized concentrations of chemicals, but at times no adequate histochemical explanation is available for certain differences in radiodensity. Such differences not only increase historadiographic contrast, but often prompt further investigation. X-ray studies of unstained sections are described later.

Some soft tissues, such as the brain, yield unsatisfactory historadiograms unless x-ray stains are used to increase tissue absorption and hence image contrast. X-ray stains, or stains which are known to increase the radiopacity of certain tissue components are as yet limited in number.

Few histological stains have been adequately tested for their possible historadiographic application. The simple immersion of histological sections in solutions containing heavy metals has been tried. This probably reflects mere mechanical trapping of metallic deposits in tissue spaces, rather than any reliable chemical interaction. However certain tissues when immersed in a colloidal suspension of thorium dioxide (Thorotrast) show an apparently selective absorption or staining of some of their components. Bohatirchuk (1953, 1961) for example reported selective absorption of Thorotrast by the white matter of the spinal cord, and also by the metastases of Brown-Pearce carcinoma in the liver. This he ascribed to an affinity between positively charged groups of the tissue for negatively charged thorium particles.

The classic heavy metal stains, such as those of Golgi and Cajal, are useful in historadiography. Microradiographs of nervous tissue so stained show good image contrast, and also strikingly demonstrate patterns of neuronal and dendritic organization. Nervous and reticular tissues reputedly show a selective affinity for silver salts, but this is not confirmed by microradiography, which shows that skin and tendon also possess a strong affinity for silver (Mitchell, 1950, 1951).

Histochemistry, owing to its better understood and more precisely controlled reactions, is the most promising approach to the production of artificial contrast for historadiography. In histochemistry, as applied to light microscopy, there is a whole group of reactions that employs metals not primarily for their high atomic weight, but more often because of their highly colored compounds or appropriate chemical actions. Recently it has been realized that the atomic weight of these metals gives this group of reactions great potential for electron microscopy, where increased beam absorption and improved image contrast is desired at sites specific to the particular reaction chosen. Similarly, some of these histochemical reactions can be applied to tissue undergoing examination by historadiography. A wide variety of biological applications suggests itself under the term x-ray histochemistry (Bell, 1969).

If such a histochemical reaction is applied to a tissue, the distribution of a tissue inclusion of high atomic number can be readily demonstrated by x-ray methods. Moreover the strong wavelength dependence of x-ray methods makes it possible to assess both the quantity and nature of the inclusion. In all cases however the result can only be as reliable as the information available on the histochemical reaction itself. A histochemical technique, for example, may give consistent results and be used with assurance to demonstrate certain components, and yet some doubts surround its actual chemical mechanism. In such a case the technique may perhaps still be usefully employed in x-ray microscopy to demonstrate the same reactions or components, even though the product may not be quantitatively interpreted.

One method applicable to historadiography is that of increasing a naturally occurring but undetectably small metal component of the tissue. Another method is that of adding a completely new metal to a tissue component or product which is naturally of too low an atomic weight to be detected. The reaction in either of these cases, as in light microscopy, may simply attach the metal to the component under study, or it may deposit the metal as a reaction product to mark the site of a specific activity. The Prussian blue reaction is an example of the first method, since it may be used to increase the concentration of iron where the original iron is in the ferric state, to give a final product of ferric ferrocyanide (Boyde and Switsur, 1963).

The use of lead or cobalt reaction products to mark the site of acid and alkaline phosphatase reactions in tissues is an example of the second method, since it introduces a metal where none occurs naturally. These are classical histochemical techniques and have been widely used in light and electron microscopy. Recently they have been successfully applied to x-ray microscopy (Bell, 1969), and used to demonstrate the capillary and neuronal patterns in the cortex of the human brain (Saunders, Bell and Carvalho, 1969).

One possible advantage of x-ray histochemistry is that fewer steps may be necessary in a given procedure. For with x-ray methods the very presence of the metal in the beam suffices, and the subsequent color reactions are unnecessary, unless of course the same specimen is to be used for comparison of historadiographic and histological appearances. Reduction in the number of steps likewise favors greater accuracy, especially if quantitative estimation is contemplated. Where the identity of a final reaction product is in doubt, x-ray analysis of either the specific absorption or emission type should be able to locate and measure the different elements present. The staining process itself may also be tested in this way at different stages of the procedure, to determine the nature and completeness of the various conversion steps and reactions. The Gomori technique for alkaline phosphatase was so studied by Hale (1962) and discussed by Cosslett and Switsur (1963).

The embedding and sectioning of mineralized tissues such as tooth and bone for microradiography is a specialized technique that presents certain difficulties. Nevertheless microradiographic surveys can be quickly made of the deposition of mineral during growth and development, and also the distribution of mineral in mature calcified tissues and various pathological conditions. Basically, formalin-fixed specimens of tooth or bone are first embedded in methyl methacrylate, and cut into thick sections with a milling machine. The sections are then ground down between glass plates to the desired thickness (50μm to 200μm). Various modifications of this technique have been described for the study of calcified tissues (Scott, Nylen and Pugh 1960), bone (Sissons, Jowsey and Stewart, 1960a,b; Hallen and Röckert, 1960; Molenaar, 1960; Bohatirchuk, 1963, 1966; Belanger, Robichon and Belanger, 1963), and teeth (Röckert, 1958; Dreyfuss, Frank and Gutmann, 1964; Engfeldt, Bergman and Hammarlund-Essler, 1954).

Ground sections of bone may be microradiographed by the method described by Engfeldt and Engström (1954). According to Engfeldt (1958) the absorption of x-rays in the 1 to 3 Å wavelength range is dependent on the presence of bone salts, so that microradiographs obtained with such radiation indicate both the amount and distribution of bone salt in the section. Softer x-rays, in the 8 to 12 Å range, have been used to study the organic part of bone.

Thinner sections can be prepared from decalcified bone or teeth. These yield interesting historadiographic results. Decalcification is best performed with a chelating agent, since tissues so prepared show a minimum of artifact (Culling, 1963). A pH5 solution of Versene, which is the disodium salt of ethylene diamine tetra-acetic acid (EDTA), is a powerful inorganic chelating agent that produces good demineralization of teeth (Nikiforuk and Sreeby, 1953). The end point of decalcification can be determined by micro-radiography, after which the sample is dehydrated, paraffin embedded, and sectioned. If desired the decalcified tooth can be sectioned longitudinally with a scalpel. This provides a convenient preparation for microradiography, since the pulp is supported by side walls of dentine. If the pulpal vessels have been injected with a radiopaque, such a preparation also gives a good microangiographic picture of dental vascular anatomy (Saunders, 1957a,b).

Finally, some of the *in vivo* techniques of histology can be extended to historadiography. The phenonomena of vital staining and phagocytosis offer yet another method of introducing radiopaque material into tissues. For example, the intravascular injection of Thorotrast or Umbrathor into the living animal some time before death (2 to 24 hr) is followed by its reten-tion by the reticuloendothelial system. Historadiography can then be used to demonstrate the reticuloendothelial elements of the liver and spleen (Bohatirchuk, 1961). The intracellular accumulation of barium sulfate in the wall of a duodenal ulcer has been demonstrated by historadiography following surgery (Röckert and Zettergren, 1963).

2.C.5. Equipment for Historadiography

Various types of equipment may be used for historadiography, but much depends on the kind of tissue being examined and the best radiation for its study. Mineralized tissues such as sections of tooth or bone can be studied with a standard x-ray diffraction tube such as used for crystallography. A Philips Norelco diffraction tube, fitted with a copper target (0.3 mm focal spot) and beryllium windows, gives good results. Better results are obtain-able with the more versatile Philips PW unit. This is a fine focus (0.4 mm) x-ray diffraction tube equipped with a variable transformer to give a con-tinuous voltage range from 2 to 50 kV. At an anode voltage of 10 to 50 kV the wavelength range varies from 1.8 to 0.4 Å. A copper target is useful, but other target metals are obtainable. Unfortunately a vacuum camera can-not be attached to such tubes, unless the camera itself is sealed with a thin plastic (Mylar) window at its tube end.

The Hilger microdiffraction unit, which is essentially a demountable, continuously pumped, microfocus tube of the Ehrenberg-Spear type, gives excellent pictures of bone. This tube is easily fitted with a vacuum camera, and if carefully focussed gives a focal spot of about 20μm diameter. Because it is supplied with interchangeable (oil cooled) targets of various metals (e.g. Cu, W, Mb, Cr), radiation of different types is easily obtainable.

Unfortunately few sealed-off x-ray tubes are wholly suitable for histo-radiography of soft tissues. The sealed-off, low voltage x-ray tube designed specifically for historadiography by Combée *et al.* (Combée, 1955; Combée and Engström, 1954; van den Broek, 1957; Recourt, 1957b; Combée and

Recourt, 1957) gives excellent results with sections of soft tissues. It became commercially available in the Philips contact microradiography unit (CMR 5). This miniature, air-cooled tube, which is fitted with a tungsten target and very thin beryllium window (50μm thick), can be operated in the 1 to 5 kV, 1 to 5 mA range with a maximum dissipation of 10 watts, and yields soft x-rays of about 4 to 10 Å in wavelength, at 3 kV.

The Philips contact microradiography unit (CMR 5) produces very sharp historadiographs because of the small effective focus (0.3 mm) of the target, and relatively large target-specimen distance (1.5 cm) as compared with the small specimen-film distance. A resolving power of 0.5μm or better may be obtained with high resolution plates. The unit is simple to operate. A microradiograph is obtained by placing both the specimen and recording film within a small cylindrical vacuum camera, which is pushed up close to the tube window prior to making the exposure. The camera supplied with the unit provides too small a field (8.5 mm diameter), and is difficult to load without damaging the specimen or scratching the emulsion. Small discs of film (e.g. Gevaert-Lippmann) can be easily cut with a punch for this camera, but it is difficult to cut circular plates of this size.

Better results can be obtained by fitting the Philips CMR 5 unit with a wide field vacuum camera, that will accept both large histological sections and standard size plates (e.g. high resolution 2 × 2 in; spectroscopic 1 × 3 in). Such a camera (Fig. 10) is preferably made out of clear plastic to permit inspection of the section during vacuum pumping, since air evacuation sometimes displaces the section. The camera chamber should be about 7.5 cm long, and be secured to the x-ray tube by four screws with an O-ring interposed. The grooved door of the chamber is fitted with a large O-ring to give a vacuum seal, and accepts a plastic drawer that is large enough to accept a standard microradiograph plate. This camera provides an image of diameter 2 cm at a tube-plate distance of 7.5 cm. The vacuum pump connection and air admittance (bleeder) valve, which are situated on the side of the camera, should be of small bore to eliminate flutter of the tissue section during pumping.

In operation, the tissue section is positioned on a plate on the camera drawer, which is then slid into the camera chamber. Vacuum pumping is then started and continued for one minute or longer, after which the pump is cut off to eliminate vibration during the x-ray exposure. On completing the exposure, the air admittance valve is opened to permit removal of the camera door and the exposed plate. A camera of this type is essential for historadiography, since it eliminates absorption of soft x-rays by air, and thus increases image contrast while reducing exposure time.

Various experimental x-ray tubes suitable for high resolution microradiography of soft tissue have been described. Some indeed are not difficult to make, such as the midget x-ray tube developed by Engström and Lundberg

(1957), which is a continuously evacuated tube constructed from two automobile spark plugs.

The Cosslett-Nixon projection x-ray microscope is a particularly versatile instrument for historadiography, because it can be used to examine both mineralized and nonmineralized tissue by either the contact or projection method. Under optimal conditions the focal spot is virtually a point source (0.1μm to 0.2μm in size), hence this instrument has a better resolving power than any other x-ray tube. This, combined with the fact that it provides primary x-ray magnification, makes this tube particularly useful for histological studies. Magnification is varied by simply altering the target-specimen and target-plate ratio, hence the closer the section is brought to the point source the larger the image.

The great depth of field provided by the point source results in all parts of the specimen remaining in focus. Consequently thick sections can also be studied and stereographs obtained by shifting the specimen laterally between successive exposures. The degree of lateral shift (x) required is related to the total magnification (m) and the average interocular distance (65 mm), whence $x = 65/m$. Actually the degree of shift can be quickly determined by trial and error.

Among other advantages, the point source of the x-ray microscope facilitates the selection of a small area of tissue for detailed study, since a limiting aperture can be placed directly over the target. A microscopic area of tissue can thus be examined and recorded either by plate or an appropriate x-ray counter.

The transmission target used in the projection x-ray microscope is simply a small disc (4 mm diameter) punched from a thin metal foil (4μm to 12μm thick), hence it can be readily changed for historadiographic experiments. Various target metals and operating voltages can be used to produce different types of x-radiation. Study of the periodic table and data on emission voltages assist in the selection of the optimal target metal and x-ray wavelength for the problem in hand. The selection of a target metal is influenced also by the nature of the tissue, whether or not the section has been stained, and the presence of foreign elements in the sample.

The operator of an x-ray microscope should bear in mind that the general radiation of an x-ray tube is a continuous band of wavelengths, which begins at a critical minimum wavelength and increases rapidly to a region of maximum intensity of somewhat longer wavelength, beyond which the intensity gradually decreases. The total intensity of the beam will depend on the voltage and current impressed on the x-ray tube, and also on the atomic number of the metal used as a target. If the voltage applied is sufficiently high, sharp high peaks of characteristic radiation are superimposed upon the continuous band. These peaks are characteristic of the target material and consist of isolated wavelengths that occur above certain voltages. Historadiography is usually performed with continuous radiation, but there are occasions when characteristic radiation can be used to advantage.

Targets of copper, silver, gold, molybdenum, tungsten and aluminum have proved very useful in projection x-ray microscopy, but it is to be expected that other metals will find particular histological applications. Transmission targets under certain conditions produce a spectrum that consists essentially of characteristic radiation. These conditions depend on the operating voltage, the absorption energy of the target material, and the target thickness (Bessen, 1957). When these conditions are not met, as for example by using higher voltages or thinner targets, more of the transmitted radiation is continuous. Studies of the radiation from an aluminum transmission target 4μm thick, showed that the characteristic curve was much stronger relative to the continuous spectrum. The characteristic radiation (Kα) of aluminum has a wavelength of 8.3 Å. Soft radiation of this type is particularly useful for historadiography of soft tissues, and can be obtained by operating the x-ray microscope in the 5 to 10 kV range. Further experimental studies of transmission targets and their spectra are desirable in the interests of improved historadiography.

Some histological preparations, such as sections of mineralized tissues, can be examined over the projection x-ray microscope under atmospheric conditions. Sections of nonmineralized tissues generally require to be studied with soft radiation to secure good contrast. Hence a vacuum camera must be fitted to the microscope to eliminate the absorption of soft x-rays by air. Since such a camera is threaded to the target assembly and lies within the magnetic field of the objective lens polepiece, it must be constructed of a nonmagnetic material such as brass. The interior of this funnel-shaped camera is stepped at 1 mm intervals to accept brass rings used to support tissue sections, and to give known target-specimen distances. A surrounding platform, grooved for an O-ring and vacuum seal, supports a plate chamber. This chamber is constructed of clear plastic to permit inspection of the specimen, and is fitted with a vacuum pump connection and air admittance or bleeder valve. The plate holder within this chamber may be adjustable to permit variation of the target-plate distance. Figure 12 shows a camera of this type.

In practice, a tissue section mounted over a brass ring is lowered into the stepped lower part of the camera, and the target-specimen distance noted. The plate chamber is then loaded under red safelight conditions, positioned over the specimen, and vacuum pumped for several minutes. The vacuum pump is then switched off to eliminate transmitted vibration. After the x-ray exposure has been completed the air admittance valve is opened and the plate removed for processing.

High magnification can be obtained by positioning the specimen 1 mm from the target, but if the specimen is placed any closer it may suffer heat damage with consequent loss of definition. It is often wiser to use lower magnifications ($\times 2$ to $\times 10$) to secure better beam intensity and shorter exposure times, thereby also reducing the risk of mechanical and electrical instabilities. The x-ray microscope should be carefully focussed beforehand with the vacuum camera *in situ;* this is done by placing a fine test grid (1500 silver mesh; 3μm bars) on the target and a small fluorescent screen within the lower part of the camera. Focussing may be difficult at low voltages (5 to 7 kV) owing to low beam intensity, unless the eye is properly dark adapted and use is made of peripheral vision. Test pictures of the grid should be taken with the camera before commencing historadiography.

2.C.6. Applications

Historadiography appears to have valuable potential for the study of both morphological and histochemical problems. The complexity of earlier equipment limited its development, but there is now a growing appreciation that it represents an important histophysical technic for which special methods of preparing tissue must be developed.

The bibliographies on microradiography prepared by the x-ray Division of the Eastman Kodak Company (1955) and Ely (1963) provide a ready reference to the literature. Some papers illustrating the type of information obtained by the contact and projection methods of historadiography will be described here, together with results obtained in this laboratory.

Basic studies of healthy and diseased tissue were reported by Lamarque and Turchini (1936 to 1942), who used a demountable x-ray tube equipped with a tungsten target. Operated at low voltages (3 to 4 kV, 100 mA) with a thin (5μm) aluminum or lithium filter interposed between the target and histological section, this tube provided remarkably detailed historadiographs of sectioned skin, muscle, prostate and thyroid gland. The absorption characteristics of various tissue components were described, and they stressed the principle of performing historadiography before and after the extraction of certain histochemical substances, such as iodine.

The order of resolution obtainable by contact historadiography is illustrated by microradiographs showing the distribution of elastic tissue in the wall of blood vessels, or the mass distribution in the cells of a sectioned hair follicle, or the process of cell division in the onion root tip (Engfeldt, Engström and Datta, 1955; Engström, Lundberg and Bergendahl, 1957; Greulich and Engström, 1956).

The highest resolution (0.2μm) obtained in contact studies of biological material was achieved by recording thin sections (5μm) of formalin-fixed rat skin, skeletal muscle, and decalcified incisor, at low voltage (1 kV). These historadiographs showed considerable cytological detail, such as the fine intercellular connections or tonofibrils of stratified squamous epithelium, which have diameters considerably less than 1μm. Historadiographs of rat skeletal muscle demonstrated the fibers and myofibrils as well as focal accumulations of highly absorptive material (presumably contractile proteins) corresponding to the arrangement of the characteristic striations. These accumulations at the cross bandings had diameters of less than 0.5μm and were separated by intervals of approximately 0.2μm (Engström, Greulich, Henke and Lundberg, 1957). High resolution microradiography at 8 to 10 Å of a 3μm section through Carnoy-fixed hypoglossal nerve cells has resolved structures down to 0.2μm (Hydén, 1960).

Greulich (1960), who compared the histological and historadiographic appearances of a variety of tissues, showed that historadiography can demonstrate some histological structures more clearly than optical microscopy, as well as add information that other techniques do not provide. His contact studies of thin sections (5μm to 8μm), exposed in the 1.5 to 2.0 kV range, showed for example the clarity with which an actively secreting osteoblast embedded in a spicule of cartilage could be demonstrated, together with its surrounding newly formed bone matrix. Other microradiograms demonstrated differences in dry mass content of adjacent nuclei in surface cells of the rat jejunum, and showed that the striated border of the intestinal epithelium of the young rat is composed of a high concentration of structural protein. As he points out, such findings encourage further investigation of their function in the organism.

Istock *et al.* (1960), Randaccio (1964), and Shackleford (1965) all used the soft x-ray and contact technique to study stained and unstained histological sections. As Shackleford observed, the interaction of soft x-rays with organic compounds is poorly understood and many organic components may be transparent to light but opaque to x-rays. In general, histological structures containing large amounts of scleroprotein, (e.g. dermal keratin and collagen) proved the organic components densest to soft x-rays. He also pointed out that while many glycoproteins react equally strongly to histochemical stains for carbohydrate, that they may show a strikingly different x-ray absorption picture. Strong PAS reactivity might or might not therefore be reflected in the microradiograph, depending on the type of glycoprotein or polysaccharide examined. For example, the surface of the rabbit tongue epithelium is strongly PAS positive and opaque to soft x-rays, whereas mucous acini, which are also strongly PAS positive, may be relatively transparent. It was also noted that cell nuclei often exhibit striking differences in x-ray opacity, although stained sections gave no indication of this phenomenon.

Historadiography has also been applied to plant histology, thin plant structures often requiring little or no preparation other than the removal of water to improve image contrast. Salmon (1957, 1961) for example showed that contact microradiography could be used to demonstrate the distribution and fluctuation of crystalline structures in many plant tissues, such as the calcium oxalate crystals that occur in the epidermal cells of the bulb in Allium and Hyacinthus. Dietrich (1960) used contact microradiography to study the plant cell, comparing each element of the stained cell with its corresponding microradiographic image. He concluded that the state of the cell membrane, cytoplasm, nucleus and nucleolus, could be more accurately depicted by microradiography than by the usual cytological technics. Recently (1966) he provided a striking cytological analysis of cell division in Lilium Candidum and Lister Orata. His microradiographs show with almost diagrammatic clarity the changes which occur in the nucleus and chromosomes during the various phases of cell division.

Historadiography has also been used in pathology. Fitzgerald (1957) studied mass changes in pancreatic acinar cells following the administration of ethionine to rats. X-ray absorption revealed that in the normal gland there is considerable cytoplasmic, relatively little nuclear and a high nucleolar mass concentration. Contact microradiographs taken three to four days after the animals had received ethionine showed degenerating acinar cells with a loss of lobular and acinar patterns. All the cells showed a marked decrease of cytoplasmic mass, and loss of the sharp delineation between cytoplasmic and nuclear mass concentrations seen in the normal. Microradiography also revealed fairly good restitution of the pancreatic acinar cells if ethionine was stopped at about ten days. But in most instances there was some variation from the normal, both cytologically and in terms of concentration of mass.

Fitzgerald (1957) also used contact microradiography to compare the mass concentrations of normal and carcinomatous cells of the human uterine cervix. A densitometer traverse over the normal and cancer image areas indicated differences in mass concentrations and suggested the possibility of applying the x-ray absorption method to cancer research.

Leborgne *et al.* (1957) used contact microradiography to study malignant tumors of the breast. They demonstrated the general parenchymal and stromal pattern of several types of cancer, as well as cell nests, areas of necrosis, and a type of calcification that occurs in intraductal carcinoma. Gros and Girardie (1964) likewise used historadiography to study breast tumors and compared histological and microradiographic appearances. They briefly described the absorption characteristics of some tissue elements, and also the method of taking microradiographs on either side of the calcium absorption edge (3.06 Å) to identify calcareous deposits.

Historadiography was used by Amprino and Engström (1952) to make a direct comparison of the histological structure and quantitative distribution of mineral salts in bone. Ground sections of bone (20μm to 50μm thick) obtained from various animals including man were examined with soft x-rays. The wavelength band employed (2.5 to 3 Å) corresponded to the maximum

absorption of calcium on the short wave side of the K-absorption edge
(3.06 Å). A reference system in the form of a step wedge of cellophane
foils was recorded alongside the historadiographic image. The absorption
power of certain bone structures was measured by photometry and expressed in units of the reference system, so permitting the calculation of the
relative content of calcium salts in different structures. They thus demonstrated that the calcium content of primary periosteal bone is always higher
(5% to 20%) than the secondary bone of Haversian systems, and that there
was a density increase from the periphery towards the central canal within
individual Haversian systems.

Subsequently Engfeldt (1958) asserted that the absorption of x-rays in
the 1 to 3 Å wavelength range was from the practical standpoint solely dependent on the presence of bone salts. Hence microradiograms of bone obtained with such radiation indicated both the amount and distribution of
bone salt in the section. By using considerably softer x-rays (8 to 12 Å)
and decalcified sections, he was able to determine the distribution of dry
mass in the organic substance.

Wallgren (1957a) used microradiography to study the rate and extent of
bone mineralization in the human fetal femur, and showed that full mineralization is achieved in the primary periosteal bone during a four to five
week period in development. The microradiographic appearance of normal
bone tissue at various ages was studied by Sissons, Jowsey and Stewart
(1960a,b). They provided a striking demonstration of bone formation and
bone resorption in the cortical bone of the midfemoral shaft in individuals
ranging between two and a half and ninety-three years of age. Briefly, their
microradiographs showed the rapid turnover of growing bone tissue during
childhood, the remarkable lack of either bone formation or resorption in
young adults, and extensive bone resorption with little bone formation in
old age.

Historadiography has provided important information on the pathological processes of various bone conditions such as Pagets disease, osteogenesis
imperfecta, osteogenic sarcoma, strontium uptake, osteomalacia and rickets
(Engfeldt et al., 1952, 1954a,b, 1955; Bohr and Dollerup, 1960; Bohatirchuk, 1963).

Contact microradiography has also been widely used in experimental
studies of cartilage and bone. Blackwood (1965) for example, in a study of
cellular differentiation in the mandibular cartilage of rat and man, used microradiography to demonstrate the mineralization that occurs in the deeper
layers of the cartilage. Fyfe (1960a,b) used microradiography to study the
effects of intermittent pressure, and also artificial torsion, on growing rabbit
tibiae. His microradiographs of pressurized tibias showed increased vascularity, widening of the hypertrophic zone of the epiphyseal cartilage, and
irregularity of the bony trabeculae. Gradual torsion of the growing tibia was

seen to produce twisting of both the cartilage cell columns and the new trabeculae. These striking pictures were taken with a Hilger microfocus x-ray diffraction tube (Fig. 29). Göthmann (1960a,b) used microradiography to study the normal arterial pattern of the rabbit tibia and vascular changes associated with fractures and their union, while Okawa and Trombka (1956) studied the vascular anatomy of the rabbit bone marrow.

Contrary to Bohatirchuk's (1963) belief, the x-ray microscope has been used for bone research, and far from presenting technical difficulties provides simplicity of operation with a wide choice of radiation and improved resolution. The first study of bone by projection x-ray microscopy (Hewes *et al.*, 1956) revealed that the diploic veins within the tables of the dog skull could be demonstrated in about five minutes in contrast with the days required by conventional technics. Jackson (1957) examined sections (50μm thick) of developing rat bone by projection x-ray microscopy and so demonstrated the epiphyseal cartilage, new bone, and initial laying down of the apatite.

Histological sections of mineralized tissues such as calcified cartilage, bone, developing teeth and dermoid teeth (Röckert and Saunders, 1958) provide contrasty projection micrographs, even when decalcified. Such micrographs provide a striking demonstration of the epiphyseal and metaphyseal growth zones of young bone such as seen in the rabbit tibia (Saunders and van der Zwan, 1960). The concentric Haversian systems with their lamellae, entrapped osteocytes, and central canals, as well as the columns of proliferating and maturing cartilage cells with their calcifying intercellular matrix, can all be clearly shown (Fig. 30). The metaphyseal area distinctly shows the calcified trabeculae, osteoblasts lining the dark marrow spaces, and the areas of new bone deposition. High power views show details such as the lacunae that contain the osteocytes, and the collagen fibrils which separate the rows of maturing cartilage cells.

Historadiography has been widely used to study the structure and mineralization of the developing and mature tooth. Applebaum (1932, 1938, 1948a,b) was the first to use x-rays to study undecalcified ground sections. He demonstrated the lamellae and Schreger bands within enamel, and Owens contour lines in the dentine. Amprino and Camanni (1956) used historadiography to analyze mineral distribution in relation to the microscopic structure of hard dental tissues. Röckert (1955, 1956, 1958) who summarized dental studies between 1931 and 1958, demonstrated variations in the x-ray absorption of the Retzius and Schreger bands in tooth enamel, and the transparency of dentinal tubules to x-rays.

Van Huysen (1960) using historadiography noted radiopaque rings indicative of peritubular calcification in young coronal dentine. Later Dreyfuss *et al.* (1964) studied dentinal sclerosis by microradiography, and showed that the gradual closure of the dentinal tubules with calcified mate-

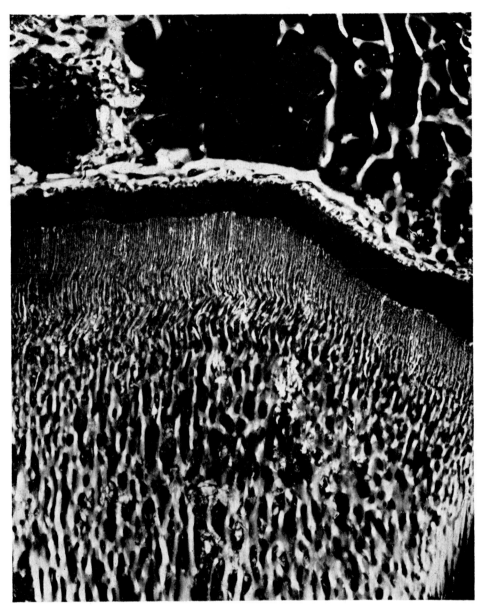

Figure 29. CMR of undelcalcified sagittal section of rabbit tibial upper epiphysis and metaphysis. The epiphysial cartilage is radiolucent. The metaphysial trabeculae are interspersed with a few Micropaque-containing vessels, visible as far as the proximal ends of the intertrabecular spaces. The living tibia had undergone torsion in opposite directions on five successive days, but no torsion for two days before sacrifice. Hence the bends visible in the secondary trabeculae and the straightness of the primary trabeculae. Mag. × 17.

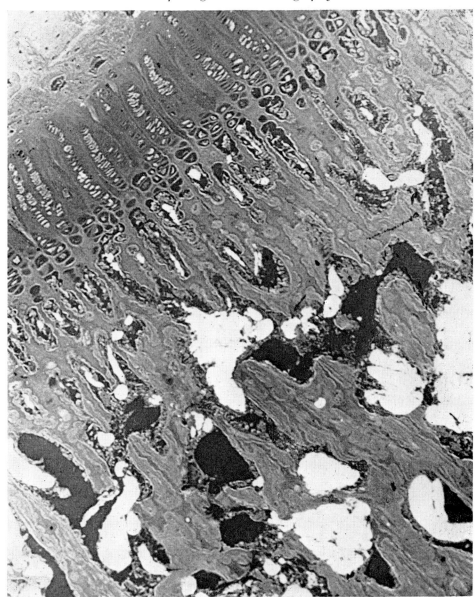

Figure 30. X-ray projection micrograph of a section of the epiphyseal growth zone of a rabbit tibia, showing young and calcified cartilage, osteoblasts along the marrow spaces, and blood vessels filled with contrast medium. Mag. × 230.

rial is a specific reaction of dentinal tubules to physiological and pathological stimuli (e.g. aging, abrasion and caries). Dreyfuss and Frank (1964) also used microradiography to study the lamellar and homogeneous structure of human cementum, which are both found in cellular and acellular cementum.

The tissue changes in dental caries have also been studied by micro-radiography (Thewlis 1940; Applebaum 1948a,b; Bergman, Hammarlund-Essler, and Lysell 1958). Recently Wyckoff and Crossant (1963) pointed out that new information can be obtained about healthy and carious dentine in human teeth by using monochromatic instead of heterochromatic x-rays for microradiography, such as the K radiation of aluminum, titanium, etc.

Ways of applying projection x-ray microscopy to the histology of un-stained, freeze-dried, animal and human tissues were explored by Saunders and van der Zwan (1960). Many tissues such as the skin, kidney, thyroid, sebaceous and salivary glands were found to give contrasty historadiographs with soft aluminum radiation even when unstained.

For example, projection x-ray micrographs of unstained human skin showed the stratum corneum, with markedly x-ray absorbent keratinized strands lying at the skin surface. Deep to the stratum corneum could be seen the layers of polyhedral cells of the stratum spinosum, whose irregular intercellular boundaries appeared as white areas of x-ray absorption. Marked absorption was also evident in the basal region adjacent to the dermis.

Historadiographs of unstained human scalp showed the cellular organization of the inner and outer root sheaths of the hair follicles with almost diagrammatic clarity, owing to their different x-ray absorption patterns (Fig. 31). From without inwards it was possible to identify the vitreous membrane, cells of the outer sheath, as well as Huxley's and Henle's layers. The manner in which these layers undergo conversion to soft keratin could also be clearly seen.

The sebaceous glands related to the hair follicles, although unstained, also provided interesting historadiographs. High power studies showed the gland lobules and their cells with striking clarity. The cell membranes were markedly x-ray absorbent and hence stood out in contrast to the accumulation of dark, radiolucent, lipid droplets within the sebaceous cells (Fig. 32). Some of the sebaceous cells could be seen to contain discrete droplets of lipid, whereas others showed the cell nucleus surrounded by confluent droplets, which gave the cell a black appearance. Other historadiographs showed lipid packed cells at the center of the gland acinus surrounded by the debris of broken down cells. The manner in which the sebaceous duct entered the hair follicle was also demonstrable.

Unstained sections of rat kidney when examined with the projection x-ray microscope showed the cellular structure of the renal corpuscles and the cortical labyrinth. The interlobular arteries, vascular pole and capillary tuft of the glomeruli could be readily identified, as well as Bowman's capsule and space and its continuity with the proximal convoluted tubule. The glomeruli contrasted markedly with the surrounding and more x-ray absorbent proximal convoluted tubules (Fig. 33). The reason for this ra-

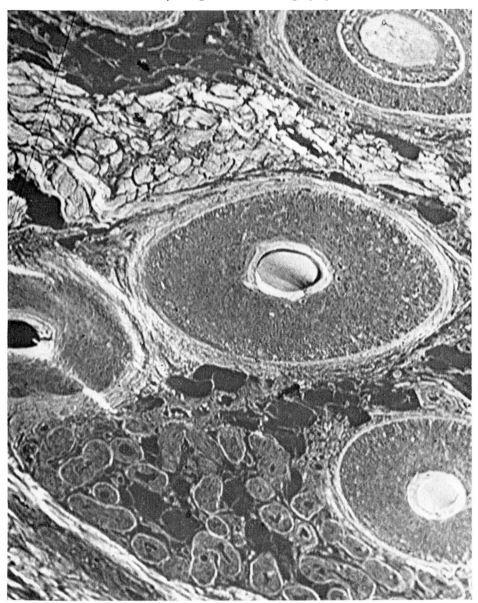

Figure 31. X-ray projection micrograph of an unstained section of human scalp showing hair follicles cut at different levels. The cellular organization of the inner and outer root sheaths is clearly shown. Mag. × 245.

diopacity differential is unknown, but is possibly related to the selective absorption that occurs in these tubules. Projection x-ray micrographs also showed that the distal and collecting tubules could be readily identified by virtue of their numerous and radiolucent nuclei.

Oderr (1964) used a specially designed microscope of the projection type

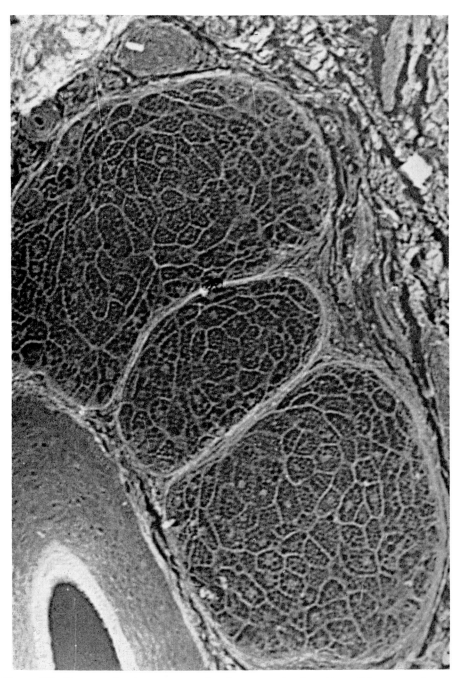

Figure 32. X-ray projection micrograph of an unstained section of human sebaceous gland showing its lobules and lipid droplets within the sebaceous cells. Mag. × 340.

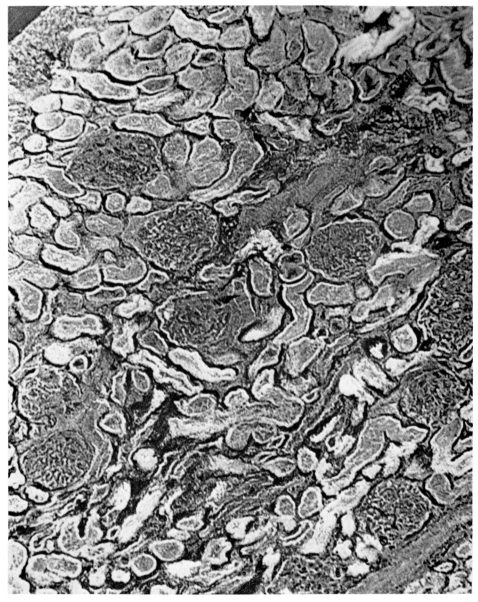

Figure 33. X-ray projection micrograph of an unstained section of rat kidney showing the renal corpuscles and cortical labyrinth. Note the capillary tufts of the glomeruli and markedly x-ray absorbent proximal convoluted tubules. Mag. × 270.

(Oderr and Dauzat, 1964) to study the internal architecture of the human lung, prepared by inflation-fixation using only alcohol and air. Sections of lung (0.5 to 2 mm thick) revealed a lace-like pattern of interlocking respiratory units. High power studies of the respiratory units showed that the air sacs or alveoli were arranged in spiral tiers, forming helices about the

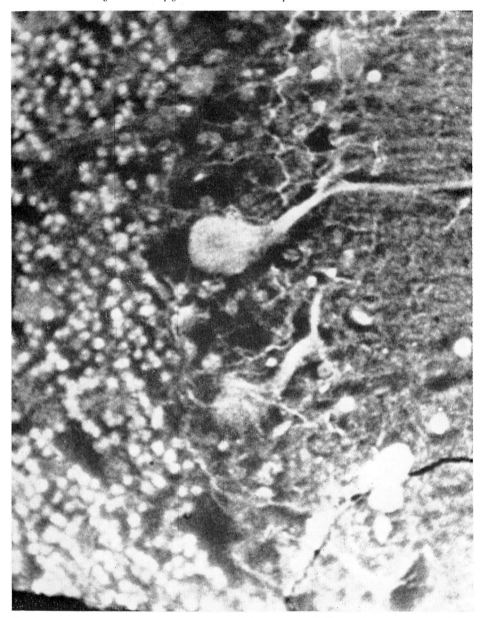

Figure 34. X-ray projection micrograph of human cerebellum showing the Purkinje cells and their dendrites, as well as the small round cells of the granular layer. Mag. × 363.

alveolar ducts. Gas exchange was calculated in the light of the geometric figures so discovered. Other lung studies were performed by Cunningham (1960), who used the projection method to examine pathology sections of various lung diseases such as asbestosis, and among other details demon-

Figure 35. Rat brain stem, 20μm parasagittal paraffin section. Adaptation of Gomori's lead stain for acid phosphatase activity (pH5). Contact microradiograph. Mag × 19. (See also Figs. 36 and 37.)

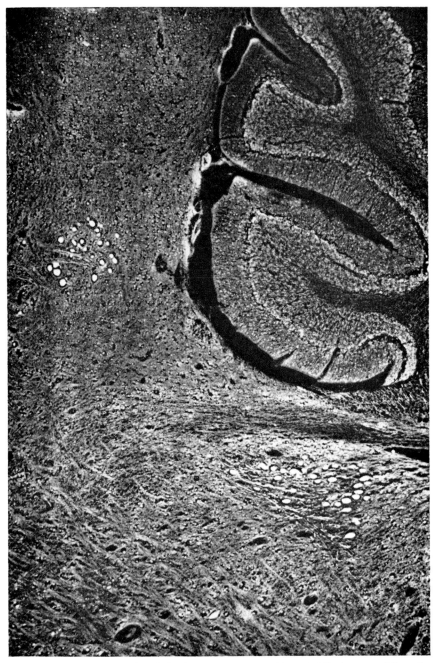

Figure 36. Rat brain stem, detail from section seen in Figure 35, showing folia of cerebellum, and two groups of cells of the mesencephalic nucleus of cranial nerve V (heavily stained with lead and therefore intensely radiopaque). Contact microradiograph. Mag. × 48.

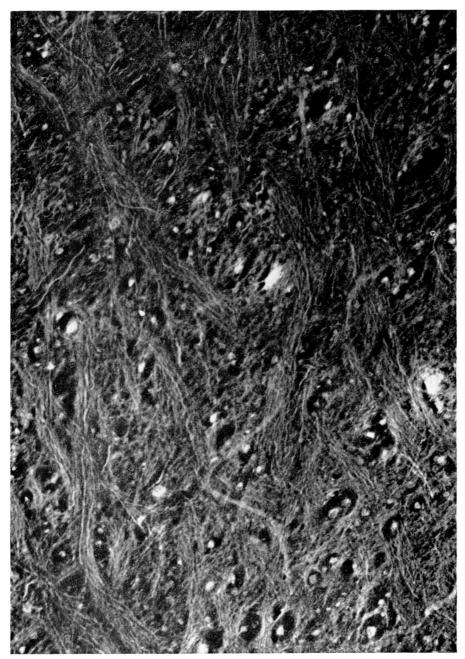

Figure 37. Rat brain stem, further detail from section in Figure 35, showing nerve cells scattered in the woven pattern of nerve fibers. Contact microradiograph. Mag. × 200.

Figure 38. Human brain, hypothalamus, 35μm frozen section. Large and small nerve cells intensely stained with lead via an acid phosphatase reaction (pH 5). Vessel fragments lightly stained with lead via an alkaline phosphatase reaction (pH 9), modified from Gomori. Contact microradiograph. Mag. × 49.

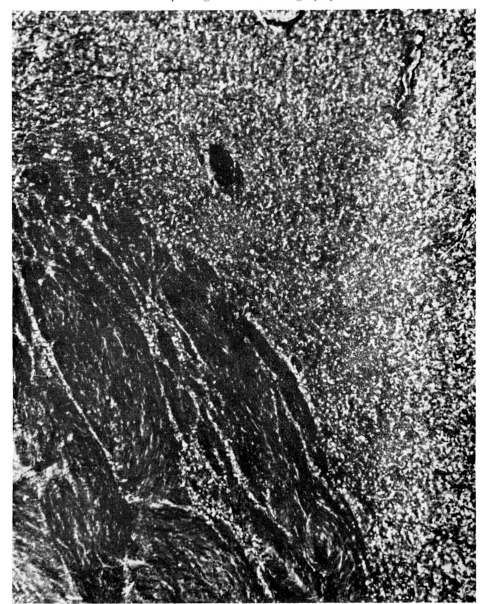

Figure 39. Human brain, hypothalamus, 35μm frozen section. Acid phosphatase reaction (pH 5). Contrast intensely lead stained cellular elements (white, radiopaque) and unstained fiber bundles (dark, radiolucent). Projection microradiograph. Mag. × 30.

strated bronchial cartilage undergoing calcification, and bronchial cilia with the characteristic phospholipid line at their base. The projection method has also been used to demonstrate the location of tin within a tin miner's lung (Saunders, 1957c), and also the characteristic perivascular

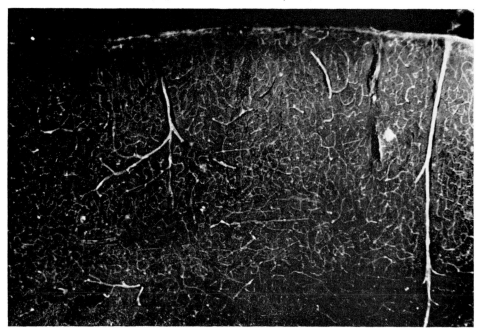

Figure 40. Human brain, sensory cortex. 100μm frozen section. Vessels stained with lead using alkaline phosphatase reaction (pH 9). Vertical arterioles, horizontal arterioles, and capillaries are obvious; a few palely stained horizontal venules can be discerned. Contact microradiograph. Mag. × 36.

disposition of iron particles from a case of pulmonary siderosis in an electrical welder (Angervall, Hansson, and Röckert, 1960).

The contact and projection methods of historadiography have been used to study sections of nervous tissue stained by metallic impregnation following the classical techniques of Golgi and Cajal. For example, historadiographs of the cerebellar cortex stained by the Cajal method clearly show the pear-shaped Purkinje cells with their dendrites branching within the molecular layer, and subjacent to them the small round cells of the granular layer (Fig. 34). But while historadiography can rapidly and accurately record the general and detailed features of tissue sections prepared in this way, such staining techniques are both time-consuming and capricious.

Recently it has been found that certain of the modern precisely controlled histochemical technics, or modifications of these, can be effectively used to produce radiopacity in a desired location for study by historadiography. For example, use of the phosphatase reactions has made it possible to demonstrate various collections of nerve cells (or nuclei) and fiber patterns within the brain stem of the rat, as well as within the human hypothalamus (Bell, 1969) (Figs. 35 to 39).

Histochemical reactions have also been used to increase the contrast of nerve cells and capillaries in the human brain, to permit their simultaneous

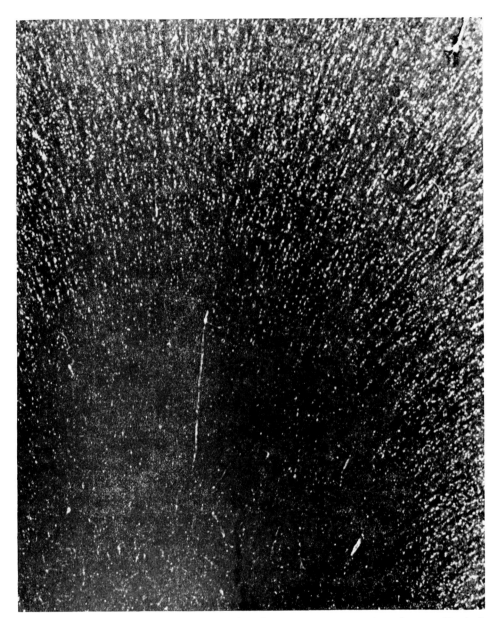

Figure 41. Human brain, motor cortex. 35μm frozen transverse section of gyrus. Vertical columns of nerve cells stained with lead using acid phosphatase reaction (pH 5). Contact microradiograph. Mag. × 53.

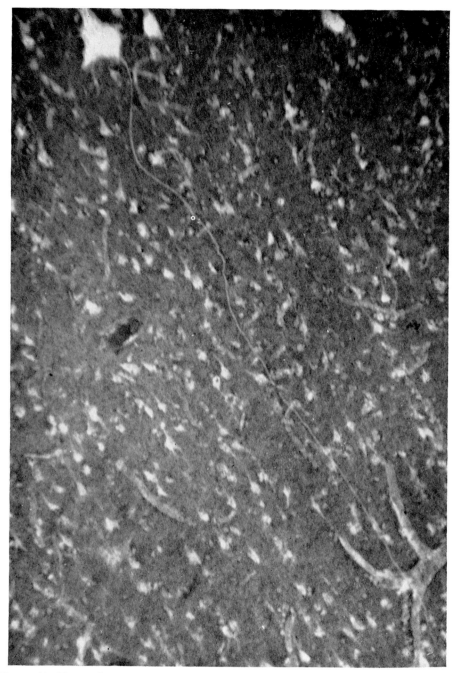

Figure 42. Human brain, motor cortex. 35μm frozen section. Nerve cells stained with lead by the acid phosphatase reaction (pH 5). The long axonal process of the largest nerve cell can be followed for some distance, past a small blood vessel lightly stained with lead by a second reaction for alkaline phosphatase activity (pH 9). Projection microradiograph. Mag. × 200.

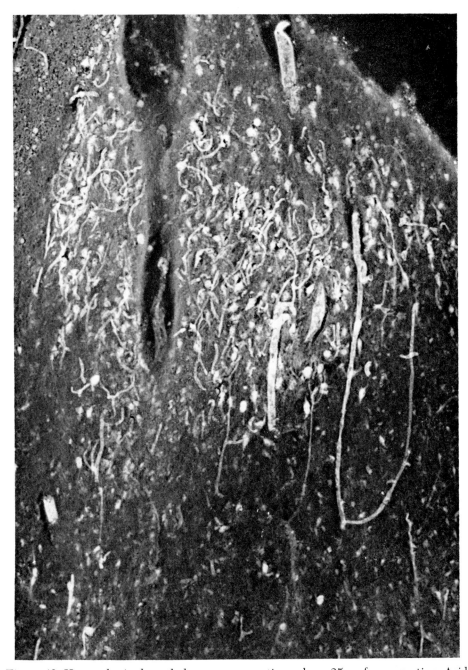

Figure 43. Human brain, hypothalamus, supra optic nucleus. 35μm frozen section. Acid phosphatase activity of nerve cells and alkaline phosphatase activity of blood vessels simultaneously stained with lead. A pH of 7 was used, halfway between the optima of the two enzymes (ph 5 and 9). Contact microradiograph. Mag. × 90.

demonstration by contact or projection x-ray microscopy (Saunders, Bell and Carvalho, 1969). No injection method can guarantee complete filling of a microcirculatory bed, hence it was decided to selectively stain the cortical capillary vessels specially for x-ray microscopy. A modification of the classic Gomori stain for alkaline phosphatase activity was used, resulting in the deposit of a black, radiopaque, lead sulfide within the vessel walls. Because the small arteries and veins in the brain exhibit a different degree of phosphatase activity there was consequently a difference in their radiopacity. The arteries stained more darkly and hence were more x-ray absorbent; thus the arteries could be readily distinguished from the fainter and less x-ray absorbent veins (Fig. 40).

Since the majority of nerve cells contain the enzyme acid phosphatase, another Gomori technique was used to bring about the deposition of lead sulfide at the site of activity of this enzyme, so making it possible to secure an x-ray picture of the nerve cell population. Incubation times however had to be increased to obtain sufficient metal deposit to produce a useful x-ray image, thereby sacrificing the delicate localization of enzyme activity for which the techniques were originally developed. X-ray micrographs taken of such sections demonstrated that the elementary organization of the human motor cortex consists of nerve cells oriented vertically to the cortical surface (Fig. 41).

After first recording the cortical blood vessels, and then the nerve cells alone, the next step taken was to evoke the acid and alkaline phosphatase reactions in the same brain section, with either a compromise pH or a sequence of two reactions at different pH's, so that the vessels and neurones could be studied together. This combined technique made it possible to secure x-ray pictures of the nerve cells and their processes, and also illustrate the intimate and important nerve cell—capillary (neuronovascular) relationship (Figs. 42 and 43).

ACKNOWLEDGMENTS

The author gratefully acknowledges the help and guidance of Dr. V. E. Cosslett, Dr. W. C. Nixon and Dr. J. V. P. Long of the Cavendish Laboratory, Cambridge University, during the planning and testing of our projection x-ray microscopes. Similarly, thanks are due to Mr. Raymond V. Ely, F. Inst. E.E., Consultant in X-ray Microscopy, Harwell, England, for his advice, generosity and expert supervision during the construction of these instruments. All of these also instructed me in the essential physics and practice of x-ray microscopy—a debt that cannot be repaid.

Thanks are due Miss Rosalie Frye, B.Sc. (Physics) (Cantab.) who assisted with the installation and testing of our projection x-ray microscope, first at Cambridge and then at Dalhousie University. For improvements in instrumental techniques I wish to thank Mr. Richard James, Mr. Stanley Sieniewicz and Dr. Leonard van der Zwan.

Special thanks are due my research technician, Mr. Victor R. Carvalho, who over the past decade has spared no effort to become expert in the management and maintenance of our contact and projection x-ray equipment, develop the necessary photographic techniques for x-ray microscopy, and apply x-ray microscopy to biological, dental and medical problems. Likewise I appreciate the careful attention given by Mrs. Pat Troyer in recording scout microradiographs, attending to specimens and assisting with photographic processing and printing.

I am indebted to my research assistant, Miss Mary A. Bell, for permission to use Figures 35 to 43, derived from her neuroanatomical and histochemical researches. Similarly, I wish to thank Professor F. W. Fyfe for permission to use Figure 29, obtained from his research on bone, and Professor Arne Engström of the Karolinska Institutet, Stockholm, Sweden, for Figure 28, obtained from his general survey on contact microradiography that appeared in *X-ray Microscopy and Microradiography* in 1957 (Academic Press, New York).

I am indebted to the following for permission to reproduce X-ray micrographs that have appeared in various journals and books: Damancy & Co., Bath, England for Figure 8, which was prepared for their brochure on colored radiopaques or Chromopaques; Grune & Stratton, New York for Figure 13, that appeared in *The Peripheral Circulation in Health and Disease* by W. Redisch, F. F. Tangco, and R. L. de C. H. Saunders (1957); the *Canadian Dental Association Journal* for Figures 14 and 15 that appeared in that journal (33:245–252, 1967); the Academic Press for Figures 16 and 17 that appeared in *X-ray Optics and X-ray Microanalysis* (1963) (Eds. H. H. Pattee, V. E. Cosslett, and Arne Engstrom); the Royal Microscopical Society for Figures 19 to 22 that appeared in their journal (83:55–62, 1964); the Elsevier Publishing Co., Amsterdam, for Figures 23 to 25 and 30 to 32 that appeared in *X-ray Microscopy and Microradiography* (1960) (Eds. Arne Engström, V. E. Cosslett, and H. H. Pattee); Hermann, Paris for Figure 27 that appeared in *X-ray Optics and Microanalysis* (1966) (Eds. R. Castaing, P. Deschamps, and J. Philibert); and Springer Verlag, Heidelberg, for Figures 40 to 42 that appeared in the *Vth International Congress on X-ray Optics and Microanalysis* (1969) (Eds. G. Mollenstedt and K. H. Gaukler).

Finally I take much pleasure in acknowledging the helpful suggestions and encouragement of Dr. W. H. Feindel and Dr. Lucas Yamamoto of the Montreal Neurological Institute, which markedly influenced our approach to x-ray microscopy of the human cerebral microcirculation.

Thanks are also due Miss Lesley Gentry, the departmental secretary, for the time and care she devoted to typing and correcting the manuscript.

CHAPTER 3

MICROANALYSES BY MEANS OF X-RAY ABSORPTION AND X-RAY FLUORESCENCE

X-RAY MICROSCOPY has become a medical tool which is capable of analyzing different components of the body tissues. Due to its nondestructive nature it also enables the investigator to make several analyses of the same specimen. Some problems may be solved by x-ray microscopy alone but many of the medical problems are such of the nature that only a combination of different analyses and tests can lead to the right conclusions. Therefore x-ray microscopical analysis has become very useful in medicine since it offers a possibility to correlate chemical data with micromorphology. The technique of x-ray absorption and x-ray fluorescence makes it possible to perform quantitative analyses in cells or sections without destroying the morphology. Generally however some sort of specimen preparation is required.

Several hundred biological papers have been published in the field of x-ray absorption and x-ray fluorescence. The papers discussed below in this chapter are chosen as examples and all have in common that they represent subjects of clinical interest or show methodological aspects of principal importance or are easy to adopt for work with clinical and/or experimental research projects.

3.A. X-RAY ABSORPTION ANALYSIS

The absorption of x-rays in a medium can be calculated from the equation

$$I = I_o e^{-(\mu/\rho)m} \tag{3-1}$$

where I is the transmitted x-ray intensity, I_o is the incident x-ray intensity, e is the base of Napierian logarithms, (μ/ρ) is the mass absorption coefficient expressed in $cm^2\ g^{-1}$ and m is the mass per unit area of the specimen in $g\ cm^{-2}$ (see Appendix, Sec. A.4).

When several absorbing components occur in the specimen, the x-ray transmission can be written

$$\frac{I}{I_o} = e^{-[(\mu/\rho)_1 m_1 + (\mu/\rho)_2 m_2 + \cdots]} \tag{3-2}$$

where the mass per unit area of each component is m_1, m_2, etc. The absorption curves for different elements at different wavelengths show sudden jumps due to electron transitions between different orbits. These sudden jumps occur always at the same wavelengths for a given element, and they are called the K, L, M . . . absorption edges of the element (see Fig. 44).

The equations above refer to monochromatic radiation. Generally it is difficult to obtain such radiation but the composition of the specimen often permits polychromatic or "white" radiation to be used for analysis. Bone, for example, contains a minor part of organic constituents. Wallgren (1957b) has calculated "that between 0.5 and 3 Å the mass absorption coefficient of apatite is roughly ten times that of water or protein. In this wavelength range, therefore, the major part of the x-ray absorption in a bone sample is caused by the mineral compartment, and the microradiographic images formed by radiation of this wavelength will thus demonstrate the distribution of mineral elements in the sample." For quantitative work absolute values of local concentration ρ (i.e. mass per unit volume) can easily be calculated when the absorption is measured over a spot with known thickness, t, since the measurement of absorption determines m and $\rho = m/t$.

In practice a step wedge of a material with similar absorption properties is often used as a reference system. This system is placed close beside the specimen and exposed at the same time.

Another way of performing quantitative analyses on mineralized tissues is by using a curved crystal which makes it possible to select x-radiation of different wavelengths. By measuring the absorption of x-rays on each side of the K-absorption edge, e.g. Röckert (1958) calculated the amount of calcium present in the specimen. This will be discussed in more detail later in this chapter.

With soft x-radiation, e.g. in the wavelength region 8 to 13 Å, the absorption in typical soft tissues depends only on the mass of the specimen. This has been shown by Engström and Lindström in 1950. Their calculations made it possible for many research workers not only to determine the mass, e.g. in a single cell or in micro areas in tissue sections, but also to analyze the amounts of organic components extracted from the tissues. In this way analyses of lipids and RNA have been performed. The determination of the dry mass is first made in the intact specimen. After extraction of lipids or RNA the determination is repeated and the difference represents the amount extracted (Hiroaka and Glick, 1963; Hydén, 1959).

For problems of interest in clinical medicine, x-ray absorption has mainly been used either on mineralized tissues or on organic tissues for mass deter-

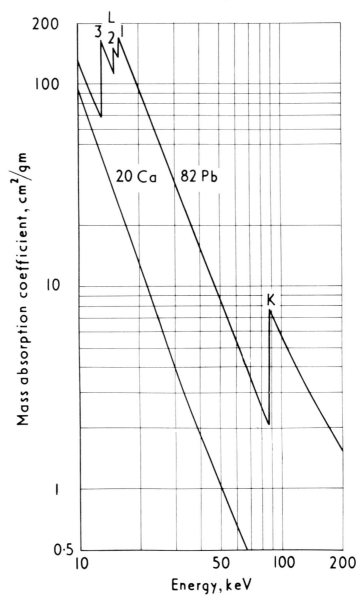

Figure 44. Variation of mass absorption coefficient of lead $(Z = 82)$ and calcium $(Z = 20)$ with energy of incident radiation. (After Omnell: *Acta Radiol*, 1957.)

mination. In order to show the possibilities of the method some examples of this type of investigation will be discussed in this chapter.

3.A.1. Specimen Preparation

For *mineralized tissues* where the minerals are to be analyzed, grinding of the material is the best way to prepare the tissue for x-ray absorption studies. For analysis of the content of minerals in bone, dentine and cementum the optimal thickness for x-ray absorption analysis with contact microradiography is within the thickness range 75μm to 100μm in order to produce optimal contrast on the film. For recording of the transmitted radiation with counters the intensity of the beam used will determine the best thickness of the specimen. It is of course easier to prepare a thick specimen than a thin one. During the grinding of the mineralized tissue the organic matrix on the surface will be destroyed. Sometimes, although more seldom, a bigger part of it in the section will be destroyed by the embedding procedure. Generally, however, it is of minor importance and can for many problems be entirely neglected. If corrections for the organic constituents are desired this can be done in several ways depending on the tissue studied. Many times decalcification of a ground section is possible and measurements of the absorption of the organic matrix in the ultrasoft x-ray region will provide the information wanted. As mentioned, the organic components in the surface layers generally will be destroyed during grinding. Absorption measurements on decalcified ground sections therefore will be reliable only if the sections are so thick that destruction of the surface layer becomes relatively unimportant. Decalcification of bigger tissue blocks followed by weighing can only be accepted in investigations where mean values are sufficient. Generally there are great local variations in the distribution of the organic matrix, e.g. in compact bone and dentine. In dental cementum the corresponding distribution is much more even.

The ideal way of preparing ground sections is of course without an embedding material. This is often possible if smaller parts of the interior of a section are to be analyzed like central parts of dentine or central parts of a bone section. If the marginal parts of a specimen are to be analyzed, e.g. cementum, they will be destroyed by most grinding procedures if an embedding material is not used. The ideal embedding material has the same hardness as the tissue. This is often difficult to obtain but methyl methacrylate and araldite are the most commonly used embedding materials for mineralized tissues. One of the best ways of obtaining consistently good embedding results with methyl methacrylate was described by Ericsson in 1965. He embeds his tissue blocks inside a methyl methacrylate tube in a simple way by using azo-diiso-butyronitrile as catalyzer (for details, see Ericsson, 1965).

The absorption of the embedding material has to be considered in each case depending on the type of tissue studied, but for many determinations of minerals it is negligible. The absorption depends on atomic number and the absorption of the constituents of methyl methacrylate, C, N, O and H, is minute.

Many grinding methods have been presented in the literature and they fulfil different requirements depending on the type of problem dealt with.

Many factors must be considered in the choice of grinding method, e.g. demands of capacity, planarity, plane parallelity and cost.

In principle three types of grinding methods have been developed:

1. Grinding by hand against a fixed grinding surface.
2. Grinding by hand against a rotating grinding surface.
3. Grinding against a rotating grinding surface with the specimen mounted on a fixture.

The first two types generally are cheaper but have a lower capacity than the third type. Better plane parallelity and capacity for grinding a series of specimens are some of the advantages of the third type. (For detailed information and survey of the literature cf. Atkinson, 1952, Hallén and Röckert, 1960; Ericsson, 1965; Sundström, 1966.)

In studies with projection x-ray microscopy where the morphological information may play the most important role it is not always necessary to prepare the specimen specially, as the point-focus of the x-ray source will produce a sharp image of all parts of an irregularly shaped specimen (see Röckert *et al.*, 1965).

For determinations of *dry mass*, microtome sections of organic tissues can be used. In paraffin embedded sections the paraffin has to be removed by chloroform. The use of frozen-dried material is often preferable as no further treatment of the material is necessary. The specimen is then sectioned frozen in a cryostat and the sections are immediately frozen-dried under vacuum at $-80°C$. The morphological orientation in such a specimen may however be difficult. A third way of preparing the material for mass determination is by using microdissection for the isolation of single cells with fine needles under the binocular microscope preferably by hand without a micromanipulator which would require a fluid working medium.

3.A.2. Thickness Determinations

In order to obtain certain quantitative data for local areas in tissues it is necessary to know the volume of the analyzed piece. The area analyzed is determined by the aperture of the photometer or counter or by the collimation of the radiation. The third dimension is the thickness of the specimen in the analyzed area. If semiquantitative estimations over larger section areas are made it is important to know that densitometric variations on the microradiograms are not due to thickness variations but to true elementary variations.

For quantitative determinations mineralized sections can be broken just at the measured point and the edge thickness measured in a profile microscope. The disadvantage of this method is its destructive nature. Soft sections or isolated cells as well as mineralized tissues can be measured with an optical-mechanical device which in principle works so that a shadow of a hair is focussed on top of the specimen in a microscope. A similar focussing is made beside the specimen on the plane supporting glass. The movement of the microscope controlled by the micrometer screw is measured with a microcator. This measurement is very accurate in tenths of a micron provided the contact between the specimen and the plane supporting glass is perfect. For further details see Hallén (1956).

3.A.3. X-ray Absorption Determinations in Mineralized Tissues

The x-ray absorption techniques have been used in combination with other techniques in many of the examples discussed below in order to obtain a maximum of significant information from the material. All investigations discussed here have in common the purpose of micromorphological correlation to data concerning the calcium or mineral content. In many of these x-ray microscopical investigations additional information was obtained from autoradiography, x-ray diffraction and tetracycline labelling.

Studies on the formation and structure of human fetal bone (Wallgren, 1957) were performed with contact microradiography in combination with x-ray diffraction and polarized light microscopy. Fetal femurs from twenty-six individuals ranging in age from 6.5 weeks to full term were studied. According to visual microscopical observations of the microradiograms, the youngest femur which showed evidence of primary ossification was a specimen from a 9.5-week-old embryo. The cartilage models were not completely surrounded by the newly formed primary periosteal collar and there were many discontinuities along its circumference apart from foramina nutritia. Two types of bone seem to build up the collar: An inner layer which is fairly homogenous, well-mineralized and containing few osteocytic lacunae, and an outer layer which is more cell rich and less mineralized than the other. Visual studies of the microradiograms of fetal femurs of various ages show how the primary periosteal collar remains intact and completes the surrounding of the bone marrow. "The original innermost layer attains a level of mineralization higher than that of the rest of the collar. Using this hypercalcified layer as a structural landmark, it is seen that the collar maintains its structural integrity throughout the greatest portion of the intra-uterine life of the bone." From the microradiograms the periosteal and endosteal bone growth was evident. The endosteal bone growth showed a less mineralized pattern and a slower rate in the appositional growth in comparison to the periosteal bone.

The periosteal bone later showed a structure of concentric series evenly spaced but discontinuous. There was a difference in structure of the periosteal bone compared to the primary collar. "At its first appearance, each layer is made up of a rather cellular, lacunar type of bone, the mineral content of which is quite high. During later development, each of these highly mineralized and cellular structures acts as a central core about which are

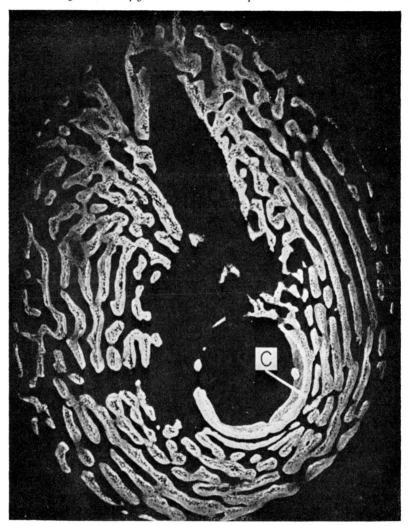

Figure 45. Microradiogram of the mid-diaphyseal cross section of a femur from a 24-week-old fetus. C = the primary periosteal collar (after Wallgren: *Acta Paediatr*, 1957).

applied new layers of less cellular and less mineralized bone." Later the spaces between the cores become obliterated by the formation of new peripheral coatings of bone (see Fig. 45).

Shortly before birth the cores and their surrounding bone become replaced by newly formed Haversian systems with a relatively low mineral content.

The start of endochondral bone formation and mineralization can be seen microradiographically as tiny spicules between the hypertrophied cells. The spicules gradually grow and their degree of mineralization becomes higher than that of the surrounding bone.

After microscopic examination of the microradiograms, Wallgren classifies the different types of bone according to their mineral content as (a) newly formed bone possessing a rather low and inhomogenous degree of mineralization, (b) older bone, which constitutes by far the greatest share of the tissue, and which exhibits a high and homogenous level of mineralization and (c) hypermineralized bone and calcified cartilage which appear to act as centers about which new bone is laid down. The quantitative results were obtained by the determination of I and I_o and the thickness of the specimen. The mass absorption coefficient $\left(\frac{\mu}{\rho}\right)$ is constant. Any changes in the value of the linear absorption coefficient (μ) reflect changes in the specimen density (ρ), i.e. the degree of mineralization. By means of the slide-rule device of Lundberg and Henke (1956) calculations were made of the linear absorption coefficient from the observed densitometric and thickness measurements. The densitometric determinations were performed in a microphotometer. The density was kept within the range in which photographic response was linearly related to x-ray intensity (0.5 to 0.7). The quantitative determinations were expressed as mean linear absorption coefficients in *mature fetal bone,* in periosteal bone and hypermineralized bone and calcified cartilage together with similar figures for adult bone. In fetal periosteal bone the density ρ is about 16 percent lower than in adult bone. The hypermineralized bone and calcified cartilage however showed a mineralization level 5 to 6 percent higher than that of the adult.

In *developing fetal bone* cross sections were studied of mid-diaphysis from the femurs of individuals representing different stages of fetal growth. The linear absorption coefficient showed an increase with increasing fetal length initially, but after the fifteenth week the degree of mineralization reached a steady level (Fig. 46).

Corresponding measurements were performed on longitudinal sections of fetal femurs at different distances from the epiphyseal tip.

Figure 47 shows the mineralization in the distal epiphyseal end of the primary periosteal collar. Values of the linear absorption coefficient are plotted against bone age as determined by the growth rate of the distal epiphysis.

X-ray diffraction of the same material showed among other things a pattern characteristic of calcium-hydroxy-apatite. The long dimension of the apatite crystallites increased until at full term, the crystallites were about 220 Å long, about the same as in adult bone. The initial growth of the long dimension is very similar to the rapid increase in mineral content, so that the rate of increase in mineral content appeared to be a function of the rate of increase in crystallite size. The factors that cause the crystallite growth to cease seem to be properties of the organic matrix which also seem to determine the crystallite orientation.

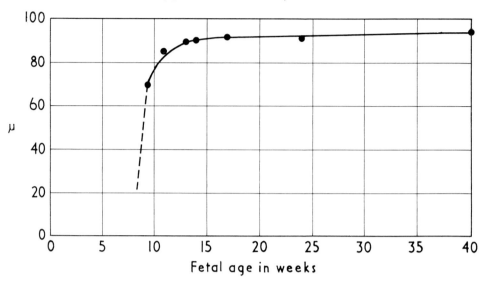

Figure 46. Mineralization of the primary periosteal collar as measured on cross sections from the mid-diaphyseal part of the femur. Values for the linear absorption coefficient for each sample are plotted against fetal age. The rapidity with which a high and nearly constant level of mineralization is attained is clearly demonstrated. (After Wallgren: *Acta Paediatr*, 1957.)

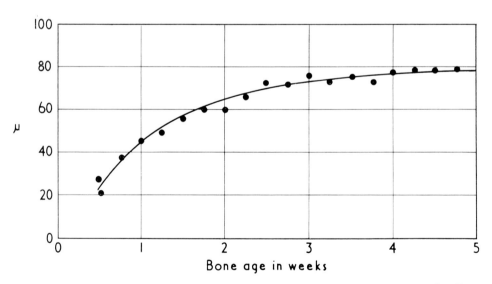

Figure 47. Mineralization in the distal epiphyseal end of the primary periosteal collar. Values of the linear absorption coefficient are plotted against bone age as determined by the growth rate of the distal epiphysis. During the period of most rapid increase in mineral content, the values shown here are in good agreement with those plotted in Figure 46. (After Wallgren: *Acta Paediatr*, 1957.)

X-ray microscopy has also been a useful tool in studies of the fate of *bone transplants and bone implants*. It is possible to obtain samples of human material during surgery, especially if repeated operations are necessary in the same area. Due to the very small amounts required for analyses there are practically never any problems of obtaining a sufficient amount of material.

Holmstrand in 1957 has combined contact microradiography with autoradiography and x-ray microdiffraction on samples from transplants and implants of fifty-six rabbits and two dogs. Cortical bone transplants from tibia were turned 90° and moved to another place in the same bone. Cancellous bone was taken from iliac wing and moved to a prepared bed in the tibia. The microradiograms of the specimens were studied both visually in the optical microscope and quantitatively with a photometer in tissues ranging from one to sixty-four weeks after the operation. From the visual examination it could be concluded "that the manner of healing from cortical transplants and implants was similar, but in their resorption and substitution with new bone within pre-existing Haversian canals they differed. The relative resorption-substitution rate was in the following order: autogenous, homogenous and frozen bone, while for cooked and glycerol ashed bone the rate was very slow." For autogenous cortical transplants there was a difference in the resorption rate of the transplants in young and old rabbits. Younger rabbits showed a considerably larger number of resorption cavities. After operation (53 to 64 weeks) parts of the cortical autogenous transplant had not been absorbed. The callus had not reached the high degree of mineralization which could be seen in the host bone and in the remaining parts of the transplants.

During the entire period of observation of the rebuilding process of autogenous cancellous transplants in the callus, there was formation of Haversian systems containing different amounts of mineral salts, and the majority of the Haversian canals were oriented longitudinally in the leg. Peripheral parts of the callus formation of lamellar bone showed a formation of lamellar bone resembling circumferential lamellae.

The healing of cancellous transplants and implants took place by callus occupying the spaces between the trabeculae which were in turn resorbed. Rebuilding processes in the callus resulted in the operated area assuming the appearance of cortical bone. The relative resorption rates were about the same for autogenous, homogenous and frozen bone, but rather slow for os purum and extremely slow for calcined bone.

As in Wallgren's work (1957b) Holmstrand's quantitative determinations were expressed as linear absorption coefficients (μ). The quantitative evaluations showed an average increase in the mineralization of about 2 percent in ten weeks, and the average time for "full" mineralization of the callus in the rabbit series was ninety weeks. No significant difference in the mineral content of the callus for autogenous and homogenous transplants was obtained. The mineralization of unresorbed parts of the transplants showed major variations. Apart from rebuilding sites in the host bone close to the transplant, the average content of mineral salts in the evenly mineralized areas of the host bone did not undergo any changes as compared with determinations made in the unoperated leg.

From the x-ray microdiffraction investigation on the same material Holmstrand could deduce the independence of mechanical forces in the ultrastructural organization of the bone replacing a transplant. The average crystallite length was 245 Å.

The autoradiographic investigation revealed a direct relationship between the uptake of Sr^{90} and the resorption rate of a transplant, due basically to a specificity in the ultrastructural organization.

X-ray microscopical studies of *human* autogenous bone grafts were performed by Breine, Johansson and Röckert (1965). The x-ray microscopical technique was combined with ultraviolet fluorescence analyses after tetracycline administration. In twenty-five cases of *cheilo-gnato-palatoschisis* the defect in the hard palate and alveolar process was filled with autogenous cancellous bone from the tibia. A sample of bone from the donor site was taken for microscopic investigation. This operation was performed at about six months of age (5 to 15 months).

In seventeen of the cases biopsies from the grafts were taken at the next stage of repair five to nine months later. Nine of these cases were given 0.75 g oxytetracycline (Terramycin, Pfizer) up to four times in intervals of two weeks starting one to two weeks after the primary operation. Contact microradiograms were made using the wave length region 2.5 to 5 Å (Be-filtered tungsten radiation). Control material for x-ray microscopy was obtained by autopsy from the palates of sixteen subjects without cleft within the age group four to sixteen months, all of whom died of acute diseases without signs of metabolic disturbances affecting the skeleton.

Clinically autogenous bone grafting to primary clefts gave a hard palate and alveolar ridge. Macroscopic clinical x-ray examination showed that atfer 2.5 months the graft seemed well incorporated in the surrounding bone. All of the x-ray microscopical pictures of biopsies from the graft five to nine months after the transplantation had an appearance in the microscope very similar to the normal healthy hard palate and quite different from the microradiograms of the tibia (see Fig. 48).

The variation in mineralization within each section of the biopsies was of the same degree as in the control material, thus demonstrating the same distribution of the de- and remineralizing process. This was interpreted as a sign of normal metabolic activity and an adaptation to the surrounding tissue demands and an indication that no dead material was filling the cleft.

The ultraviolet fluorescence studies showed that two weeks after grafting the tetracyclines were already incorporated. In cases where several dosages were given about two weeks apart several fluorescent lines at about equal distance illustrated the speed and evenness of the mineralization process.

In an investigation like this the surgeons who worked out a satisfactory and functional operative method with clinically good results could use x-ray microscopy as a tool to check the biological correctness in their clinical work. Tetracycline analysis in ultraviolet light then becomes a very suitable method for clinical work of this kind where the use of radioactive isotopes is not wanted.

The nature and behavior of the mineral phase in *healing of fractures* was studied by Nilsonne in 1959. He used both projection x-ray microscopy and

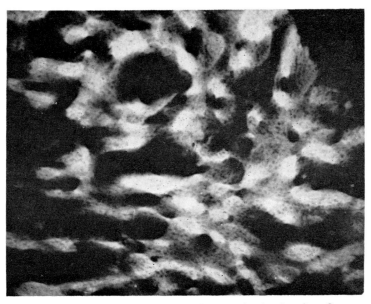

Figure 48. Microradiogram of part of a transplant in the hard palate five months after transplantation. Patient's age at the time of biopsy was one year (orig. mag. × 64).

contact microradiography in combination with x-ray microdiffraction and autoradiography. His material consisted of rats and dogs. In the anaesthetized rats, closed fractures were obtained as diaphyseal fractures by breaking the right tibia between the fingers. The clinical after-effects of the injuries appeared to be small, and after only a day or two the leg was back in normal use. Three series of rats were studied: (a) twenty rats, two of which were sacrificed each week after the day of fracturing; (b) five rats which were killed at set intervals during a period of twenty-one months; (c) nine rats sacrificed in close succession during a fourteen day period commencing immediately after the fracturing.

The study on dogs was performed on ten animals after breaking the anaesthetized dog's femur against a blunt edge with a heavy hammer. "In this way oblique to transverse fractures were obtained, no crush or comminute fractures were observed." After twelve to sixteen days the fractured leg was in a "clinically" full function. The animals were sacrificed at fourteen days to twenty-five months after the fracturing.

In both the x-ray microscopical methods used, the recordings were made on films and the quantitative evaluations were performed with a photometric procedure. Nilsonne has expressed his quantitative results in linear absorption coefficients. For the autoradiographic investigations the rats had been given a carrier-free $Sr^{90}Cl_2$ solution intraperitoneally three days before the animal was killed. By comparison of the auto- and microradiogram of the same specimen it was possible to correlate the localization of the isotope with the distribution of the mineral salts in the callus region.

In the diaphyseal fractures of the tibia in rats the first traces of mineral salt in the callus could be seen as early as on the fourth day after the fracture. They appeared as small bone salt deposits in the periosteal cuff. The periosteal reaction often takes place along the whole length of the fractured bone. After one week the periosteal cuff is sparsely mineralized, "but after three weeks the bone salt accumulation has greatly increased, particularly near the fracture ends." A varying degree of mineralization was seen in the periosteum. Resorption cavities could be seen at the fracture ends. The intermediately situated callus showed trabeculae which decreased in size and number from the fracture ends. Even after ten weeks the mineralized callus was not continuously spanned at all points between the fracture ends. At the fourteenth week a mineralized connection was seen between the two periosteal callus halves. Lamellar bone with Haversian systems of a low degree of mineralization appeared at this time. In the further development a continuous building and rebuilding of the bone trabeculae was observed.

The autoradiograms showed an intense uptake of Sr^{90} in the least mineralized parts of the callus. The isotope was especially concentrated in the periosteal callus cuff. Even in the endosteum there was a lesser isotope accumulation, which increased towards the fracture ends. As for the callus it was observed in general that the higher the degree of mineralization, the weaker was the uptake of isotope.

For the quantitative measurements, the problem arose of how to detect the earliest mineral deposits. This was accomplished by the use of CrK radiation, which has a higher absorption in the bone salt than CuK radiation. The quantitative data expressed in linear absorption coefficients showed that at the fracture ends there was a temporary demineralization during the first five days, followed by a remineralization (Fig. 49).

The increase in degree of mineralization in the first five weeks in fracture ends, periosteal callus and intermediated callus resembles the corresponding curves by Holmstrand (1957). The curves are shown in Figure 50.

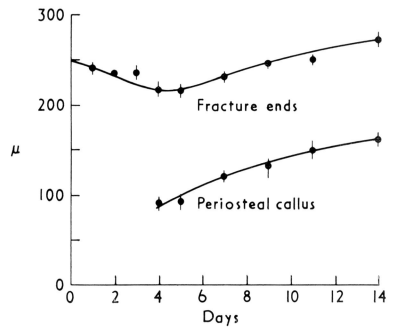

Figure 49. Mineralization in the early stages of fracture repair in the tibia of rats as represented by the linear absorption coefficients (μ) of the bone salt for CrK$_a$ radiation. (After Nilsonne: *Acta Orthop Scand*, 1959.)

Nilsonne found that "an analysis of the measurements for the periosteal callus shows that this mineralized faster in the part nearest the fracture ends than in the part near the callus center. From the dynamic point of view this indicates a nonuniform process of mineralization, that is to say the

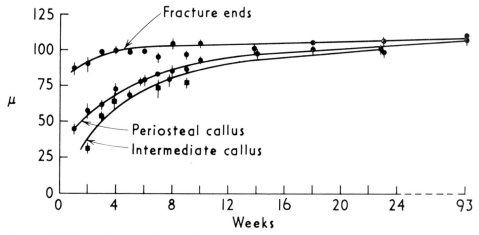

Figure 50. Mineralization of the callus region in the tibia of rats up to a fracture of 93 weeks as represented by the linear absorption coefficients (μ) of the bone salt for CuK radiation. (After Nilsonne: *Acta Orthop Scand*, 1959.)

mineralization quotient between the different levels in the periosteal callus cuff varies at different times."

The diaphyseal fractures of the femurs in dogs showed a mineralization pattern in principle not differing from that found in rats. Normally the dog's skeleton has a more inhomogenous distribution of bone salt compared to rats. The mineralization starts mainly in the periosteal region. The unequal distribution of mineralization was seen even when the rebuilding of the fracture was in its final stage.

The quantitative evaluation of the experiments in dogs showed the quickest increase in mineralization during the first three months. From one month's fracture age onwards the mineralized callus structures were of such dimensions that the inhomogeneity factor mentioned above was negligible. After eleven to twelve months the degree of mineralization reached the level of the fracture ends. The fracture healing had to run for twenty-five months until it reached the degree of mineralization of normal bone.

The x-ray diffraction studies showed that hydroxyapatite occurred in every stage of the development of the callus. The crystallites generally showed a preferred orientation corresponding to the longitudinal direction of rebuilding of the fractured region.

Investigations of a clinical population show that about 10 percent have microscopic *otosclerosis* and about 1 percent have clinical manifestations in the form of hearing loss. The knowledge about the pathogenesis of otosclerosis is very limited. Röckert, Engström, Hallén, Herberts, Lidén, Nordlund and Shea in 1965 have used x-ray microscopy not with the idea of putting forward new hypotheses but rather to make every effort to assemble basic facts. One of the sectors where basic information is lacking is the mineralizing process in otosclerosis. In the investigation mentioned, both contact microradiography and projection x-ray microscopy were used as well as fluorescence microscopy in ultraviolet light after tetracycline administration. In material like this the projection x-ray microscopy has the advantage that no specimen preparation at all is necessary.

The material studied consisted of twenty stapes and 140 fragments of stapes from 123 patients, all operated on for otosclerosis. All the specimens had been obtained from operations where stapedectomy had been performed and where the diseased area had been removed together with its surrounding healthy bone tissue. The patients were one hundred women and forty-three men, between twenty and eighty years old. All of them had shown clinical signs of otosclerosis. As controls, nonotosclerotic stapes from fifteen cases of autopsy material were studied.

For the UV-fluorescence investigation sixty patients had been given 1 g Tetradecin Novum (Astra) in single doses, one week to one month before surgery. Sixty-three patients had not been given any drug at all of this kind, thus serving as a control group.

The x-ray microscopical pictures of healthy parts of the stapes showed a sponge-like internal appearance. Evidently a large amount of organic matrix is present in the footplate. Its porosity gives room for a widely branched network of vessels. One can easily see how the big- and medium-sized vascular channels in the healthy footplate are mainly localized in its peripheral torus.

A larger variation in mineralization was generally seen in the otosclerotic parts. Sometimes, however, a solid, evenly mineralized footplate could be seen. Microradiograms of the crura normally showed two or three vascular channels which generally ran parallel to the crura. The head of the stapes showed the greatest individual variation in mineralization. Some parts were completely demineralized. It is well known that arthrosis deformans occurs quite frequently in this joint (see Fig. 52 and compare with Fig. 51).

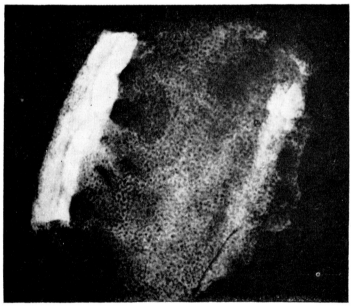

Figure 51. X-ray microscopical picture of part of a healthy human stapedial footplate. Note the mineral content of the limbus region and the vascular channels. The porous appearance of the mineralized parts is also clearly visible (orig. mag. × 70).

The material did not allow any classification of the otosclerosis into different groups, neither based upon the x-ray microscopical picture nor on the age of the patient's history. It was also difficult to estimate whether a less mineralized part in an otosclerotic focus was the result of demineralization or just a poorly mineralized area from the beginning.

The tetracycline technique did not give any information in this material. No difference in incorporation was found between the material treated and the control material. It was also impossible to find any correlation between

the activity of the disease, estimated from case history and the local findings at operation, on one hand, and the degree of fluorescence on the other.

Quantitative contact microradiography has been applied to studies of
osteoporotic bone and compared to normal bone by Jowsey *et al.* (1965).
The investigation also included labelling techniques using yttrium-91 (Y^{91})
and tetracyclines. By the giving of repeated tetracycline dosages to the same
patient and then by studying the fluorescence in the sections in ultraviolet
light, this technique gave information about the bone formation. Y^{91} appeared in high concentrations in areas where resorption is known to occur.

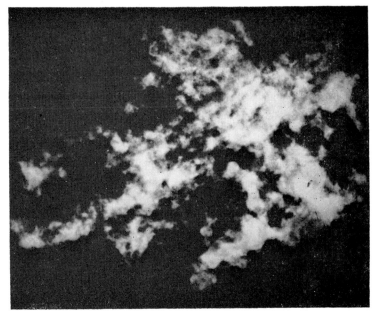

Figure 52. Microradiogram of an otosclerotic fragment, showing an advanced disorgani-
zation of the mineralized component (orig. mag. × 80).

In one investigation the correlation was studied between the surface-
length of bone formation as indicated by microradiographic criteria and by
tetracycline labeling. In another investigation the correlation was studied
between the surface-length of bone resorption obtained by microradiographic criteria and by uptake of Y^{91}. In the first study samples were taken
from the femoral metaphysis of ten people between the ages fifty-six to
eighty-five. In the second investigation samples were taken from the metaphysis and diaphysis of two monkeys six months and five years of age
respectively. In both studies a straight line was obtained at an angle of
about 45° when tetracycline data or autoradiographic data were plotted
against the microradiographic data.

In another investigation the authors used the quantitative microradiographic technique in order to study bone formation and bone resorption in clinical material. Normal bone was obtained from autopsy material from patients who died as a result of traumatic accidents, cardiac infarction, or cerebral hemorrhage. The other group of material was obtained from patients who were clinically diagnosed as having osteoporosis. The osteoporotic criteria were a roentgenographically visible ballooning of the intervertebral bodies and fractures of the femoral neck. Any patient who had been in bed for four days or more was excluded from the study as bed rest reduces the amount of bone formation. "Specimens were taken from midshaft of the femur, the neck of the femur, the lumbar vertebra, the anterior iliac crest one inch above the anterior superior tuberosity, the posterior iliac crest half an inch anterior to the ilium, and the fifth to eighth rib at least three inches from the sternum. In most cases, samples were taken from only two or three sites and in most cases of osteoporosis only a single sample was taken from the femur or the anterior or posterior iliac crest." The material ranged from about fifteen to ninety years of age, the osteoporotic material being obtained of course only from the older subjects.

This investigation showed that normal bone formation and resorption decreases greatly from fifteen to twenty-five years of age. Then they stay at a fairly constant level or increase a few percent continuously toward the older ages. In the rib, the normal resorption does not show the early decrease in the younger groups. In general, early decrease in both formation and resorption in normal bone is of the magnitude of about 3 to 20 percent with the highest value for formation in femur midshaft. In all the types of bone sample studied there was no significant difference between the figures for osteoporotic bone as compared to normal in respect to formation. In the resorption studies, however, each osteoporotic group showed more resorption than the control group. The authors found a significant difference ($P < 0.001$) in normal versus osteoporosis in figures expressed as 8.5 ± 3.1 and 15.7 ± 5.5 percent respectively per unit length of intratissual bone surface. However, in Cushing's syndrome, after corticosteroid therapy, and after immobilization, bone formation decreased.

The method just described for osteoporotic studies has certain disadvantages as a diagnostic tool in clinical medicine. A relatively large piece of bone is needed as small specimens do not give representative results. Generally bone biopsy is too painful for the patient. Biopsy of the rib is relatively painless, but "this bone does not provide satisfactory specimens for study because there is a consistent difference between resorption and formation that results in a slow decrease in cortical thickness after the age of twenty-five, and the differences between normal bone and osteoporotic bone are small."

Röckert in 1958 determined the *calcium content in cementum of teeth*. A Cosslett-Nixon x-ray tube was used. Areas no larger than $75\mu m^2$ were analyzed in the specimens. By the use of a curved pentaerythritol crystal in two positions the transmitted beam could be monochromatized on each side of the K-absorption edge of calcium. By adjustment of the crystal, either of two wavelengths of the beam (3.14 Å and 3.00 Å) could be diffracted into a proportional counter (Fig. 53; cf. also Sec. 5.C.3). From the measured ratio of the intensities of the transmitted beams on each side of the K-absorption edge, the calcium content was calculated from the equation

$$\frac{T_1}{T_2} = e^{-m\,[(\mu/\rho)_{\lambda1} - (\mu/\rho)_{\lambda2}] - m'\,[(\mu/\rho)'_{\lambda1} - (\mu/\rho)'_{\lambda2}]} \tag{3-3}$$

where

T_1 and T_2 are the measured transmissions at the wavelengths λ_1 and λ_2,
$\quad m =$ mass/cm² of calcium,
$\quad m' =$ mass/cm² of the matrix,
$(\mu/\rho)_{\lambda1} =$ mass absorption coefficient of calcium at wavelength λ_1,
$(\mu/\rho)_{\lambda2} =$ mass absorption coefficient of calcium at wavelength λ_2,
$(\mu/\rho)'_{\lambda1} =$ mass absorption coefficient of the matrix at wavelength λ_1, and
$(\mu/\rho)'_{\lambda2} =$ mass absorption coefficient of the matrix at wavelength λ_2.
(Equation 3-3 is deduced in slightly different form and discussed in Sec. 5.A.3.)

The investigation was performed on teeth from eighteen monkeys (Macacus rhesus) of various ages ranging from less than one year to six to seven years. Ground sections were used with a thickness varying between 10μm and 100μm. The calcium contents were measured at three different levels in the cementum of the root.

For the entire material (236 measured points) the mean obtained for the calcium content in the cementum was 0.32 mg Ca/mm³ or, expressed in percent dry weight of the entire cementum, 23% Ca, assuming a value of 2.0 for the density of cementum and a water content of 32 percent. (The optimal working range for the method was from 3.10^{-9} to 4.10^{-10} g of calcium in the area measured.)

In the material as a whole the values varied within a very wide range: 0.10 to 0.83 mg Ca/mm³. A comparison between deciduous and permanent teeth did not show any noticeable difference in calcium content. No systematic difference between different parts of the cementum could be detected in this material with the method used.

The variation in the same age group, i.e. the biological variation, in many cases was larger than the variation between different age groups.

The mean of the calcium content for each kind of tooth, disregarding age, showed only a small variation. In other words, for the group of animals investigated there was no correlation between position in the mouth and the calcium content of the cementum.

In thirty-five old deciduous teeth where the root was progressively resorbed the amount of calcium in the cementum did not deviate from that of
the permanent teeth. There was no systematic difference between the regions in the cementum of old deciduous teeth. This indicates that the demineralization occurring during the resorption seems to be local and not to
be associated with a general decalcifying process.

A comparable contact microradiographic investigation on cementum
from monkeys (Macaca cynomolgus) was performed by Ericsson (1965)

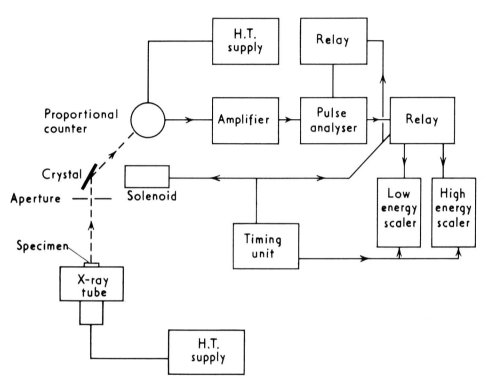

Figure 53. Block diagram of the equipment used for x-ray absorption measurements at
two wave lengths, one on each side of an absorption edge. (After Röckert: *Acta
Odontol Scand,* 1958.)

using a photometric recording principle. He compared teeth with and
without antagonists. The mean value for the mineral content in cementum of the control side was 43.4 percent by volume. There was no
significant difference in the mineral content between the cementum on the
experimental side of the jaw as compared to the control side. "On the other
hand, the mineral content was significantly higher in the cervical cementum
than in the region in the middle of the root and at the apex. Similarly, the
mineral content in the cementum from the middle of the root was significantly higher than that in apical cementum."

3.A.4. X-ray Absorption Determinations of Dry Mass

In 1950 Engström and Lindström published a method for the determination of the mass of structures in single cells by using microabsorption measurements of x-rays. Their theoretical calculations have been very valuable to many research workers especially in the field of quantitative cytology. The method provides measurements *in situ* in a smear of cells or in a microtome section of biological tissue as well as in parts of or in the entire body of an isolated cell. The masses to be determined were of the magnitude 10^{-8}–10^{-13} g.

In their quantitative calculations they assume that in average protein the percentages of the principal elements are: 51% C, 16% N, 24% O, 7% H and about 2% S. Their calculations showed that the wavelength region 7.65 to 12.64 Å is a favorable region where the absorption of other elements in the usual physiological concentrations is considerably less and negligible. The theoretical background is discussed in Section 5.A.2. Rather large variations in the proportions of the elements C, N and O do not affect the final absorption result. They also showed that "Na, Mg, P, S, Cl, K, Ca and Fe may occur in concentrations which are considerably higher than their percentages in biological material in general with the exception of bone tissue, without disturbing the mass determination by x-ray absorption measurements. Other elements with atomic numbers both higher and lower than 30 occur in biological material in such small concentrations that they do not disturb the measurements." Hydrogen however, could be expected to interfere. Engström and Lindström's calculations showed that hydrogen produces a negative error, and the sum of the errors from the other elements is positive, all together generally less than 5 percent. By using a contact microradiographic technique with a step wedge of cellulose nitrate foils, Engström and Lindström could perform quantitative measurements on tissue sections. The wavelengths produced by their equipment were within the calculated optimal region, i.e. about 8 to 13 Å, in the soft x-ray region. The microradiograms showed the mass distribution within the sections and within the cells and could be assessed quantitatively by photometric recording.

This technique has been used by many authors for various biological problems. One of its disadvantages has been the time-consuming procedure which is involved, especially in the photometric evaluation. It has also been necessary, in order to obtain reproducible results, to develop the x-ray films with a standardized procedure.

In order to overcome these disadvantages Rosengren (1959) developed an apparatus based upon Engström and Lindström's theoretical calculations but without using any film. His method is especially suited for measurements of an entire specimen, e.g. whole dissected cells rather than parts of tissue sections. In his method the intensity of the transmitted radiation is measured with a proportional counter (instead of film as in contact micro-radiography). Rosengren's small x-ray tube was operated at about 3 kV; the radiation passed through a 9μm aluminum filter and gave a modal wavelength of about 8.3 Å. In the use of this type of wavelength region, there is the disadvantage of a slight, almost negligible inaccuracy as compared to the use of radiation giving maximum contrast (15 to 20 Å). However, the direct transmittance measurements are not affected by photographic and densitometric errors and will therefore be so much more accurate that the total errors of the mass determination can be kept within reasonable limits. By comparing the transmittance of x-rays through an aperture with and without the specimen the mass of the entire specimen can be obtained by means of Equation 5–37, which is derived and discussed in detail in Section 5.C.1.

It is important that the specimen should not be too thick as too low a transmission would cause considerable errors (see Sec. 5.C.1). Schematic diagrams of the arrangement used by Rosengren are shown in Figures 54 and 55.

Björkerud (1966) has used the mass-determination method of Rosengren in a study of the correlation of succinate oxidation and the mass of micro samples of human arterial intima. Flakes of intima about 500μm in diameter, including the internal elastic membrane, were dissected out from the cystic artery. The samples were obtained from cholecystectomies uncomplicated by inflammation or icterus (Björkerud and Rosengren, 1966). The results of the mass determination are interesting in themselves. In material from eleven patients of varying ages (3 men and 8 women) there ocurred a significant increase of the mass per unit area in the intima with increasing age. The determinations of the succinate oxidation were performed on these small samples using the micro diver technique of Zajicek and Zeuthen (1956). The results could be expressed in consumed microliters O_2/hr/μg. The aim of this investigation was to study the succinoxidase activity as an expression of the metabolic activity of the intima. The results showed a significant decrease of the succinoxidase activity in the intima after fifty years of age among women. The succinoxidase activity occurs in the mitochondria and it is likely that the findings indicate a severe reduction in the energy producing capacity of the intima of women older than fifty years. These findings are especially interesting in view of the known difference between men and women in the development of atherosclerosis. Men show a gradually increasing amount of atherosclerotic change in their arteries from the age of forty-five and upwards. Women on the other hand generally show a delay of ten to fifteen years, until after the menopause, before the atherosclerotic changes rapidly increase (Mitchell and Schwartz, 1965). Another finding that points towards the same factor was reported by Oliver and Boyd (1959) who showed that the relative resistance towards coronary

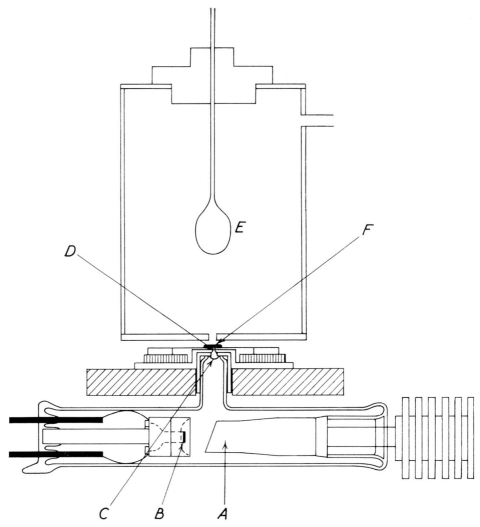

Figure 54. Arrangement for direct absorption measurements in the ultrasoft x-ray region, showing x-ray tube, specimen mount and proportional counter. A = anode. B = filament. C = 9μm Al window. D = specimen stage with aperture. E = high voltage electrode. F = specimen (after Rosengren: *Acta Radiol*, 1959).

infarction among women before the menopause disappeared after bilateral oophorectomy.

Another example will be mentioned where studies of the mass of materials of clinical interest have been performed. The variation of mass among cells from human uterine cervical carcinoma *in situ* was studied by Fitzgerald (1956). He used the technique described by Engström and Lindström. He found in his material no substantial indication "that there is a great increase of mass either cytoplasmic or nuclear in the carcinoma *in situ* cells of human uterine cervical cancer." This finding will reduce the value

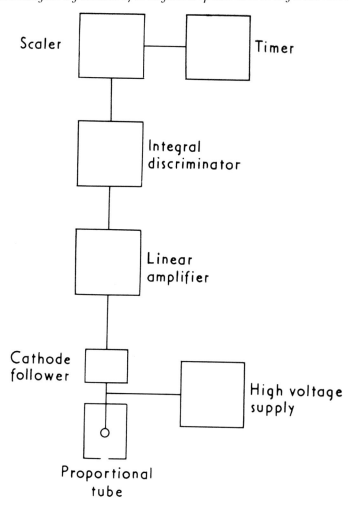

Figure 55. Block diagram of the electronics used for the equipment shown in Figure 54. (After Rosengren: *Acta Radiol*, 1959.)

of the soft x-ray absorption method as a diagnostic tool for this type of material.

3.B. X-RAY FLUORESCENCE MICROANALYSIS

For the microanalysis of biological tissues, the x-ray fluorescence technique has been used only during the last decade. Although the fluorescence intensities are very low compared with those obtained by electron excitation in the microprobe, the method has the advantage in biological work that there is no danger of specimen damage caused by the exciting beam. It is possible to determine the total mass of an element in a biological structure of nonuniform thickness, provided that at any point the thickness does not

exceed that at which absorption effects become appreciable (for explanation, see Sec. 5.D.3). For example, in the determination of potassium, for
thicknesses up to about $5\mu m$ to $10\mu m$ dry tissue, the measured fluorescence
intensity is directly proportional to the mass of element present. Due to the
relatively low intensity of the fluorescent radiation absorption in air between the specimen and the counter becomes important. A slight vacuum of
the order of 5 to 10 mm Hg or registration in an He atmosphere will reduce
most of that negative effect. Any possible disturbing fluorescent radiation
from the components of air is likewise reduced. The necessity of a moderate
vacuum is an advantage of the x-ray fluorescence method as compared to
electron probe analysis where the requisite high vacuum may be intolerable
in some work.

The main problems to be overcome in the practice of the technique are
the collimation of the primary beam by an aperture system which must
produce little or no scattered radiation, and the efficient collection and discrimination of the radiations excited in the specimen. For the latter, the
alternatives of a nondiffracting proportional-counter system, and a curved
crystal spectrometer are available. The limiting factors which determine the
sensitivity of the method are twofold; first, the intensity of scattered radiation of a wavelength close to that of the fluorescence line; and second, the
intensity in the fluorescence line. Thus, even though a crystal spectrometer
may give a much better rejection of scattered radiation, its low efficiency
will partly offset this advantage (see Sec. A.7). All the results of the investigations which will be dealt with in this chapter were obtained from a
nondiffractive system.

In all the following investigations, a Cosslett-Nixon type of tube served
as x-ray source. It has the advantage of giving a high x-ray intensity over a
small spot. Over the target two electron microscope apertures were placed,
$100\mu m$ and $200\mu m$ in diameter. Over the upper aperture a thin mylar foil
coated with iron, Fe_2O_3, was glued. The specimen was then placed on top
of that, on another thin mylar foil. The mylar foils contain H, N, O and C
and do not disturb the measurements.

When the x-ray beam hits the specimen, most of the energy passes straight
through as the transmitted beam. A minor part excites secondary radiation,
i.e. x-ray fluorescence. The wavelengths of this radiation depend on the
elements present in the specimen and the intensities are directly proportional
to the amounts of the constituent elements. A sector of the fluorescent
radiation was picked up by a gas proportional counter which, together with
a multichannel pulse height analyzer, was able to sort out the different
wavelengths.* Specifically on an oscilloscope a curve with several peaks

---

* Better performance is now available by replacing the gas counter with a solid-state
silicon detector (see Sec. A.7).

was seen, each peak representing one element and the height of the peak corresponding to the amount of that element present in the specimen. An electrical typewriter connected to the pulse height analyzer automatically transformed the curve and its peaks into figures. The peaks of some elements may come close to others. If the peaks are superimposed, it frequently happens that one of the two elements occurs in negligible amounts in biological tissues.

A certain high tension on the tube is necessary to produce x-rays of energies high enough to excite the element to be studied. For the investigations of potassium, 20 kV has been used.

The disposition of instrumental components is shown in block-diagram form in Figure 56.

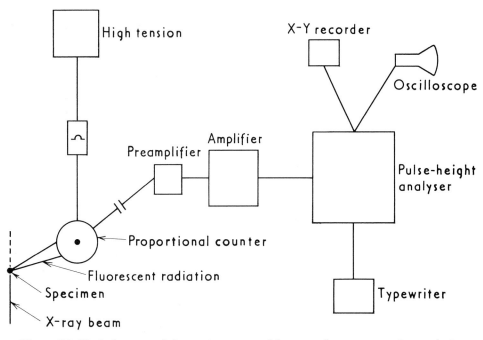

Figure 56. Block diagram of the equipment used for x-ray fluorescence microanalysis.

In order to decrease the absorption in air of the fluorescent radiation the specimen is surrounded with an evacuated brass box which has a hole for the window of the proportional counter. As mentioned before a moderate vacuum (pressure about 10 mm Hg) is enough. Evacuation also removes most of the argon. Argon produces a peak fairly close to potassium so that their interference is considerably reduced by the evacuation around the specimen. An iron foil placed over the aperture is used in order to correct for variations in the primary beam, so that the results could be expressed as ratios K/Fe.

In order to obtain absolute quantitative results one may calibrate with a reference system containing a known amount of potassium. This was prepared in the following way: A solution with a known concentration of KOH was mixed with 50% I^{131}. From the mixture a microdrop was placed on a mylar foil over the 200μm aperture on the x-ray tube. After the intensity of x-ray fluorescence was recorded the foil with the drop was moved into a scintillation counter. From the known radioactivity per microliter of the solution the volume of the drop was easily calculated and from that, the absolute amount of potassium in the drop. By repeated experiments a standard curve was obtained, in the shape of a straight line. Ratios K/Fe were plotted along the y-axis and absolute amount of potassium along the x-axis. From this standard curve a measured K/Fe ratio then could easily be converted into an absolute amount of potassium. In favorable cases the sensitivity of the method is about 10^{-11} g.

The statistical error for the amounts of potassium in a single nerve cell was estimated to be approximately 4 percent.

3.B.1. Specimen Preparation

As has been mentioned already, for potassium analyses it is necessary to use fairly thin specimens, around 5μm in thickness. Most of the investigations performed so far have been made on isolated cells. When a cell is dissected out and placed on a mylar foil it generally becomes flat like a pancake. Then there is no risk of excessive thickness. If a specimen is too thick, up to a certain point a correction factor for absorption may be calculated provided the thickness has been accurately measured.

The dissection technique is extremely important for potassium analyses. All glassware must be made of potassium-free pyrex glass. The dissection medium must not contain any interfering elements in large amounts. If it does they must be removed from the surface of the cell before it dries on the mylar foil. The dissection technique can be checked in this respect, e.g. by performing test dissections in radioactive solutions. The fluid left on the specimen could be assayed in a scintillation counter. For the dissection of nerve cells the dissection temperature turned out to be important for the determination of potassium (Hamberger and Röckert, 1964). At 37°C the nerve cells showed a tendency to lose their potassium gradually with time but the glial cells showed a tendency to take up the same amount during the first hour after the death of the animal. This was compared with dissection at 4°C and 20°C where the loss of potassium was greater.

In general, all preparative techniques for single cells or for microscopic parts of tissues must be designed to avoid changes in the content of the element to be studied, e.g. by leakage from the cell or contamination from the dissection medium. Modifications of preparative technique for each element and specimen are often necessary. For details, the reader is referred to the various investigations mentioned below, as well as to Haljamäe and Röckert (1969) and Röckert (1969).

3.B.2. Applications

X-ray fluorescence microanalysis has been used for qualitative and semi-quantitative analyses in several investigations. The *tissue reaction to barium sulphate contrast medium* has been studied by Röckert and Zettergren in 1963. Peculiar granulomatous lesions in the abdomen had been observed on several occasions. Correlation of clinical and pathological data in such cases suggested that the granulomatous changes were caused by barium sulphate, which after x-ray investigation had leaked through a peptic ulcer in either the gastric or the duodenal wall into the peritoneal cavity. In all the cases studied, the granulomatous changes were chiefly located in the greater omentum. Since the lesions differed from barium granulomatas previously reported in the light-microscopic literature, x-ray microscopy and x-ray fluorescence microanalyses were performed of the phagocyting cells in these granulomas. After staining with hematoxylin-van Gieson the cytoplasm of the cells was greyish and showed fine granules which clearly refracted the light. No birefringence could be seen in polarized light. Quite a number of the cells were perivascularly located. The cells were sometimes densely packed and sometimes separated by fibroblasts, lymphocytes or plasma cells (see Fig. 57). Contact microradiography of sections from the same area, with x-rays 1.2 to 4 Å in wavelength, showed a number of cells

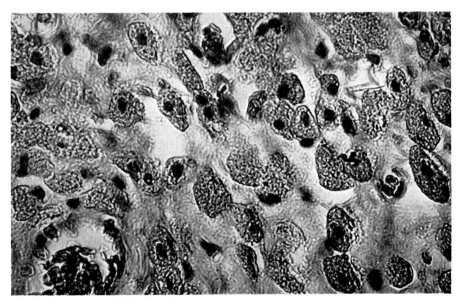

Figure 57. Cells containing cytoplasmic barium granules seen in transmitted light. From the subserosal connective tissue in the gastric wall.

appearing with white cytoplasm due to the high x-ray absorption. The nuclei were relatively nonabsorbing and appeared as black spots in the cytoplasm (see Fig. 58).

For x-ray fluorescence microanalysis, about five of the cells were isolated and put on top of the aperture of the tube as described above. The peaks on the oscilloscope corresponded to the emission peaks of barium. A comparative investigation made with a small drop of Mixobar, the same contrast medium used for the clinical gastric x-ray examination, produced peaks identical in position with those from the cells (see Fig. 59). The only difference was the height of the peaks, due of course to the larger amount of material analyzed. This confirmed that the cytoplasm of the cells contained barium contrast medium.

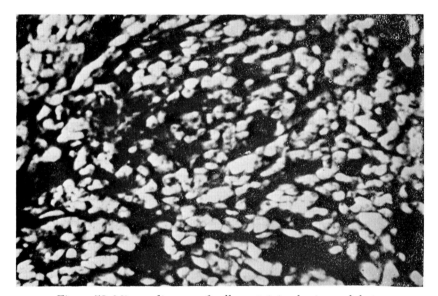

Figure 58. Microradiogram of cells containing barium sulphate.

In order to get a rough idea of how long the barium granulomas might be detectable after their development, white rats were given 0.8 cm^3 barium sulphate (Mixobar) intraperitoneally. A week after the injection granulomatous spots were seen in the fat tissue of the greater omentum and in the peritoneum. Microscopically the spots were all like barium granulomas in man. Seven months after the injection small greyish plaques were seen in the peritoneal cavity. The refringence and the granules of the cells in this instance were less distinct than those from fresh granulomas and the microscopic picture showed fewer granules in the cytoplasm.

X-ray fluorescence microanalysis was used in the development of a method for measuring quantitative relative *microscopic changes in vascularity* as standard methods are not accurate enough. With this method it

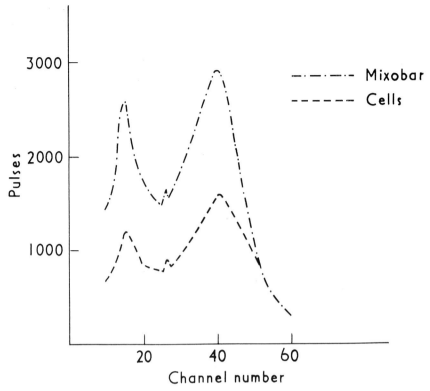

Figure 59. X-ray fluorescence curves as seen on the oscilloscope. Top curve from the contrast medium; bottom curve from the cells. (After Röckert and Zettergren: *Acta Pathol Microbiol Scand*, 1963.)

was possible to measure changes of vascularity *in vivo* within circular areas 50μm to 750μm across. The method is applicable to surface areas or membranes.

Changes in the vascular bed in the web of the living frog's foot, in areas 200μm across, were examined by determining changes in the amounts of iron present under different conditions (Röckert, 1965). The total amount of iron in the area analyzed was measured for a period of one-half minute. Most of the iron analyzed was due to the hemoglobin present in the erythrocytes, and variations in the serum iron were relatively unimportant. The experiments were performed in air with the specimen fixed in relation to the proportional counter. A peak representing argon in the air was used as a reference. Expression of the results in ratios Fe/A made them independent of variations in intensity of the primary beam. (In this case the use of an iron foil as a reference was of course unnecessary.) When a small drop of iron solution of known concentration and volume was used as a standard, the results could be expressed in grams per unit area. Model experiments revealed measurable differences between the values at room temperature and during exposure to a stream of cold air. No systematic investigation of the subject has been performed yet.

X-ray fluorescence microanalysis of intracellular potassium in nervous tissue has been performed *after X-irradiation at increased oxygen partial pressure* (Röckert and Rosengren, 1966). Several authors have pointed out the possibility of improving the response in radiotherapy by increasing the concentration of oxygen surrounding the cells. The anoxia in some malignant tissues may reduce the sensitivity to x-irradiation quite considerably. With clinical material especially good results have been obtained in the head and neck region with irradiation at a pressure of three atmospheres absolute, 100 percent oxygen (Churchill-Davidson, 1964).

Nervous tissue contains nondividing cells, nerve cells, as well as dividing cells, glial cells. In many respects these two types show a connection for example reflected in their respiratory enzyme activities, content of nucleo-protein, and the rate of loss of potassium during dissection at 37°C. Intra-cellular potassium has turned out to be a very sensitive indicator of partial or total cell damage as potassium is mainly located intracellularly. This response comes much sooner than light microscopical changes (Hamberger and Röckert, 1964).

White rabbits were given 3000 R over the lateral vestibular nucleus and compared with normal nonirradiated rabbits as well as with a group exposed to pure oxygen at four atmospheres absolute in a pressure chamber which had also received 3000 R in a single dose from a 3000-Curie Co^{60} source. The irradiated rabbits showed a maximum response after four to six days with a potassium loss of 52 percent in the nerve cells of the irradiated area and a loss of 48 percent for glial cells compared to the control group. There was no significant difference for either nerve or glial cells when they were irradiated under increased oxygen partial pressure. Irradiation with dosages smaller than 3000 R showed a potassium loss in proportion to the dose given. The lack of difference between the populations irradiated under pressure and not was probably due to the fact that the "healthy" brain has an optimal oxygen supply. The damage to the sodium pump is fortunately not in-creased at high pressure in healthy nervous tissue. This is the base on which a successful hyperbaric radiotherapy will be built. A corresponding in-vestigation on tumor cells has not yet been performed. The results men-tioned will make it possible to verify objectively the irradiation effect on tumour cells under the same conditions. This type of investigation may also serve as a model for similar radiobiological research on other tissues, ex-pressing the efficiency of the combination of hyperbaric oxygen and radio-therapy.

The x-ray fluorescence microanalysis of potassium has been carried out on isolated *muscle fibers* in order to increase the knowledge of the patho-physiological mechanism behind *hemorrhagic shock,* as well as obtaining better information for safer postsurgical treatments in regard to parenteral therapy (Haljamäe, 1967; Hagberg, Haljamäe and Röckert, 1967). Earlier methods are based on biopsy analyses of larger pieces of muscle. This involves a crude approximation because of the presence of fat cells, connective tissue, vessels and nerves. For routine clinical use, analyses of potassium in serum have been the only way of obtaining information about the body electrolytes. This determines the status of the tissue electrolytes only very approximately. Therefore a biopsy technique has been developed for obtaining parts of muscle fibers. A needle is inserted in the muscle and a cylindrical piece of tissue about 2 to 3 mm long is taken out. This can be performed on any patient and causes a minimum of trauma. The potassium analyses then give specific information about the intracellular reactions of the muscle cells.

The experiments were performed on anesthetized dogs which were bled to a blood pressure of 45 mm Hg and kept at that level for two hours and fifteen minutes. Their normal blood pressures were 160 to 180 mm Hg. A piece of muscle was removed, from which single muscle fibers were dissected out by hand under a binocular microscope over a period of time of forty to fifty minutes. The fibers were placed on mylar foils and allowed to dry. The intactness of the fibers was checked by phase-contrast microscopy. A criterion of the intactness was the presence of cross-striation. Samples were taken before bleeding and after the blood had been returned into the animal. By the use of the x-ray absorption technique of Rosengren for mass determination, all the potassium analyses could be expressed as potassium per unit weight of muscle fiber.

The results from near 400 specimens from seven dogs showed a 40 to 50 percent reduction of the intracellular amount of potassium after irreversible hemorrhagic shock as compared to before the shock. In cases of induced reversible shock the drop of the intracellular amount of potassium was about 25 percent and after a mild reversible shock only a 10 percent reduction was obtained.

The possible future clinical use of this technique is obvious. Some sources of error however must be considered. The dissection technique is very important. A cut muscle fiber shows a loss of potassium at the cut ends but has a constant potassium content in its middle part. Therefore only the central parts of these pieces were analyzed. As the dissection medium, oxygenated Ringer solution was used. As the Ringer solution contains potassium it is very important to suck away the excess of fluid around the specimen when it has been placed on the mylar foil. The sucking is done with small capillary pipettes. Control tracer-radioactivity measurements of the potassium then left on the surface of the specimen have shown the amount to be negligible. It is also necessary to check that no change of pH occurs in the dissection medium.

An attempt to analyze the fluid in dental enamel by x-ray fluorescence microanalysis was made by Bergman, Lidén and Röckert in 1966. The enamel of an erupted tooth is in contact with fluid on all sides. The inside of the enamel is in contact with the dentinal fluid, the outside with the saliva or the gingival fluid. The loosely bound water with its dissolved molecules and ions probably occupies about 2 percent by volume of the enamel and serves as a connecting link between the outer and inner fluid environments of the enamel.

The relative osmolality of the saliva in relation to blood would indicate a constant flow of water from the tooth surface to the pulp. This direction may, however, be reversed if the hydrostatic pressure of the blood is sufficient, or if concentrated sugar is ingested. On the basis of the relative concentrations one would expect that calcium would diffuse from the blood to the saliva, and the phosphates in the reverse direction. Sodium should migrate to the saliva and potassium to the blood.

There seems to be reason to assume that the most important pathways for molecules and ions through the enamel are those parts of the enamel prisms where the crystallites are less densely packed and not so well oriented. These parts of the prisms are usually called enamel sheaths and interrod substance.

The material studied consisted of eight premolars from patients ten to fourteen years of age. The teeth were fully erupted, and were extracted for orthodontic reasons. After extraction the enamel surface was cleaned with a tooth brush under running water for thirty seconds, and in redistilled water for fifteen seconds. It was then dried with filter paper and air-blasted. After covering the fissures with wax, a polythene bag was mounted around the crown and sealed 1 mm from the cementum with wax. The root and the cervical area were then covered with wax. During these procedures the enamel was never touched with defiled material. The whole manipulation required about four minutes, and the tooth was then kept in an exsiccator for twenty-four to forty-eight hours. A fluid slowly collected in the bag, and was visible under the microscope. The microsamples were taken in two ways. In some cases centrifugation was used, and the fluid was then forced via a hole in the polythene bag and a glass funnel into a capillary. In the other cases the microsamples were obtained under the microscope by using thin glass capillaries which were allowed to perforate the bag. The fluid then entered the capillaries as a result of capillary attraction. The volume of the fluid was determined by calibrating the capillaries using radioactive iodine (I^{131}) with a known radiation per microliter. The analyzed drops were of the magnitude of a few thousandths of a microliter. The (Ca+K) values varied between 8 and 40 mg/100 ml. The K/Ca ratio is about 30 in saliva and only about 2 in serum. The mean value for the enamel fluid was only 0.7 indicating a relatively high calcium content compared to potassium. The mean value for the calcium content was 1.5 mM which is in the neighborhood of unstimulated parotid saliva. The limited number of analyses did not allow definite conclusions. A systematic error may be present. The results indicated, however, that the enamel fluid of young teeth contains appreciable amounts of calcium and potassium. The absolute concentrations of these elements showed a great variation but their ratio is relatively constant. Calcium might be of importance for the posteruptive maturation of enamel and for its resistance against caries.

The results are preliminary but a more extensive investigation may be of clinical value since nutritional variation may occur and may be affected by differences in composition of the enamel fluid.

The investigations mentioned in this chapter are examples of the versatility of the x-ray microanalytical methods. These are of particular importance considering that clinical diagnostic demands are requiring more and more at the cellular level.

ACKNOWLEDGMENTS

I wish to express my sincere thanks to Mrs. Ingrid Lundberg and Mrs. Britt-Marie Otterstedt for all the help when preparing the manuscript and to authors and journals for allowing me to publish pictures and diagrams:

Figure 44. Professor Å. Omnell and *Acta Radiologica.*
Figures 45, 46 and 47. Docent G. Wallgren and *Acta Paediatrica.*
Figures 49 and 50. Docent U. Nilsonne and *Acta Ortopaedica Scandinavica.*
Figure 53. *Acta Odontologica Scandinavica.*
Figures 54 and 55. Docent B. Rosengren and *Acta Radiologica.*
Figure 59. *Acta Pathologica et Microbiologica Scandinavica.*

CHAPTER 4

MICROANALYSES BY MEANS OF
THE ELECTRON PROBE

ELECTRON-PROBE MICROANALYSIS is a technique for detecting and measuring the amounts of chemical elements within microregions of a specimen. The physical fundamentals of the technique are outlined in Chapter 1 and elaborated in Chapter 5. In this chapter we shall first discuss the medical work which has been done with the method. For the reader who wants to carry out his own studies we shall then consider certain essential practical matters, namely specimen preparation, the danger of damage caused by the probe, and histological identification of the analyzed microvolumes.

Within the broad field of biological applications, the definition of "medical studies" is rather arbitrary. Nevertheless, simply to save space we shall avoid the description of nonmedical investigations, merely citing them where they provide an important background.

4.A. MEDICAL STUDIES

4.A.1. Concentrations of Foreign Matter

Electron probe analysis is well suited to the identification of foreign particles which sometimes appear in *lung pathology*. In a study by Robertson *et al.* (1961), tin was positively identified in individual phagocytosed particles in the lungs of tin smelters (Fig. 60). They found that silicosis from free silica dust is prevalent among tin miners, but in the plants where tin is bagged, the silica particles are washed away and the condition suffered by the baggers, as well as the smelters, is stannosis.

A more widespread industrial hazard is asbestosis. In many industrial processes which contaminate the atmosphere in an obvious way, the workers are also exposed incidentally to asbestos fibers. Thus asbestos fibers, as well as amorphous iron oxide and corpuscles similar to asbestos bodies, have been found in the lungs of shipyard flame cutters and welders who are exposed to high concentrations of iron oxide. ("Asbestos bodies" are asbestos fibers coated with an iron-rich layer or with iron-rich rings deposited by a reaction of the host's tissue.) In cases of mesothelioma it was essential to

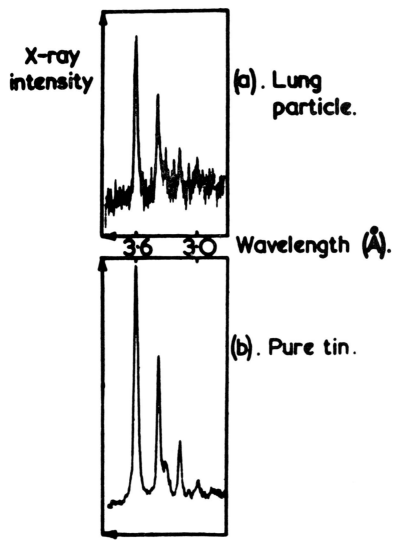

Figure 60. X-ray spectral data identifying tin in a foreign particle in human lung. (From Robertson *et al.*, 1961, courtesy of *The Lancet*.)

determine whether the corpuscles were genuine asbestos bodies or pseudoasbestos, ferruginous bodies. Von Rosenstiel and Zeedijk (1968) were able to remove the deposits at least partly from the fibrous cores by micromanipulation and etching, and they then proved by electron probe analysis that the stripped cores were amphibole (iron-silicon) asbestos, while other fibers which had not been coated in the host were chrysotile (magnesium-silicon) asbestos.

Microprobe analysis of the asbestos fibers is technically complicated not only because of the overlying deposits but because the fibers are often much less than 1μm in diameter, too thin for accurate use of the conventional quantitative theory for thick specimens (Section 5.E.7.a). Quantitative analysis of the individual fibers has been discussed in detail by Andersen (1967). He studied lung sections fixed in formaldehyde, dehydrated in alcohol, embedded in paraffin, cut at 5μm, deparaffinized in xylol and mounted on plastic supports. Some of the sections were ashed, which concentrated the interesting elements by removing volatile components and often exposed bare fibers to the probe.

A quite extensive study of asbestos and asbestos-like particles in pulmonary tissue has been carried out by Berkley *et al.* (1967), using a wide variety of techniques including electron-probe analysis. They confirmed that a large fraction of the observed fibrous particulates was genuine asbestos, and established that amosite and possibly crocidolite were the main mineral types in their samples.

Banfield *et al.* (1969), using solely the electron microprobe, have analyzed both asbestos bodies and anthracotic pigment (mainly coal dust), in both lung and hilar lymph node tissue. The x-ray analyses of the asbestos bodies revealed the presence of the elements usually associated with these objects, but in addition there was an unusual finding: concentrations of titanium in both the asbestos bodies and the pigment bodies. In the case of one patient, who was a painter, the titanium may have come from the titanium oxide which is often used as a pigment in paints, but the origin of the element in most of the specimens is obscure.

Technically, the work of Banfield *et al.* is important as one of the first biological microprobe studies to use transmission-scanning microscopy (explained in Sec. 5.E.1) as a means to improved visualization of the tissue sections during probe analysis.

Accretions in the *kidney*, caused by foreign material, have been the subject of extended study by Galle. He has used the combination of electron microscopy with x-ray microanalysis (Sec. 5.E.1) to examine ultrathin sections prepared for electron microscopy in a conventional way. The gold salt aurothiopropanol, which is used for human therapy, was given to rats subcutaneously (Galle *et al.*, 1968). High doses caused acute tubular nephritis while repeated smaller doses produced lesions only in the proximal tubular cells. Deposits appeared first in the mitochondria, which later expanded and were expelled into the lumen. Microprobe analysis unequivocally demonstrated a high concentration of gold in these deposits (Fig. 61).

After subcutaneous injection of the well-known nephropathenogen uranyl nitrate, uranium was found to be localized in "needles" in the cells of the proximal tubule (Galle, 1966, 1967a), while the glomerular lesions were found to be relatively unimportant.

Lead poisoning also affects the cells of the proximal tubule, producing mitochondrial lesions and cytoplasmic and nuclear inclusions. Probe anal-

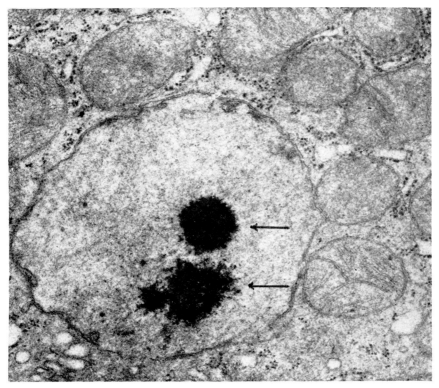

Figure 61. Renal deposits found after the administration of aurothiopropanol. Electron-probe microanalysis demonstrated a high concentration of gold in these deposits. (After Galle *et al.*, 1968.)

ysis positively demonstrated iron in the cytoplasmic inclusions, but the composition of the nuclear inclusions does not seem to have been established (Galle and Morel-Maroger, 1965; see also Gueft *et al.*, 1964, and Carroll *et al.*, 1970).

Deposits were seen in the renal biopsy of a patient with chronic nephritis. The microprobe analysis of these deposits revealed large amounts of calcium, phosphorus, sulphur, and magnesium (Galle *et al.*, 1968), and the source was then identified as a powder, taken for many years for stomach troubles, containing magnesium carbonate, magnesium trisilicate and calcium bicarbonate.

Calcification in the kidney was studied after injections of four different agents into rats intraperitoneally (Galle, 1967a, 1967b). Oxalate produced large crystallites in the cytoplasm of the tubular cells, and microprobe analysis proved that the cation in the crystals was calcium. Phosphate ions led to small crystalline deposits in the same site, characterized by a constant ratio of calcium/phosphorus. Injection of calcium itself produced mainly "noncrystalline" deposits in the hyaline substance of the basal membranes in the proximal tubule (as seen in Burnett's syndrome), with variable proportions of calcium, phosphorus and sulphur. Vitamin-D intoxication led to basal-membrane deposits containing a fixed ratio of calcium to phosphorus.

Gueft *et al.* (1964) have analyzed the renal tubular casts found in autopsy tissues of patients with liver disease after doses of bunamiodyl (an iodine-based oral radiographic contrast medium). The probe data showed that iodine had accumulated excessively in the kidney (presumably as a consequence of impaired function of the liver).

In a more general recent study, Carroll *et al.* (1971) have sought, found and analyzed concentrations of heavy elements in tissue sections of many human organs, with special attention to possible *differences between malignant and "normal"* tissue. They report the existence of localized regions of high metallic concentration in samples of thyroid, lung, breast, stomach, colon, ovary, and rectal tissue, each region containing one or more of the elements Mg, K, Ca, Ti, Cr, Mn, Fe, Ni, Cu, Zn and Ba. Such regions were found in both malignant and "normal" samples, but the frequency of occurrence appears to be greater in malignant material.

The foregoing studies have all shown the effectiveness of electron-probe x-ray microanalysis for chemically characterizing small individual entities seen in the optical or electron microscope. A problem which is more formidable technically as well as more profound biologically is to trace a chain of events which begins with a gross implant and culminates in a diseased tissue containing no dense deposits as landmarks. For example, one wants to follow the fate of corrosion products when inflammation occurs around metallic prosthetic devices, or one wants to understand the process leading to malignant tumors after the injection of cobalt powder into rats (Heath, 1960). Microprobe assays of metallic elements have been carried out in the neighborhood of prosthetic devices (Mellors and Carroll, 1961; also J. Scales, unpublished, and D. Mears, unpublished) and in the neighborhood of implants of cobalt or nickel chips or powder (T. Hall, unpublished). Usually one may find detached extracellular bits of the implanted material; there is an extremely narrow (submicron) border around the implant or the bits where metallic elements are "detectable" in lowered concentration (perhaps artefactually due to x-ray fluorescence—see Sec. 5.E.7.a); and in the bulk of the tissue, even nearby or at sites of reaction, the metallic concentrations are not detectably elevated. This is not surprising since the pathological effects may sometimes be due merely to mechanical intrusion, and even when there is evidence for specific chemical reactions (Heath, 1960), the effects may be due to elemental concentrations far below the limit of detectability. However, the quoted studies were all carried out several years ago, and more might be discovered with modern microprobes of higher sensitivity, or with combined electron microscopy-microanalysis where conceivably intracellular inclusions might be localized to cell organelles.

4.A.2. General Pathology

In *Wilson's disease,* copper is accumulated extraordinarily in certain tissues. So much copper is stored in the eye that the outward appearance changes. Tousimis and Adler (1963; see also Tousimis, 1971) have shown by microprobe analysis that this copper is sharply localized to Descemet's membrane of the peripheral cornea, precipitating in bands subjacent to the corneal epithelium (Fig. 62).

The accumulation in liver in Wilson's disease, though not as striking, is still extraordinary and presumably more significant. Goldfischer and Moskal (1966) sought to show that the copper is localized in the lysosomes of the

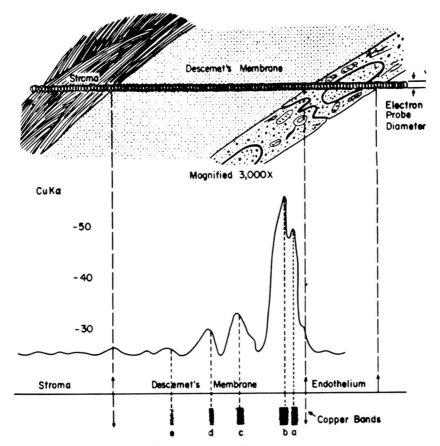

Figure 62. The graph shows the variation in the intensity of copper x-rays, recorded as the electron probe was scanned along a line across the Descemet's membrane and adjacent tissues in the eye. (From Tousimis and Adler, 1963, courtesy of the Histochemical Society and the Williams and Wilkins Company.)

hepatic cells. They examined 7μm tissue sections mounted on glass slides, with the microprobe run at 30 kV. As they realized, under these conditions the spatial resolution is too poor for the clean analysis of individual lysosomes because of diffusion of the electrons in the tissue and electron-backscatter from the glass. (Better spatial resolution could have been obtained by the use of low kilovoltage, or with thinner specimens mounted on thin supports, as discussed in Section 5.E.5.) However, they incubated the sections in a Gomori medium to deposit lead at sites of acid phosphatase activity, which is believed to be especially concentrated in the lysosomes. The probe study showed a close correlation of the distribution of copper and lead, giving relatively direct evidence for localization of the copper at sites of high acid phosphatase activity, and presumptive evidence that the element is localized in the lysosomes.

Marshall (1967) has studied pathological accumulations of *iron in the liver* in hemosiderosis, hemochromatosis, and among the Bantu people of South Africa.

The adults of the Bantu population all have abnormal amounts of iron stored in the liver because of the use of iron cooking utensils. Figure 63 shows a section of such a liver viewed in various ways. Deposits, which are not prominent in the optical microscope but are quite clear in the back-scattered-electron image, are shown unequivocally in the x-ray image to be sites with a high content of iron. The image formed by white x-rays shows the distribution of total dry mass in the section (cf. Sections 5.E.1 and 5.E.3); since the deposits do not show especially high total mass, it is clear that not only the total amount of iron, but also the iron *weight-fraction* is elevated at these sites. The quantitative analysis of the individual deposits, according to the method described in Section 5.E.7.b, gave a mean weight-fraction of iron of 3 percent.

Iron-rich deposits are more prominent in hemosiderosis and hemo-chromatosis (Fig. 64). The quantitative analysis of these deposits gave an iron weight-fraction of 20 percent with a standard deviation of 11 percent in hemosiderosis, and 27 ± 9 percent in hemochromatosis (the difference was not significant statistically). However, the highest observed weight-fractions were 51 percent in siderosis and 49 percent in chromatosis. These fractions are high enough to preclude ferritin (iron weight-fraction 26%) as the storage form; almost pure hemosiderin is indicated. In chromatosis the entire tissue is visibly discolored and the mean iron content is higher than in siderosis, but this seems to be due to a larger number of deposits and to a higher iron content in the tissue between the deposits, rather than any difference between the individual deposits themselves in the two conditions.

The retention of *copper in cirrhotic liver* has been studied under experi-mental conditions by Russ and McNatt (1969). They induced cirrhosis in rats by carbon tetrachloride inhalation and added copper acetate to the

LIVER (4)

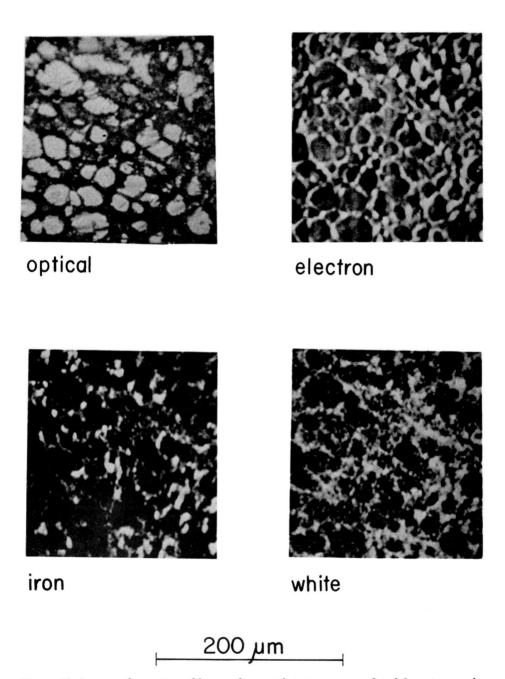

optical   electron

iron   white

200 μm

Figure 63. Images of a section of human liver, rich in iron accumulated from iron cooking utensils. Optical-microscopy image and three scanning images, based on backscattered electrons, iron x-rays, and continuum x-rays. (From Marshall, 1967.)

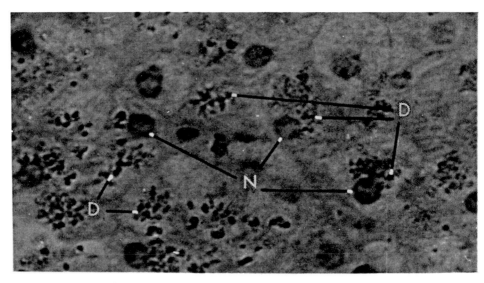

chromatosis

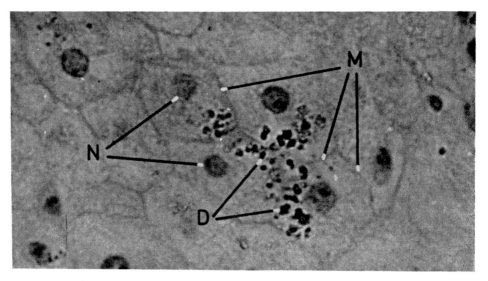

siderosis

Figure 64. Iron-rich deposits in hemochromatosis and hemosiderosis, seen by optical microscopy. D: deposits. N: cell nuclei. M: cell membranes. (From Marshall, 1967.)

animals' drinking water at a level of 0.2%. By combining two special microprobe techniques, transmission scanning microscopy (explained in Sec. 5.E.1) and nondiffractive x-ray spectroscopy with a solid-state detector (Appendix, Sec. 7), the authors were able to determine the distribution of the copper burden on an exceptionally fine scale. In their scanning images, they were able to identify cytological entities such as mitochondria, glycogen bodies, and regions of endoplasmic reticulum, and they were able to determine separately the amounts of copper associated with each.

Nephrocalcinosis has been observed in patients with *adenomas of the parathyroid gland.* Galle *et al.* (1968) have simulated this effect by the injection of parathyroid hormone into animals. Microprobe analyses revealed calcium, phosphorus and sulphur in the concentric deposits seen in renal basement membranes; calcium, phosphorus, sulphur and magnesium in the deposits in the tubular lumen; and calcium and phosphorus in the aggregates of "fine needles" seen in the cytoplasm of cells of the proximal tubule. The aggregation of needles in these cells seems to be specific to hyperparathyroidism.

Carroll and Tullis (1968) have reported microprobe observations of elevated amounts of titanium and zinc in *leucocytes of human lymphoblastic lymphoma.* They analyzed individual cells in smears of whole blood, of bone marrow, of suspensions of isolated cells, and of centrifugal concentrates of these suspensions. The specimens were smeared onto single-crystal wafers of highly polished silicon, which is relatively free of impurities. It is very difficult to control contamination sufficiently in a study like this, where extremely small amounts of element are detected; unambiguous identification of the cells is a further difficulty. Problems have arisen in an effort to repeat the data (Beaman *et al.,* 1969), so that there must be some question of their validity in spite of the precautions taken and the apparent difference between normal and lymphoma-derived cells in the original observations.

Disturbances of *protein metabolism* in the protein-deficiency disease *kwashiorkor* have been studied in easily accessible tissues, specifically hair (Sims and Hall, 1968a) and skin (Sims and Hall, 1968b, and to be published). In *hair* the microprobe assays were limited to comparisons of density (dry mass per unit volume) in control and disease-affected specimens, by means of the method described in Section 5.E.3. Pairs of transverse sections were cut from control and kwashiorkor hairs embedded side by side to assure equal section thickness. After mounting each pair on a very thin nylon support, each section was scanned in turn individually in the microprobe, and the total count for a band of quantum energies in the x-ray continuum was recorded. For fixed operating conditions and in a fixed time, the total count is proportional to the total mass in the scanned field. Hence when the scanned field is large enough to enclose each section totally, and after the background from the nylon support is measured separately and sub-

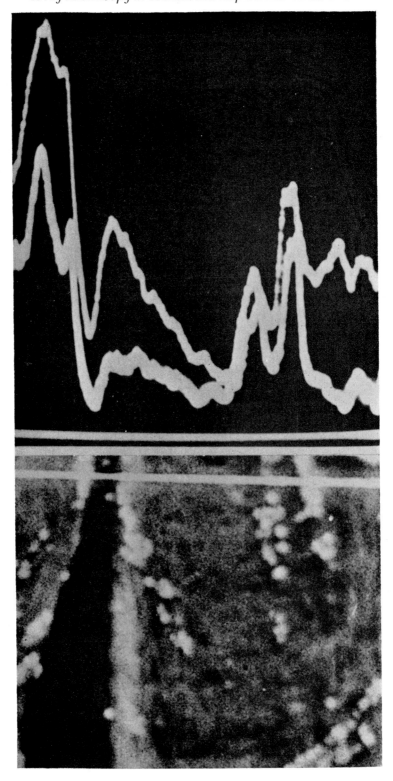

tracted, each pair of counts gives the ratio of the total masses of the two sections. The densities of the sections could also be compared after the areas of the sections were measured from photomicrographs, since each density was proportional to total mass per unit area (the two sections being equal in thickness).*

Differences between the control and the kwashiorkor specimens were manifest in the larger areas of the control sections, and in the greater swelling produced in kwashiorkor hairs upon immersion in water. The greater swelling suggests weakened stability of the cross links between polypeptide chains. However, the microprobe analyses showed no consistent differences in density of the sections.†

In the studies of *skin*, microprobe data were obtained from transverse sections along lines of scan extending through the various strata of the viable epidermis and the corneum. Again the x-ray continuum intensity was used as a measure of dry mass per unit area (almost entirely protein in this tissue), but in addition the intensity of sulphur characteristic radiation was recorded as a measure of the mass of sulphur per unit area (for instrumental theory and practice, see Sec. 5.E.7.b). The sulphur content was regarded as an index of the incorporation of sulphated amino acids, presumably largely into keratin.

The linear scans of the sections of normal skin show a very consistent pattern (Fig. 65). Just before the scan moves from viable epidermis into corneum one sees an abrupt increase by a factor of two to three in the dry mass per unit area and a parallel increase in the amount of sulphur per unit area. The closeness of the correlation is shown when the two signals are recorded simultaneously on a pen-recorder (Fig. 66). Thus there is a large increase in the *amount* of sulphur per unit volume, but no increase in sulphur *weight-fraction* (i.e. sulphur mass/total dry mass). When combined with

* As noted in Section 5.E.3, the scanning mode can be used not only for a measure of the total mass within a scanned field, but also for a direct measure of mass per unit area averaged over the field. However, the resulting average refers to the *entire* scanned field, and for sections of hair the rectangular field of scan cannot be fitted closely to the circular or oval sections. It was therefore better to measure density by dividing the total masses by the optically measured areas.

† The density is not uniform over an individual section of hair, but varies with distance from the axis of the hair. The density determined from a field scan is the average density for the entire section. Local densities in 1μm spots also were compared by static-probe counts. Again there was no apparent systematic difference between the control and the kwashiorkor sections.

A very recent microprobe study of hair pathology has been reported by Brown *et al.* (1971).

Figure 65. X-ray signals recorded during a linear scan across a transverse section of normal skin. The scan goes across corneum and adjacent viable epidermis, with a tear in the section between them. The sulphur x-ray intensity (upper trace) and the continuum x-ray intensity (lower trace) vary in parallel.

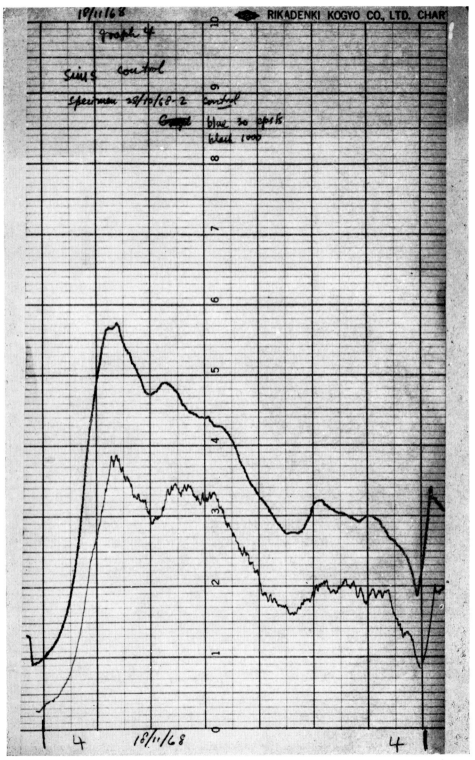

Figure 66. Like Figure 65, but with the x-ray intensities recorded on a moving chart. In this normal skin there is a close parallelism between the sulphur signal (lower trace), and the continuum signal (upper trace).

other data (Sims and Hall, 1968b), these results prove that adjacent to the corneum there is a zone of synthesis of protein (and not merely a compression with loss of water). In the pathological skin the pattern is erratic. Some of the linear scans give the normal pattern, but a displacement of the probe of only a few micrometers from a normal line may give a linear scan like Figure 67 where the overall synthesis of protein adjacent to the corneum seems normal, but there seems to be a failure to incorporate sulphated amino acids, presumably signifying a disruption of the synthesis of keratin.

Because of the variable pattern revealed in the linear scans, a different technique was used to characterize the diseased tissue quantitatively. Counts were recorded during 100-second field scans to give *average* values for selected fields, each rectangular field being in the range of 10μm to 50μm on a side. Each field was restricted either to corneum or to the adjacent epidermis. In the control sections the data reiterated the pattern consistently seen in the linear scans: The sulphur per unit volume and the total dry mass per unit volume in the corneum were both much greater than in the adjacent epidermis, by a factor in the range 2 to 3. In the sections of diseased tissue, although the dry mass in the corneum was again two to three times greater than in the adjacent epidermis, the corneal sulphur value was approximately the same as in the adjacent epidermis. This was true even in fields within which some of the linear scans gave the normal pattern, so the field-scanning technique confirmed quantitatively that in kwashiorkor there is, on overall average, a substantial failure of keratin synthesis adjacent to the corneum.

Microprobe studies of *zinc in the prostate gland* have been carried out by Zeitz and Andersen (1966; see also Andersen, 1967). The medical interest is in the possibility of exploiting the extraordinarily high concentration of zinc in the prostate to concentrate a chelating agent like dithizone (diphenylthiocarbazone) as a possible therapy for benign hypertrophy of the gland (Philips, 1961; Schrodt *et al.*, 1964). Microprobe analyses of normal rat glands readily confirmed the histological distribution of zinc indicated by staining methods (special concentrations in the basal cytoplasm of the epithelial cells), and showed that there were micron-scale spots of extremely high zinc concentration with proportional elevations in the amounts of calcium and magnesium. The degree of chelation of zinc by injected dithizone could be determined by the simultaneous assay of zinc and sulphur since dithizone contains sulphur enough to show up well against the endogenous level. Thus it was confirmed that usually only one fifth of the zinc in the gland was chelated, and it was learned that a still smaller fraction of the zinc was complexed at the foci of very high zinc concentration.

Microprobe assays of *iodine in thyroid follicles* have been reported by Robison and Davis (1969) and by Zeigler *et al.* (1969). Robison and Davis were concerned with the question of the degree of variation of iodine con-

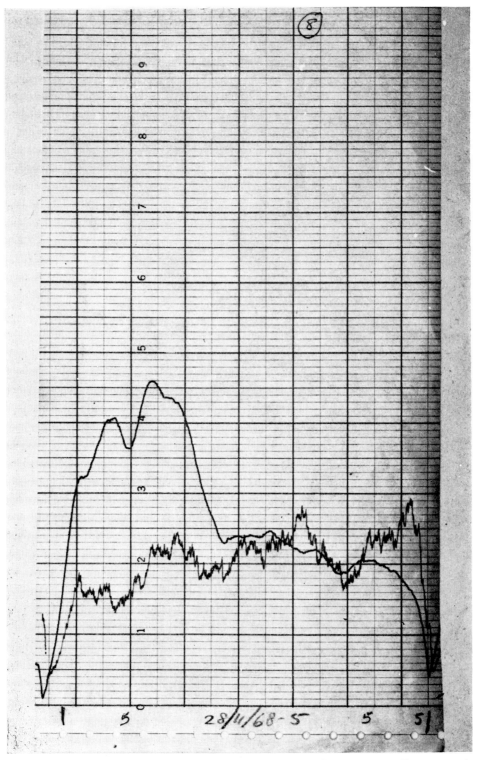

Figure 67. Like Figure 66, but in pathological skin. The sulphur content (lower trace) fails to follow the rise in total dry mass (upper trace) at the edge of the viable epidermis and in the corneum.

centration among follicles, and they showed in rats that there is very little variation within a group of follicles in one animal, although there is large variation between individual animals. Zeigler *et al.* were concerned with the correlation between iodine concentration and their "thyroid activity index," and with the effect of a diet supplemented with iodine. In their experimental animals (sparrows), it was found that the thyroidal iodine concentration was inversely proportional to thyroid activity, and directly proportional to the amount of iodine ingested. Like Robison and Davis, they found uniform iodine concentrations for follicles within one animal, with no difference observed between peripheral and central follicles.

Iodine-labelled antigen has been localized histologically by microprobe analysis by Clarke *et al.* (1970). In experimental studies with tuberculin-sensitized rats, they had previously observed intercellular processes formed between macrophages and lymphocytes, and they now sought to show that antigenic material was transferred along these processes. Their study benefited from the exceptional spatial resolution available with EM-MA instrumentation (cf. Secs. 5.E.1 and 5.E.5.c). Iodine counts above background were indeed obtained with the microprobe centered at various segments of the processes, although the statistical uncertainties in this pilot study seem too large to permit firm conclusions.

The microprobe has also been used for the measurement of the concentrations of a wide range of *elements in body fluids* (Lechene, 1970; Morel *et al.*, 1969). The samples of fluid are dried prior to analysis. It may seem odd to use a microprobe analyzer, which is intended above all as a tool for mapping elemental distributions, for the assay of a specimen like an aliquot of fluid, within which each constituent element is presumably uniformly distributed. However, the technique has important advantages: very high sensitivity, making it possible to analyze very small samples; the relative simplicity of quantitation; and the wide range of analyzable elements. Morel *et al.* have worked especially on the analysis of electrolytes in fluids obtained by micropuncture from kidney tubules, comparing electrolyte levels in rats under a variety of experimental conditions, in particular during salt diuresis.

The determination of local *electrolyte concentrations within kidney tissues* is of course extremely important for medical science, and the microprobe is therefore being applied vigorously in this field in spite of great technical difficulties. Early studies for feasibility were done by Ingram and Hogben (1968), and experimental medical studies are now being pursued by a large group (Höhling *et al.*, 1971; Kriz, 1971; Kriz *et al.*, 1971; von Rosenstiel *et al.*, 1970; von Rosenstiel *et al.*, 1971; Schnermann *et al.*, 1970). The main problem is to preserve the natural distribution of the electrolytes, i.e. to prevent the displacement of these elements during preparation of the specimens prior to analysis. Very rapid freezing of a tissue block is therefore the first step in all of the methods used in this field. Ingram and Hogben proceeded to freeze-dry their blocks and embed in vacuum and then section; they were able to show at least some preservation of electrolyte gradients across cell boundaries. In the more recent work, the frozen blocks have been sectioned and the frozen sections have been dried by sublimation in vacuum. In such preparations of rat tissues, it has been possible to see the effects of diuresis on cellular electrolyte concentrations, and to observe the differences in concentrations in different regions of the kidney, although it is clear that the preparations are not free from elemental displacements, especially of sodium. A major remaining problem for quantitative analysis is probe damage (see Sec. 4.B.2), and a more general problem stems from the fact that the measurements are of elemental mass-fractions in dried tissue, whereas the quantities immediately relevant biologically are the original concentrations of the electrolytes in their solvent (water). Efforts are under way now to measure the electrolytes in frozen sections which are held at such low temperatures that the water is retained in the form of ice, even during the microprobe analysis (see Sec. 4.B.1.d; and Echlin, 1971). This radical procedure should give the best preservation of the electrolyte distributions and should come close to measuring directly the true electrolyte concentrations, but the feasibility of the method is not yet established.

Microprobe analysis has been used in an elegant procedure for the localization and identification of fixed anionic groups, and the measurement of their concentration, at the *articular surface of uncalcified human cartilage* (Maroudas, 1970a,b). After dissection and prior to dehydration, the cartilage was equilibrated with dilute $CaCl_2$, under conditions where free electrolyte was excluded from the tissue and all of the retained calcium was bound to the fixed anionic groups. The microprobe assay of the calcium then revealed the distribution of these groups. Further, from the simultaneous assay of sulphur and the determination of the Ca/S ratio, it was possible to estimate the proportions of chondroitin and keratin sulphate, and to conclude that hyaluronic acid could not be present superficially on the articular surface.

4.A.3. Early Stages of Calcification

One of the fundamental biological processes is mineralization, manifested normally in growing bone and teeth and pathologically in aging arteries. The process must include early stages which should be called prestages of mineralization proper, when calcium and/or phosphate concentrations are built up, at very small sites at least, before any genuine mineral appears. In the view of one school (Glimcher and Krane, 1962 and 1964), during the prestages calcium is bound to phosphate groups which themselves are bound to collagen. A contrary viewpoint held by many scientists (Urist, 1966) is that calcium is first accumulated by binding to some amorphous component of the soft tissue matrix, perhaps to sulphated mucopolysaccharides, or to tissue proteins like elastin. Electron-probe microanalysis can contribute critically to knowledge of this subject, because the heterogeneity of the tissues calls for highly local analyses, and because the processes may be understood largely in terms of the concentrations of chemical elements. Hence there have been many microprobe studies of tissues at early stages of calcification.

It would be out of place to review here studies of normal calcification of cartilage (Brooks *et al.*, 1962; Andersen, 1967; Höhling *et al.*, 1967 a, 1970; Suga, 1969; Carlisle, 1970; Hall *et al.*, 1971). Nor can we discuss a very interesting exploration, using a variety of techniques including microprobe analysis, of a model system mineralizing *in vitro* (Rubin and Saffir, 1970; see also Saffir and Rubin, 1970). As topics of more direct medical interest, we shall consider microprobe observations of *early calcification in arteries and in experimental epidermal calcinosis.*

Figure 68 (from Hall *et al.*, 1966) is a montage of microprobe calcium x-ray scan-images of a transverse section of human aorta, age sixty-nine years. The images go all the way from the inside to the outside of the arterial wall, and it is immediately evident that the calcium-rich plaques are present at all levels. X-ray analysis of these plaques invariably gives a calcium/ phosphorus ratio similar to that of the mineral apatite, $3 Ca_3 (PO_4)_2 \cdot Ca (OH)_2$ (see also Tousimis, 1966).

The *pre-stages* of mineralization are to be found in the areas *adjacent* to the plaques. The calcium and phosphorus concentrations in these areas are much less than in the plaques (going down to calcium weight-fractions of approximately 0.1%), so that probe analysis requires techniques which avoid background from the specimen support. Höhling *et al.* (1967b) have analyzed such areas in 5μm sections of formalin-fixed aorta mounted on very thin nylon films, using the theory of Marshall and Hall (Sec. 5.E.7.b) for quantitation. The results are shown in essence in Figure 69. In spots with very low calcium weight-fractions, one may find ratios of Ca/P well below apatitic (presumably intracellular spots containing phosphates in nucleic acid and other compounds) and Ca/P ratios far above apatitic (presumably extracellular, with very little phosphorus). At higher calcium weight-fractions, perhaps ½ to several percent, the Ca/P ratio is always far above apatitic. Finally at still higher calcium weight-fractions, the apatitic ratio is approached. (There is evidence, not entirely convincing statistically and therefore drawn as a dashed line in Figure 69, that the ratio first drops *below* apatitic, perhaps indicating a phase of octacalcium phosphate.) These observations show plainly the existence of a phase of accumulation of cal- cium which is not bound to phosphate.

The observed Ca/P ratio depends very much on preparative technique.

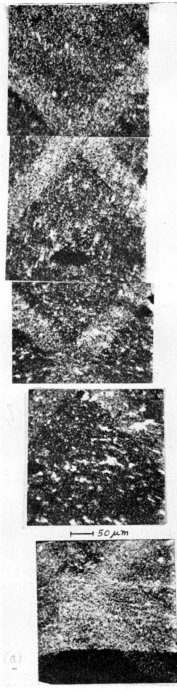

Figure 68. Calcium x-ray images showing the distribution of calcium in a transverse section across the entire wall thickness of a human aorta, age sixty-nine years. (Montage, with successive images aligned for continuity of fields. The grid pattern is due to background from a supporting nickel grid.) (From Hall *et al.*, 1966. Courtesy of John Wiley and Sons.)

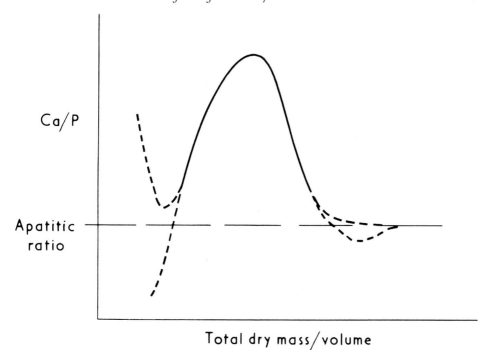

Ca/P

Apatitic
ratio

Total dry mass/volume

Figure 69. Stages of mineralization in aorta, graphically summarized. There is an accumulation of calcium, not bound to phosphate, at an intermediate stage prior to plaque formation.

In the studies of aorta, not enough frozen sections were examined to provide a good comparison with the formalin-fixed tissue. However, in other tissues there have been extensive direct comparisons of sections of chemically fixed tissue with dried sections cut from quickly frozen blocks, and with ultrathin sections cut from freeze-dried blocks after embedding in Araldite. Both in the soft-tissue matrix adjacent to developing dentine (Höhling *et al.,* 1968) and in mineralizing cartilage (Höhling *et al.,* 1970) the Ca/P ratio in the prestages is again far above apatitic in the chemically fixed specimens, but the ratio is well *below* apatitic when liquid fixatives have been avoided. Thus it seems clear that the premineralized tissue contains much phosphorus in compounds which are lost during wet preparative procedures. This does not affect the important conclusion, drawn from the studies of the chemically fixed tissues, that the calcium is not bound to phosphorus at this early stage.

How then is the calcium bound? To check the possibility that it is bound to sulphated mucopolysaccharides, some years ago Hale *et al.* (1967) tried to correlate calcium and sulphur x-ray scan-images in hardening arteries, both untreated and pretreated with hyaluronidase in order to remove one potential calcium-binder in advance. Since the sulphur is present also in presumably irrelevant amino acids, they also tried to correlate calcium and iron after applying an iron-stain for mucopolysaccharides. The results were not conclusive. But such selective chemical pretreatments to remove or stain certain components seem essential if one is to apply microprobe analysis constructively to this important problem. Chemical treatment is being combined with static-probe quantitative analyses for Ca, P and S in work now in progress (Höhling *et al.*).

Pathological calcification occurs of course not only in arteries but in many other tissues. Gabbiani *et al.* (1969) have applied the electron microprobe to the study of the role of *iron in experimental cutaneous calcinosis*. Calcification of the skin (calciphylaxis) was induced in rats by the administration of $FeCl_2$ ($500\mu g$ in 0.5 ml of H_2O) twenty-four hours after the administration of dihydrotachysterol. In the electron microscope, selective deposition of iron was seen to occur on collagen fibrils. When $FeCl_2$ alone was administered, mineralization did not occur but microprobe analysis established that local levels of calcium and phosphorus did nevertheless actually rise. Apatite formation occurred only in animals which had received the DHT as well.

4.A.4. Bone

We shall pass over the many electron-probe studies of normal bone (Mellors, 1964, 1966; Baud *et al.*, 1963; Tousimis, 1963; Burny *et al.*, 1968; Suga, 1967, 1969), and we shall discuss only two studies of direct medical interest.

Baud *et al.* (1968) have studied the distribution of *strontium* incorporated into growing bone. Mice were fed a diet containing strontium as lactate and phosphate. Microprobe analysis showed with high precision that much strontium was incorporated into the layers of bone formed during treatment, but very little strontium found its way into the older bone tissue. (For specialized technique, and some preliminary biological results, see also A. R. Johnson, 1969.)

Dihydrotachysterol (DHT) is known to affect calcium metabolism. In the electron microscope, after DHT treatment, one may see numerous osteocytes with a finely dispersed, electron-dense material in their mucopolysaccharide sheaths, surrounded by decalcification "haloes" (Remagen *et al.*, 1968). By probe analysis of ultrathin sections in the combination electron microscope-microanalyser, Remagen *et al.* (1969) have shown that in such sheaths the calcium weight-fraction may be as high as several percent, while the Ca/P ratio is, loosely speaking, in the neighborhood of apatitic (see Fig. 70). These findings are consistent with the view that in the course of the pathologically high calcium turnover induced by DHT, calcium salts are set free from the mineralized matrix around the osteocytes and are taken up in the cell sheaths.

The microprobe can readily localize histologically a number of minor

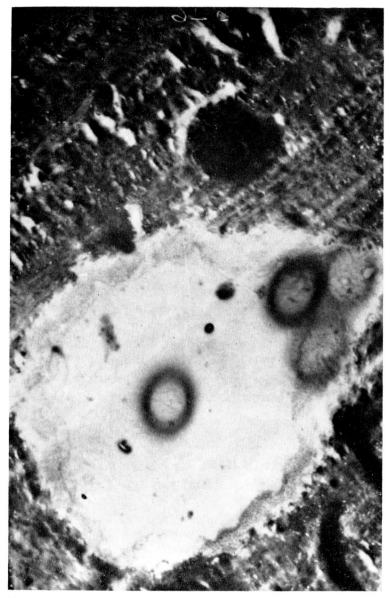

Figure 70. Microprobe analysis of an osteocyte, the cell sheath, and the surrounding mineral in an ultrathin section from an animal subjected to dihydrotachysterol. The analyzed areas can be identified by the contamination spots left by the electron probe. (From Remagen *et al.*, 1969, courtesy of Springer Verlag.)

bone elements, for example sulphur, zinc, and magnesium (Suga, 1969). A different type of analysis, quite desirable in bone pathology but reliably achievable only with great care, would be the accurate local measurement of Ca/P ratios. While it has been shown that consistent Ca/P ratios can be observed under uniform normal conditions (Mellors, 1964 and 1966), it is very difficult to get accurate *quantitative* results in diseased mineralized tissues because of the large and variable effect of x-ray absorption in the tissue. (With regard to this effect, see Sec. 4.B.1.e, especially Fig. 74; also Frazier, 1966.)

4.A.5. Teeth

The hard tissues of teeth, enamel and dentin, are mainly highly structured, inorganic, semicrystalline materials whose properties depend fundamentally on the concentrations of certain chemical elements. Not only calcium and phosphorus but also the minor elements sulphur, magnesium, zinc, fluorine, chlorine and potassium are or may be fundamental. Since local variations or local defects in enamel and dentin are reflected in altered local elemental concentrations, electron-probe analysis is a powerful tool for the study of dental microstructure and dental disease.

Here again we shall not dwell on studies dealing simply with technique or with normal material. For general and technical studies see Boyde *et al.*, 1961; Boyde *et al.*, 1963; Rosser *et al.*, 1967; Baud and Lobjoie, 1966a; Frazier, 1966. For studies of minor elements see Söremark and Grøn, 1966 (chlorine); Andersen, 1967 (carbon, nitrogen, sodium, magnesium, chlorine and, very slightly, fluorine); Suga, 1969 (sulphur, magnesium and zinc); Wörner and Wizgall, 1969 (fluorine). We shall discuss three types of medical study: assays of calcium and phosphorus in carious lesions, assays of minor elements with direct medical significance, and studies of dental fillings.

Calcium and Phosphorus in Carious Lesions

Frank *et al.* (1966) have compared calcium and phosphorus concentrations in "caries-susceptible" and "caries-resistant" teeth. They studied two groups of "resistant" teeth: unerupted healthy bicuspids from patients nine to twelve-years-old, and erupted bicuspids from patients fifteen to thirty-years-old who had neither cavities nor fillings. The "susceptible" teeth came from patients fifteen to thirty-years-old, who presented a great number of fillings and active caries. There did not seem to be any significant difference among the groups in average calcium and phosphorus concentrations, either in enamel or in dentin. (However, the dentin in the unerupted teeth was mineralized much less than in the erupted teeth.)

The same authors analyzed individual lesions in the enamel of carious teeth. Figure 71 shows a typical linear scan across such a lesion. Above most

Figure 71. Variation of calcium x-ray intensity along a linear scan across an incipient carious enamel lesion. E: enamel surface. E-S: enamel surface layer (30μm thick). R: loci of calcium decrease at stria of Retzius. C: boundary between enamel carious lesion → (to the left) and optically normal enamel (to the right). (From Frank *et al.*, 1966, Courtesy of the International Association for Dental Research.)

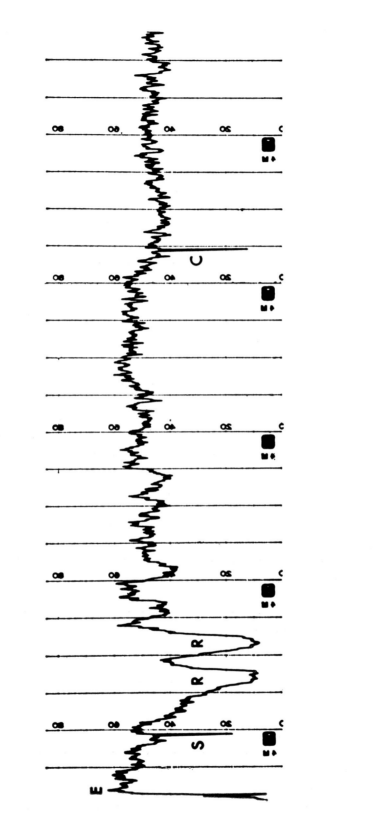

lesions there usually was an intact surface zone of the enamel, approximately 30μm thick, in which the Ca and P concentrations were normal. Both Ca and P were depleted in the underlying lesions, with depletion along the striae of Retzius greater than in the enamel at the same depth between the striae.

Individual incipient carious lesions in human teeth have also been analyzed by Frazier (1967). He reported that at the surface of a natural white spot lesion of enamel, the degree of mineralization was sometimes higher and sometimes lower than in the adjacent surface enamel. At the level of the lesion itself he found hypermineralized areas adjacent to extremely decalcified zones, with low Ca/P in the hypermineralized areas due to excess phosphorus, and high Ca/P in the hypomineralized areas due to very low phosphorus.

Andersen (1967) has assayed carious lesions in both enamel and dentin. In linear scans, he found depressed concentrations of both Ca and P in every lesion. In healthy teeth the ratio Ca/P was the same across enamel and dentin; in dentinal lesions the ratio Ca/P was reduced and in lesions of the enamel it was elevated.

Minor Elements

Frank *et al.* (1966) have observed that the concentration of *chlorine* is depressed in carious lesions of enamel in parallel with the depressions of the concentration of Ca and P. They noted further that in those cases where there was no intact surface layer of enamel above the lesion, the chlorine level was again low, in parallel with Ca and P, in the lesion at the surface.

There is a contrary comment by Andersen (1967): ". . . in the region of caries formation, Na and Cl appear to be little affected by tooth deterioration, maintaining the distributions observed in healthy enamel." However Andersen's comment seems to be more tentative and based, in the case of Cl, on few observations. He assayed *magnesium* more systematically and extensively in carious lesions, in both enamel and dentine, and found that in severe lesions the reduction in Mg was even more pronounced than the drop in Ca and P levels.

Baud and Lobjoie, 1966b, have studied the distribution of *fluorine* in the enamel of teeth which had been exposed for fifteen minutes to a 1% aqueous solution of sodium fluoride and then cleaned sufficiently to remove salts adhering to the surface. By comparing fluorine x-ray scan images of treated and untreated teeth they showed that the treatment produces a band of elevated fluorine concentration perceptible to a depth of approximately 10μm below the surface (see also Sheble and Ocumpaugh, 1971).

Saffir and Ogilvie (1967) have studied the effects of *cariostatic agents* (*fluoride* and *sodium trimetaphosphate*) on the distribution of elements in the teeth of rats placed on a cariogenic diet. In addition to a control group on the diet, they studied the teeth of rats on the same diet supplemented with either or both of the preventive agents. In comparison with the control group, caries were significantly reduced in all of the test groups. There was no apparent change in elemental microdistributions in the fluoride group, but changes were seen in the phosphate group, especially in the distribution of potassium. This element was found normally to be uniformly distributed throughout the enamel, and uniformly distributed at a higher concentration in the dentin. In the group on the phosphate-supplemented diet, the distribution in dentin remained uniform, but a gradient developed in the enamel with the concentration at the outer surface 35 percent less than at the dentin-enamel junction.

In later reports in their series of studies, Saffir and coworkers have concentrated on the localization and assay of fluorine itself and on local Ca/P measurements after the administration of the same two cariostatic agents. By an extraordinary perversity of nature, when fluorine x-rays are detected with diffracting crystals, there may be serious interference from higher order reflections of the radiations from just two other elements—namely, calcium and phosphorus! Special techniques have been developed (Saffir *et al.*, 1969, 1970a) to overcome this interference and achieve analyses with the desired sensitivity for fluorine in dental specimens. The fluorine level was observed again to drop very rapidly with distance from the surface of a tooth, much as reported by Baud and Lobjoie (1966b), and the administration of trimetaphosphate led to an increase in the concentration of fluorine at the tooth surface, and also led to an increase in Ca/P ratio at molar surfaces in contact areas (Saffir *et al.*, 1970b).

Besic *et al.* (1969) have done microprobe assays of Ca, P, Mg, Na, and Cl, in the form of linear scans, in undecalcified sections of teeth, including *mottled teeth affected by a high fluorine level* in drinking water. Their study was preliminary and demonstrated a technical capability, but the number of observations was not large enough to lead to biological conclusions.

With respect to pathological alterations in the concentrations of Ca, P and minor elements in enamel and dentin, it is important to keep in mind the warning already mentioned in connection with the hard tissues of bone: in and around small lesions in thick dense specimens, accurate quantitative analyses are impossible because one cannot correct accurately for x-ray absorption in the specimen. The difficulty is illustrated schematically by Figure 74 (Sec. 4.B.1.e). Qualitatively, the apparent pattern of changes will generally be valid, although small apparent changes may be due to altered electron penetration and x-ray absorption in or around a lesion rather than real changes in the concentration of an assayed element (cf. Boyde *et al.*, 1963; Frazier, 1966).

Dental Fillings

Teeth sometimes become discolored after being filled with an amalgam of silver, tin and mercury. Wei and Ingram (1968) have used microprobe analysis to clarify the causes of this effect. They sectioned discolored, filled teeth and performed static-probe analyses in the amalgam and in the dental substance, as well as linear scans across the boundary between them. In the amalgam itself mercury was seen to be homogeneously distributed, while the distributions of silver and tin were inhomogeneous, with high concentrations of tin in regions void of silver. Tin concentrated at the interface, and migrated consistently into enamel and dentin (with the higher concentration in the dentin). The other two elements neither migrated into the dental substance nor concentrated at the interface. Evidently the discoloration of dental substance by old amalgam fillings is due to tin, rather than mercury as had been supposed.

Bax and van der Linden (1968) have similarly studied diffusion from amalgam into tooth, but have especially looked into the effect of the zinc phosphate cement which is often used. With an amalgam containing Hg, Ag, Sn and Cu they confirm that in the absence of cement, only Sn concentrates at the interface and diffuses into the dental substance. When the cement is used, the diffusion of Sn into the tooth is blocked but instead there is a substantial diffusion of zinc.

Von Rosenstiel *et al.* (1968) have not studied filled teeth but have concentrated on the metallurgy of the amalgam itself. They have used the microprobe to identify the phases at different preparative stages of Ag-Sn-Cu amalgam. Initially copper at concentrations below 2% was found to be dissolved in the Ag_3 Sn phase, while any higher amounts of copper went into a sharply segregated Cu_3 Sn phase. After casting, instead of a simple mixing of these two phases, they observed *two* Cu-Sn phases distinguished by different copper contents; and after a final "homogenization," while the system reverted to a single Cu-Sn phase, two Ag-Sn phases could be distinguished.

4.A.6. Histological Stains

There have been several microprobe studies of histological stains. Although so far none seems directly related to medicine, this class of work must be discussed because medical applications will obviously be forthcoming.

The microprobe can directly confirm how faithfully a stain displays the distribution of its target-compound, when the stain and the target both contain identifying chemical elements. For example, the performance of the von Kossa "calcium" stain can be precisely determined by simultaneous microprobe assays for calcium and for the silver in the stain. Thus it was confirmed by Gardner and Hall (1969) that the stain localizes quite well to regions containing calcium phosphate in mineralizing cartilage, while lower concentrations of calcium in phosphate-free regions are not stained. This is just as expected from the chemistry of the stain.

Scheuer *et al.* (1967) have used Timm's method to localize copper in

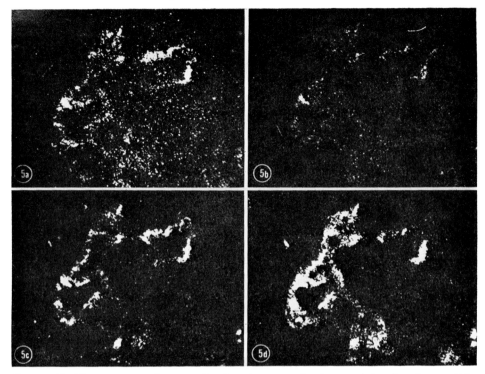

Figure 72. X-ray images showing the distributions of calcium, (a), phosphorus (b), cobalt (c) and sulphur (d) in a section of rat kidney taken to the cobalt sulphide stage of the Gomori method for alkaline phosphatase. (From Hale, 1962, courtesy of the Rockefeller University Press.)

electron micrographs of rat liver. In this method copper sulphide is first formed, and then silver is deposited at the sulphide to give granules visible in the electron microscope. The granules were analyzed in the combination electron microscope-microanalyzer (Sec. 5.E.1) with the intention of demonstrating the spatial coincidence of granules, copper and silver. However, the analysis revealed that although the granules were indeed silver deposits, no copper was present—it had been removed in the preparative procedure. Subsequent tests with radioactive copper confirmed its complete disappearance.

Stains often involve several stages of chemical reaction before a visible product is achieved. One may wonder how faithfully the spatial distribution of the original target-compound is preserved at each stage. If the intermediate stages are identifiable by chemical elements, the electron probe can be used to check the spatial fidelity of the sequence. For example, phosphatase enzymes are often localized by adding substrate and calcium so that the calcium combines with the phosphate released by the enzyme; the calcium may then be displaced by added cobalt, to give cobalt phosphate (still invisible), and finally the phosphate may be replaced by added sulphur to give a visible cobalt sulphide. The fidelity of this procedure can be checked *at each stage* by electron-probe assays for phosphorus, calcium, cobalt and/or sulphur. For example, in the x-ray scan images of Figure 72 (from Hale, 1962) one sees the localization of all four of these elements at the final stage of staining for alkaline phosphatase in rat kidney. (Better images can be obtained now, especially for phosphorus, with more sensitive, modern microprobes. See also Engel *et al.*, 1968; Libanati and Tandler, 1969.)

When a reactant is observable in electron-scan or x-ray-scan images or by static-probe x-ray assay, one may be able to use it as a stain in the microprobe with no need to achieve an optically visible product. Potassium pyroantimonate can be used in this way as a stain for the intracellular localization of inorganic cations (Tandler *et al.*, 1970). Furthermore, an additional mode of quantitative histochemistry is obviously opened up if the reaction deposits some element stoichiometrically at the site of the target compound. Thus Sims and Marshall (1966) have pointed out the possibility of local assays for nucleic acid by means of measurements of chromium in the microprobe after staining the nucleic acids with gallocyanin chrome alum.

4.B. TECHNICAL ASPECTS OF MICROPROBE STUDIES

4.B.1. Specimen Preparation

4.B.1.a. Types of Specimen

Microprobe analysis can be applied to a wide variety of specimens. The chief restriction is to dried material, since the analysis must be performed in a vacuum. Feasible specimens include thick blocks of tissue, conventional sections a few microns in thickness, "ultrathin" (electron-microscope) sections, isolated cells, and microincinerated material.

With thick specimens the analysis is restricted to a surface layer defined by the range of the incident electrons. One can study natural surfaces, or inner areas exposed by stripping of histological layers, sectioning, sawing or freeze-fracturing. The *quantitative* analysis of thick specimens is generally not to be expected unless the surface is made flat in some way (see "hard tissues" below), because absorption effects in the irregular surface of a thick specimen are incalculable. But *qualitative* analysis in thick specimens is useful, and will become popular especially now that there are commercially available spectrometric attachments for scanning electron microscopes. Because of the tremendous depth of focus of the electron-backscatter image in these instruments, they can provide qualitative analyses of irregular surfaces with good histological correlation.

Tissue sections being the most frequent form of biological specimen for probe study, we shall separately discuss below the preparation of sections of hard tissues, sections of soft tissues and ultrathin sections.

Isolated cells can be deposited for probe analysis in smears of tissues or cell suspensions, by pipette from cell suspensions, by micromanipulation or in sprays.

Microincineration, by getting rid of irrelevant organic material, can make it easier to see the nonvolatile constituents of a specimen and can sometimes reduce x-ray background. The technique is limited to the study of relatively refractory material, and x-ray background will be reduced only if the beam can be confined to the analyzed microconstituent or if the specimen support is very thin or low in atomic number. A good example is the study of asbestos fibers in the lung (Andersen, 1967; see also Sec. 4.A.1).

4.B.1.b. The Conducting Coat

Since biological specimens are usually electrically nonconducting, electrical charge may build up within a specimen during probe analysis. This can lead to high local fields and discharges which make the probe jump about. Also, the electrostatic forces may be strong enough to distort a thin specimen gradually or quickly or to disrupt a weak one.

To avoid these effects, it is sometimes sufficient with thin specimens to mount them on a conducting support—for example, isolated cells placed on conducting silicon discs (Carroll and Tullis, 1968). With thick objects, antistatic agents have sometimes been applied to the bulk of the specimen (Sikorski *et al.*, 1968). But by far the most common procedure is to apply a conducting coating to the surface in a vacuum evaporator, as is often done for electron microscopy.

In thin specimens, such a conducting coat is near enough to all points within the specimen to prevent any significant accumulation of charge. In thick specimens, even though charge may accumulate well below the surface where there is a long discharge path to the conducting coat, the probe will not be deflected because it is screened by the conducting sheet at ground potential. Hence analysis may proceed, but it is important to realize that certain other bad effects may still occur (Sec. 4.B.2).

A continuous electrical path must be provided all the way from the conducting coat to ground potential. When the specimen is mounted on a large, nonconducting support, for instance a glass slide, it may be useful to paint a substantial conducting streak (usually colloidal graphite in water—"Aquadag") from the edge of the specimen itself to the rim of the metallic holder. Further, the conducting top of the specimen must not be isolated by sides which lack a coat due to "shadowing" in the evaporator. To avoid uncoated sides, as well as isolated patches in irregular top surfaces, it is best to tilt the specimen and to rotate it during evaporation of the coating (Boyde, 1967; Boyde and Wood, 1969). Evaporator accessories for this purpose are now commercially available. A conducting streak painted down the side of thick specimens may also be helpful.

The atomic number of the coating should be low to minimize x-ray absorption. A layer of carbon 100 to 200 Å thick, or a layer of aluminum 50 to 100 Å thick, conducts electricity sufficiently without interfering seriously with optical microscopy.

Since biological specimens are poor conductors of heat as well as electricity, the second important function of the conducting coat is to help to remove the heat generated by the impact of the probe electrons. Roughly speaking, there is often a danger of overheating when probe currents greater than 50 nA are required ($1 \text{ nA} = 10^{-9}$ ampere). An extrathick coating may then be used. The effectiveness of this tactic depends on the distance from the coating to the most remote plane reached by the electrons, where the highest temperature will develop. In thick specimens, a high probe voltage would make this distance too great for effective cooling, but high voltage would be undesirable anyway since it would spoil the spatial resolution (see Sec. 5.E.5.a). For thin specimens mounted on very thin plastic films, it is most effective to coat both top and bottom so that the biggest distance to a coating is half the thickness of the specimen rather than the entire thickness. An extra-thick coating can then be very effective; it has been shown (Hall *et al.*, 1966) that probe currents up to $1 \mu A$ (10^{-6} ampere) can be tolerated with no damage evident in $10 \mu m$ sections of soft tissue mounted on very thin films and coated on both sides to a total of 1000 Å of aluminum.

When thick coatings are needed for high thermal conduction, aluminum is preferable to carbon because Al has the larger coefficient of conductivity and is more readily deposited in heavy layers of controllable thickness. Carbon is usually evaporated from the pointed tip of a carbon rod pressed by a spring against another rod, and the evaporation ceases spontaneously when the tip becomes blunt, whereas one can readily totally evaporate a small measured mass of aluminum.

Thick coatings have inescapable disadvantages. Usually optical microscopy is limited to the reflection mode and the available optical images are serviceable but not as good as with uncoated or lightly coated specimens viewed by transmission. Also, soft x-rays may be seriously attenuated in a thick aluminum coat (Fig. 73).

4.B.1.c. The Specimen Support

Ordinary glass microscope slides, which can be accommodated in the specimen chambers of several commercial instruments, are often suitable as specimen supports and are, of course, especially favorable when one wants to view the specimen by transmission optical microscopy. High-purity quartz slides may be used instead when the electrons pass through the specimen and would produce too much background characteristic radiation

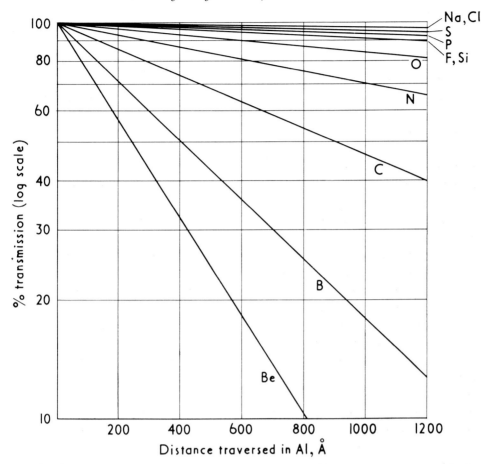

Figure 73. Transmission of the K radiations of the elements Be, B, C, N, O, F, Si, P, S, Na and Cl through aluminum.

in the many elements present in ordinary glass. Lucite or Perspex (methyl methacrylate) can be used at low probe currents, if the support must not contain silicon.

If inspection by transmission optical microscopy is not necessary, the most convenient supports are opaque materials with high electrical and thermal conductivity. For the inspection of thin specimens by reflected light, it is best for such supports to be highly polished. If the operating conditions (probe voltage and thickness, density and degree of porosity of the specimen) allow the probe electrons to traverse the specimen and strike the underlying support, the material must be selected to avoid the generation of excessive backgrounds of continuum and characteristic x-rays. Since the continuum background is proportional to atomic number, a material of low atomic number is best. Carbon blocks are cheapest and most readily obtained; beryllium has been used and gives the lowest continuum background of any thick support; and Carroll and Tullis (1968) have used single-crystal discs of silicon.

As to characteristic background, it is obviously undesirable for the support to contain much of the element of interest, and purity (absence of this element in the support) becomes very important when very small amounts must be assayed in the specimen. In extremely sensitive work one may wonder whether any thick support can be satisfactory; tiny inclusions of impurities can produce backgrounds under the probe even when the element is undetectable by bulk assay of the material. For example, such inclusions are sometimes found even in single-crystal silicon, a material which has been recommended as a support on account of its high purity (Beaman *et al.*, 1969).

Another type of specimen support is very thin plastic film, generally 1,000 to 2,000 Å thick, prepared in the laboratory. The manufacture of such films and the work of mounting specimens on them is much more laborious and delicate than the use of thick supports, and the end-product is fragile, but for certain studies the advantages of the thin supports are vital. In the first place, since they produce a very low continuum background under electron impact, much less than even a beryllium thick support (cf. Hall *et al.*, 1966), they should be used for measurements in thin specimens near the limit of detectability. Secondly, they permit the use of high probe voltages in the study of thin specimens since they do not spoil spatial resolution by electron backscatter, and the background from a very thin support does not increase with probe voltage. With high probe voltages the x-ray signal is not limited to the soft end of the x-ray spectrum. Finally, high voltages, low x-ray background and low electron-backscatter are all essential to the methods described in Sections 5.E.7.b. and 5.E.3 for quantitative local measurements of elemental weight-fractions, elemental amounts per unit volume, and total mass per unit volume in thin specimens.*

Thin plastic films can be prepared in the laboratory from Formvar or collodion by the methods well known to electron microscopists. There is some evidence that nylon films are more stable under the electron beam. To make nylon films, put DuPont "Zytel" nylon chips in isobutyl alcohol, 100 g of nylon per liter of alcohol, heat in a liquid bath and maintain a temperature of 90°C while stirring until the nylon dissolves (approximately 2 hr), cool to approximately 60°C and drop on the order of 0.1 ml of solution from a pipette onto a large surface of clean water. The nylon spreads to a thin film and may be handled from this point on like collodion. The film should be allowed to rest undisturbed for two to three minutes on the surface of the

---

* Commercial films such as 6μm Mylar, 2μm polycarbonates, or polypropylene stretched to 1μm thickness, are all too thick to offer the advantages described here. These commercial films are sometimes useful because they are readily available and easily cut to size, but in principle they offer no advantage over a Lucite slide.

water to polymerize. The stock gels on cooling and must be heated again for reuse, but dissolution then occurs much sooner and without stirring (for further details, see Brown *et al.*, 1948).

Nylon films are strong enough to cover large areas without any reinforcing grid. In fact a grid not only adds to the background generated by stray electrons, it *reduces* the stability of the preparation under high probe currents, presumably because of stresses caused by thermal gradients. However, for the study of EM specimens in combination electron microscope-microanalyzers, conventional preparations with Formvar-covered grids are almost always satisfactorily stable.

One can also drape sections across a grid without any supporting plastic film, in the hope of reducing background still further. This scheme is not often useful, partly because of the generation of background in the grid by stray electrons, but mainly because the plastic film is usually needed to stabilize the preparation. However, in the combination electron microscope-microanalyzer, a related electron microscopist's technique may be valuable; the specimens can be mounted on a plastic film containing tiny holes, and one can hope to find interesting areas of specimen suspended across the holes (Charles, 1966). Since the x-ray signals from "ultrathin" EM sections are often weak, the reduction in background at the holes can be important, but unfortunately it is far from total; perhaps half the background is found to remain, coming not from the supporting film but from stray electrons striking various things.

4.B.1.d. Loss or Displacement of Elements During Specimen Preparation

Tissues are quite commonly immersed in liquids during the preparative procedures of fixation, embedding, removal of embedding material, or staining. There is obviously a danger that such baths will inadvertently remove the interesting chemical elements or seriously alter their distribution before the specimen comes to be analyzed. The danger ranges from insignificant in some cases to prohibitive in others. We cannot enter here into a survey of the acceptability of various wet procedures in a variety of applications, but we shall summarize "dry" procedures which generally protect the integrity of the specimen much more reliably.

1. *Sections of Freeze-dried, Embedded Blocks.* A block of fresh tissue can be frozen rapidly by plunging it into a very cold bath (commonly isopentane or liquid propane). The freezing of the surface keeps the liquid out of the interior, and the rapidity of the freezing should prevent the formation of large ice crystals and keep the constituents in place fairly well. The tissue may then be dried slowly (over one or a few days usually) by sublimation in vacuum. After sublimation is complete, if sections are desired the block may be infiltrated in the usual way with one of the standard embedding media (for full details see Pearse, 1968). One should choose a medium

which is stable under the electron probe, to avoid the necessity of removing it with a solvent after sectioning. In this respect the epoxy resins are best, methacrylates are adequate in ultrathin sections, while paraffin is not stable under the probe.

This procedure can give high quality sections, and in practice there has been no evidence of the alteration of elemental distributions during in-filtration. The main disadvantage, aside from the slowness of the freeze-drying, is associated with the presence of embedding medium during analysis. Since the analysis refers to the specimen as presented to the probe, measurements of weight-fractions or total mass refer to the tissue itself plus the medium, rather than to the tissue alone, and generally one does not know in a local assay how much of the analyzed mass consists of embedding medium.

In microprobe work, the method has been applied to the study of the distribution of the diffusable electrolytes K, Cl and Na in 3μm sections of muscle and mucosa (Ingram and Hogben, 1968), and to the analysis of Ca and P in ultrathin sections of tissues at prestages of mineralization (Höhling *et al.,* 1970). An additional feature of the technique used by Ingram and Hogben was fixation of the tissue with osmium vapor after freeze-drying and before embedding. Osmium vapor can serve as both a fixative and a stain, revealing structure in optical transmission, EM, and electron-scan images. Compared to liquid fixatives or stains, a vapor is much less likely to disturb the distribution of chemical elements.

Läuchli *et al.* (1970) have developed an elegant procedure which they use for the preservation and determination of the distribution of electrolytes in botanical material. After freezing a block quickly, they freeze-substitute with anhydrous ether, and then infiltrate with their special low-viscosity epoxy embedding medium. Sections 1μm to 2 μm thick are cut with an ultra-microtome, with hexylene glycol in the microtome trough instead of water to avoid leaching material from the sections.

2. *Dried Sections of Frozen Blocks.* If one wants sections with a thickness $\geq 2\mu$m, a faster procedure is to freeze the block of tissue quickly, section the frozen block in a cryostat, and place the frozen sections, not yet dried, directly onto the specimen support. From this point the most impeccable practice is to dry the section by sublimation in vacuum, never allow-ing it to thaw. More often the section is allowed to thaw and is then dried in vacuum by evaporation or by sublimation after refreezing. At least there is no danger that constituents will be *lost* in thawing, and the danger of slight redistribution is usually not worrisome. A section takes only some minutes to dry, because it is so thin. Although fine structural detail is generally not preserved as well in dried sections of frozen blocks as in sections of freeze-dried, embedded blocks, at thicknesses $\geq 2\mu$m the quality of the cryostat sections is adequate for virtually all microprobe studies, and

the cryostat method is preferable because it is fast, there is no infiltration by liquid embedding medium, and there is no embedding medium in the end-product.

Cryostat sections thicker than 2μm have been common in histology for many years. *Ultrathin* cryostat sections, untouched by solvents and good enough for EM work, have not been routinely available although specialists have recently developed methods to produce them (Appleton, 1970, 1971; Christensen, 1969; Hodson and Marshall, 1970). As yet these methods are not widely established, so the sectioning of freeze-dried, embedded blocks is still the prevalent technique to preserve the original elemental distribution in sections which are to be viewed by transmission electron microscopy.

3. *Air-drying.* Preparation can be extremely simple for thin specimens which do not have to be sectioned, for example smears of isolated cells. Such material would be rapidly dried in the vacuum of the microprobe instrument if introduced wet, but since it deteriorates if left wet and unfixed on the specimen support, a prompt preliminary drying may be desirable. This drying can be achieved in a minute or so by exposure to a stream of warm dry air.

Andersen (1967) has directly compared microprobe assays in nucleated red blood cells prepared by air-drying, freeze-drying and wet chemical fixation. His data emphasize the danger of elemental loss in the wet methods. The same lesson is taught by the comparison of ultrathin sections prepared by conventional wet methods with thicker sections of frozen blocks (Höhling *et al.*, 1968), and with ultrathin sections cut from freeze-dried embedded blocks (Höhling *et al.*, 1970).

It must be recognized that none of the forementioned procedures can really strictly preserve the original spatial distribution of elements in a tissue. In all of these procedures fixation, by dehydration or other means, is used to stop chemical reactions and prevent deterioration, and dehydration occurs anyway in the vacuum of the microprobe instrument when not imposed sooner. But the removal of water *must* disturb the distribution of the elements which were in solution. These elements may be lost (likely in wet fixation) or deposited on adjacent surfaces (likely in the freeze-dehydration techniques). If the material is ultimately embedded, these elements *may* end up suspended in the embedding medium (not too likely, and certainly not easily controlled). Hence a procedure which would be much closer to ideal, especially in the study of electrolytes, would be to freeze the tissue block rapidly, section, and analyze the section *without dehydration*, maintaining at all times a temperature so low that deterioration and sublimation would be negligible, and the water would be retained as ice, even locally under the electron beam during analysis. Temperatures near that of liquid nitrogen, if maintainable throughout, would certainly suffice. Preliminary experiments (Echlin, 1971) have shown that this radical procedure may indeed be feasible. When and if the method comes into general use, it will be necessary to recognize the limitation imposed by the finite speed of freezing of the tissue block, and the inevitable fine-scale displacements of solute elements around the crystallites of pure ice which must form as the front of freezing advances through the block, but the preservation of electrolyte distributions should be much better than achieved with the earlier techniques.

We must consider finally an even more radical "preparative" and analytical procedure. The specimen may be contained in a special "environmental cell" during analysis, so that its environment can be controlled—held at atmospheric pressure and normal humidity if desired—while the column of the electron probe instrument remains at vacuum. Remarkable transmission scanning images have been obtained with such an environmental cell even at the relatively low column voltage of 20 kV (Swift and Brown, 1970a). The possibilities for x-ray elemental analysis remain to be established, but it seems that some types of specimen should be analyzable in a state very close to natural.*

4.B.1.e. Problems Especially Associated With Sections of Hard Tissue

The hard tissues of bone and teeth differ from other biological tissues in being predominantly mineral, while they differ from most minerals in containing organic and fluid phases.

Hard tissues are dense and have a high mean atomic number. Consequently, except in extremely thin sections, x-ray absorption within the specimen is always considerable. Accurate microprobe analysis is then possible only if the specimen surface is smooth and flat; otherwise surface irregularities produce large irregular absorption effects.

Most often, hard tissues have been prepared for probe study by embedding according to standard histological practice, sectioning with a heavy-duty microtome, and grinding and polishing according to standard mineralogical practice (for details see the references of Sec. 4.A.4 and Sec. 4.A.5). However, as remarked already, embedding entails the risk of elemental displacement unless a "dry" method is used, and polishing entails a considerable risk of contaminating the surface with abrasive, distorting the surface, or redistributing the surface material in ground form. For the production of sections Andersen (1967) therefore recommends simple cut-

---

* A major problem with environmental cells is to provide walls strong enough to hold against the pressure differential, yet thin enough to pass the electrons without too much scatter. The transmission scanning images (obtained as explained in Sec. 5.E.1) have yielded a spatial resolution better than was generally expected, perhaps because the scattered electrons do not contribute to these images, at least when the electron detector is small. However, the scattered electrons *will* of course continue to excite x-rays in the specimen, so that the wall of an environmental cell may degrade the spatial resolution of x-ray analysis much more than it degrades the spatial resolution of the scanning electron image.

ting of unembedded material followed by drying. He uses a diamond saw (diamond $<1\mu$m) for sections of approximately 100μm; for sections of a few μm ($<10\mu$m), he freezes the tissue for greater rigidity and uses a microtome in a cryostat. The resulting surface may be sufficiently flat and smooth, though not as fine as a polished one. Drying is usually desirable immediately after cutting to prevent deterioration of the organic phase, and it can be accomplished in a vacuum, or alternatively, since the material is relatively refractory, in a drying oven.

In fact, it is not to be assumed that grinding and polishing will always produce a smooth surface. When there are local variations of hardness, as occur in the hard tissues with their inorganic and organic phases, different areas may be worn away at different rates, leaving local protrusions or depressions. The final surface should always be inspected critically.

A further degree of freedom in preparation of the surface is introduced by "machining" with electron or ion beams. Boyde *et al.* (1963) have studied dental surfaces prepared by etching with a 5 kV argon ion beam. The beam was produced by a special attachment to a scanning electron microscope, so that they could watch the surface as it was treated. They did not produce extremely flat surfaces (indeed they were primarily interested in the structural implications of the observed local differences in rates of erosion). However, a surface which is seen to be rough in a scanning electron microscope may still be quite smooth enough for microprobe analysis, and the method avoids the forementioned hazards of mechanical grinding and polishing: contamination with abrasive, distortion of the structure, and redistribution of ground material over the surface.

The inhomogeneity of hard tissues, in conjunction with their high x-ray absorption coefficient, raises another serious barrier to accurate quantitative analysis. The electron probe penetrates to very different depths in pores, organic regions and inorganic regions, and there are severe local variations in the attenuation of the emerging x-rays (see Fig. 74). A partial solution to this difficulty is to remove the organic and liquid phases and infiltrate with a medium which is similar to the inorganic phase in its electron stopping-power and x-ray absorption coefficient. The inorganic phase of the tissue may then be analyzed without unpredictable absorption effects. More explicitly, if the electrons of a static probe can be confined to the inorganic tissue itself, then it will not matter if the emergent x-rays have to pass through some of the added medium; a full quantitative analysis will still be possible (see Sec. 5.E.7.a). On the other hand, if the probe electrons dissipate some of their energy in the medium, the measurement of weight-fractions must refer to the local aggregate of tissue and medium rather than to the tissue alone; however, *ratios* of elemental concentrations in the inorganic tissue will be validly obtained so long as the medium does not contain these elements.

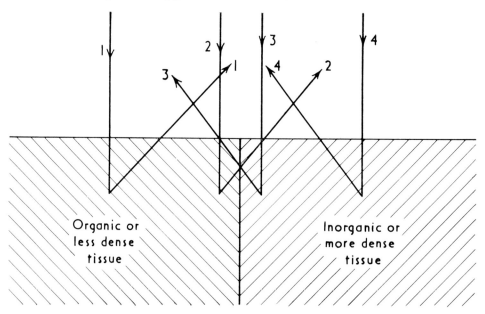

Figure 74. The uneven attenuation of x-rays in inhomogeneous hard tissue. Signal 2 is attenuated more than signal 1 and can speciously suggest lowered concentrations in the less dense phase adjacent to the denser phase. Signal 3 is attenuated less than signal 4 and can speciously suggest elevated concentrations in the denser phase at the junction. Even the signals unaffected by inhomogeneity—signals 1 and 4—are differently attenuated, and can be compared accurately only if they are known to be unaffected by inhomogeneity and the differing attenuations are taken into account (cf. Frazier, 1966).

Sodium tungstate was used as a matched medium by Boyde *et al.* (1963), but they found it unsatisfactory because of the low penetration of electrons into tungstate. Frank *et al.* (1966) seem to have been more successful with potassium iodide.

In sufficiently thin specimens, the probe electrons lose only a small fraction of their energy, x-ray absorption is negligible, and a completely different theory of quantitation can be used (Sec. 5.E.7.b). However, for specimens of hard tissue to be sufficiently thin in this sense, the thickness must be of the order of 0.1μm or less—in the range of EM "ultrathin" sections.

4.B.1.f. Problems Especially Associated With Sections of Soft Tissue

The requirement of a smooth flat surface, which must be imposed in microprobe analyses of hard tissues as well as in mineralogical and metallurgical analysis, is generally not as stringent for soft-tissue studies because of the much lower x-ray absorption coefficient. The face of a section of embedded tissue is certainly amply smooth without further treatment. Unembedded sections are generally all right when mounted evenly on flat supports. It is only when a section is grossly irregular (curled up or with projecting filaments, for example), or when long-wave-length x-rays of low penetrating power are to be recorded (roughly speaking, wave lengths longer than the 8-Å K-radiation of aluminum, atomic number $Z = 13$), that the irregular x-ray absorption associated with an irregular surface becomes important.

The difficulties of quantitative analysis in soft tissues stem from the low electron stopping power and from inhomogeneity. The situation may be summarized as follows:

In thick unembedded specimens there will almost always be substantial x-ray absorption because the radiation will come largely from deep down if a high-voltage probe is used, and only soft, poorly penetrating x-rays will be generated if a low-voltage probe is used. The inhomogeneity of dried un-embedded tissue makes it impossible to take account of absorption accurately.

In thin unembedded sections, because of the porosity or inhomogeneity of dried tissue, even low-voltage electrons may pass completely through some areas, and it is impossible to satisfy reliably the condition on which the conventional thick-specimen theory of quantitation rests, namely that the energy of the probe electrons is reduced at least down to the excitation energy of the recorded x-rays before the electrons can pass through the specimen.

Consequently, for reliable quantitative analysis, either the specimen must be embedded, or it should be quite thin and mounted on a very thin support to permit the application of a totally different theory of quantitation.

If the section is embedded in an organic plastic the electron stopping-power and the x-ray absorption coefficient will be similar in the tissue and in the medium, and the feasibility of quantitative analysis will be just as discussed in the preceding section. A "dry" embedding procedure should be used to preserve the spatial distribution of the assayed elements. The re-quirements of the thick-specimen theory can be reliably satisfied even in rather thin embedded sections, so long as the probe voltage is not much in excess of the excitation energy of the observed x-rays. Permissible probe voltages can be determined from Figure 106 (Sec. 5.E.5.a) showing the maximum range over which electrons of a given initial energy can produce a given characteristic radiation. In thicker embedded sections of soft tissue there is no danger of electrons going all the way through the specimen, but the lateral spread of the electrons is similar to the maximum range, which would be some tens of microns at high voltage, so that low probe voltages are still essential for the sake of good spatial resolution.

The alternative mode of analysis is to study thin unembedded sections so that x-ray absorption is slight, and to mount the sections on very thin supports and use high probe voltages so that the energy of the probe electrons within the specimen is nearly constant. The thin specimen theory of Section

5.E.7.b is then applicable. For soft tissues this procedure is valid, with probe electrons of 35 to 40 keV and observed x-ray wave lengths less than 8 Å, for section thicknesses up to several micrometers. So the preparative task is to mount unembedded sections of ordinary thickness on very thin supporting films.

In the author's laboratory the usual specimen support is nylon film, 1000 to 2000 Å thick, stretched over the end of a 1cm aluminum tube. The nylon film is first formed as a layer floating on water as described in Section 4.B.1.c. At this point there are innumerable ways of handling the material, but most often we transfer the film to the end of a tube by holding the tube in the water under the film, raising the tube so that the film is stretched around it, cutting away the film around the tube, and getting rid of excess water with tips of absorbent paper. Alternatively, one may have the film floating in a container with a drain at the bottom, with several of the tubes placed on a raised, perforated base under the water surface; one opens the drain, lets the film settle on the tubes, and cuts the film around each tube. The second method covers several tubes simultaneously, while the tubes are capped one-by-one in succession in the first procedure, but in both cases many tubes may be prepared from a single floating film. The specimen can be maneuvered onto the film at any time after the film has been mounted onto its tube.

To obtain unembedded sections, one can either section embedded tissue and subsequently remove the medium, or avoid embedding altogether by sectioning frozen tissue. If embedding medium is to be removed, an easily dissolved medium like paraffin is preferable.

It is not easy to obtain flat deparaffinized sections mounted with good adhesion on thin film. The following procedures have been used successfully:

1. (A. J. Hale, 1965, unpublished). Float the paraffin-embedded section on xylol to deparaffinize. Transfer the section with a fine glass rod through absolute alcohol and dilutions of alcohol into water. Scoop the floating section from the water onto the film. Dry for twenty-four hours in an oven at 37°C.

2. (R. T. Sims, 1966, unpublished). Flatten the paraffin ribbon or section by placing it on a drop of water on a warm microscope slide. Let slide cool and float section onto water. Scoop section onto filter paper and let it dry *thoroughly*. Carry the section on the filter paper through xylol, two washes of absolute alcohol, and diluted alcohols, into water. The section should float off the paper in the water. It can be sucked up onto the tip of a blunt pipette and expelled onto the film. The section may spontaneously lie flat, or else it may be teased into flatness.

3. (R. T. Sims, 1967, unpublished). Flatten paraffin section on microscope slide as above. Decant water and dry the section on the slide in an oven at 37°C overnight. Apply solutions by pipette: repeated applications of xylol

for approximately ten minutes, followed by absolute alcohol, diluted alcohols and water. (If the section is sufficiently stable, the absolute alcohol may be followed directly by water, and the turbulence when the water hits the alcohol may simplify removal of the section from the slide.) Lift the section from the water on the slide with a fine tool (a single hair from a camel's hair brush is recommended, taped to a handle). The section curls onto the tool. Put a drop of water onto the nylon support-film and transfer the section to the drop. It should flatten onto the nylon as the water is gently removed, usually by absorption into lens tissue or other paper.

4. (R. Fearnhead, 1966, unpublished). Float paraffin section onto warm water to stretch it flat. Scoop section onto nylon film, let it dry *thoroughly* in desiccator. Immerse in xylol for a minute or more; gentle agitation may be helpful. Remove xylol in vacuum desiccator.

5. (R. Höhling, 1966, unpublished). Transfer the section, held on dissecting needle or brush, through xylol, absolute alcohol and diluted alcohols. Put a drop of water on the nylon film and deposit the section on the water. Gently remove water with absorbing paper as above.

It is clear that during preparation, de-embedded sections are much exposed to liquid baths. Unless it is known that such treatment will not spoil the intended observations, it is better to study frozen unembedded sections. Such sections will generally adhere well if they are transferred directly from the microtome to the nylon film and are allowed to thaw, briefly at least. (Sometimes there is electrostatic repulsion between film and section, which may be overcome according to A. J. Hale by applying a preliminary slight evaporated coating of carbon to the film.) Adhesion is less predictable and may be a problem when the section is not allowed to thaw before it is dried by sublimation, in order to minimize the redistribution of solutes.

Figure 75 shows a typical $5\mu m$ section mounted on a nylon film.

4.B.1.g. *Problems Especially Associated With EM Sections*

Microprobe analysis can readily be applied to ordinary EM sections in the combination electron microscope-microanalyzer. The sections should be coated with carbon to remove electrical charge, as is common in electron microscopy. The problem of artefacts usually does not enter when one merely wants to characterize chemically inclusions which are seen in the transmission EM images of a given specimen. But if one wants a chemical analysis of the original tissue, including components which are easily lost or displaced in preparation, then sections should be prepared by a "dry" method—either freeze-drying of the block followed by vacuum-embedding and sectioning, or sectioning of the frozen block in a cryostat (cf. Sec. 4.B.1.d).

Difficulties can arise in the analysis of regions close to a bar of the supporting grid. Excessive background may then be produced in the grid by stray electrons, or the bar may block the path of the x-rays to the spectrometer. Therefore a very fine supporting grid is unfavorable. Patterns with 80 lines per cm or 40 lines per cm have been satisfactory. Grids of high atomic number (gold or palladium) are disadvantageous as they produce a relatively high continuum background. Copper grids are used most often and nickel may be used instead when copper is to be assayed in the specimen.

In some commercial electron microscope-microanalyzers, the quality of the EM image is very good and there is no need to survey the specimen in

Figure 75. A 5μm section of soft tissue mounted on a thin nylon film. The nylon film in this case is stretched over the end of a short length of aluminum tube; the inside diameter of the tube is 1 cm.

a separate electron microscope. In commercial units which are modified electron-probe instruments, the EM image is not so good and preliminary survey and micrography in another instrument may be necessary. Good fields should be found by preliminary survey also when sections are sent for analysis from an EM laboratory to the analyst's laboratory. In such cases the sections should be mounted on "field-finder" grids—commercially available grids with code markings—and data should be supplied so that the analyst can find the best fields promptly. This "trivial" detail can save hours of precious instrument time.

4.B.2. Damage to Specimens in the Electron-probe Instrument

Specimens may be damaged within the instrument by the vacuum or by the electron beam. The effects of vacuum are fairly well known, largely from experience in ordinary electron microscopy. There may be shrinkage or distortion, and solutes in the aqueous phase cannot remain there since the aqueous phase is removed.

The likelihood of probe-induced damage depends on the toughness and thickness of the specimen, the volatility of the elements to be studied, and the probe current needed to get adequate x-ray intensity. Except for the purpose of highly quantitative analysis, the discussion of which we defer to the end of this section, there seems to be no question of overheating and little danger of probe damage with probe currents up to about 10 nA ($1 nA = 10^{-9}$ ampere). At higher probe currents one should be aware of the possibility of several types of damage, although currents up to approximately 50 nA are often tolerated with no special precautions.

At high current there may be gross damage—curling or disintegration of the specimen—presumably caused chiefly by overheating. There are also insidious kinds of damage which may be invisible:

> 1. Volatile elements may be removed at high temperature. Losses, or at least displacements, of sodium and potassium have been observed (Hodson and Marshall, 1971; Borom and Hanneman, 1967).
> 2. Some elements may be redistributed. In thick specimens, they may be moved away from the surface to a depth beyond reach of the probe. This effect has been seen chiefly in glasses, and experimental studies have left some confusion about the cause, but electrostatic forces due to accumulated charge seem to be involved (Vassamillet and Caldwell, 1968).
> 3. In inhomogeneous hard tissue, there may be no change apparent because the hard fabric remains intact, while more labile regions are damaged. The probe may even drill small holes, which can be overlooked although they are observable under careful inspection.

Another effect of the probe is to "crack" organic vapors and deposit a layer of carbon on the surface of the specimen. This "contamination" is familiar to electron microscopists. The rate of deposition depends on the tightness of the vacuum system, the types of pumps and oils employed (most of the vapor may come from rotary-pump oil during preliminary pumping!), the type of vacuum gaskets, the degree of vapor-trapping provided, the temperature of the specimen, and the length of time since the instrument was last opened to air. Efficient trapping near the specimen can reduce the rate of deposition drastically (Ranzetta and Scott, 1966), but in fact the contamination is unimportant in most studies anyway. Generally contamination is worrisome only if carbon is to be assayed, or if the recorded radiations are oxygen-K, nitrogen-K, or similarly soft L-radiations which can be significantly attenuated in the contamination layer, or if one must record a weak x-ray signal in ultrathin sections. In the last case the peak/background ratio may be poor and a large part of the background may come from the accumulating contamination, so that it may be necessary to average backgrounds recorded before and after the peak is measured in order to correct for background adequately.

In order to minimize the damage associated with high probe currents, the following points should be noted:

1. For the effective removal of heat there should be intimate contact between the specimen and its support. For this reason thick rigid supports should usually have a polished surface, and sections mounted on very thin flexible films should adhere well.

2. Very high probe currents can be tolerated by thin sections mounted on very thin films, coated heavily on both faces with aluminum (see Sec. 4.B.1.b).

3. Reinforcing grids impair the stability of very thin supporting films, presumably because of stresses set up by thermal gradients.

4. When there are small dense inclusions in a section of soft tissue, the maximum tolerable probe current at an inclusion may be much less than elsewhere. This is because the inclusion will absorb much more energy per unit volume at a given current. Hence the current may have to be reduced at dense inclusions to avoid overheating.

5. Paraffin-embedded sections do not tolerate high probe currents; the paraffin melts and warps. For the study of embedded sections, epoxy embedding media can be used.

6. Probe damage is rarely troublesome in the study of ultrathin sections in the combination electron microscope-microanalyzer. The *dissipation* of heat in an ultrathin section is quite effective since there is a short path to the conducting coating at the surface, while the *absorption* of energy from a high-kilovoltage probe is reduced in proportion to the thickness. Furthermore, it is not possible to put high probe currents into the very fine probes which are often needed in the study of EM sections. For these reasons, conventional preparations on grids are found to be stable in spite of the grids (point 3 above). The main risk occurs not during actual probe analysis, but during preliminary inspection under ordinary conditions of electron microscopy.

Since probe damage may occur in many ways, in any given case it is difficult to guarantee the absence of damage in advance. It may be important to observe the x-ray intensities immediately on a count-rate meter as the probe is brought onto fresh areas, and to watch for any changes in the readings. Amounts of individual chemical elements can be monitored through the intensities of their characteristic x-rays, and the stability of the matrix can be monitored in thin sections of soft tissue according to the method of Section 5.E.3.

When accurate quantitative analysis is intended, probe damage raises serious problems. The electron beam may remove considerable fractions of the elements O, H, C and N from organic specimens, so that the total dry mass may be considerably reduced in the electron-irradiated volume. Since the local mass-fraction of an element is the ratio of elemental mass to total mass, the reduction in total mass must lead to a proportionate error in the assay of elemental mass-fractions. At present there is not much knowledge about mass-loss effects. Lineweaver (1963) has presented data on the removal of oxygen from thick inorganic specimens (glasses); Stenn and Bahr (1970) and Bahr *et al.* (1965) have extensively studied mass loss from organic specimens under conditions of electron microscopy; and Höhling *et al.* (1972) have measured the loss of mass from 5μm sections of kidney under various operating conditions in the conventional microprobe. Stenn and Bahr argue that most of the loss results not from rise in temperature but from the direct rupture of chemical bonds and consequent loss of the gaseous products, and the data of Bahr *et al.* show that such loss can occur extremely quickly, at a very low electron dose, of the order of 10^{-10} coulomb/μm^2. This means that the damage would be done already in less than a second of irradiation with a microprobe even if the probe current were kept as low as 1 nA (an excitation much too low for x-ray analysis)! Clearly, while it is important to monitor the stability of the x-ray signals with time and to confirm that they do not change, the appearance of constancy under normal methods of observation is no guarantee that the specimen has not been altered—all of the changes may have occurred too quickly to be observed. The data of Höhling *et al.* suggest that under suitable operating conditions (thin specimens with adequate coatings, low probe currents, and thin-film specimen supports selected for stability) the mass-loss effect may be kept small, but it is urgent to obtain information soon with special, highly sensitive x-ray monitoring arrangements to pin down the effects which occur already at very low dose.

4.B.3. Histological Correlation

During microprobe study the analyst must be able to view the specimen well enough to put the probe where he wants. In most modern instruments the specimen can be viewed by an optical microscope while it is under the probe, so that the histologist can often have his accustomed view of stained specimens. Optical microscopy is most effective if the specimen can be mounted on a rigid, transparent support (Sec. 4.B.1.c) and if a thin conducting coating is sufficient (Sec. 4.B.1.b).

Electron-scan images (Sec. 5.E.1) are also available. The contrast in these images can be improved and particular histological features can be accentuated by the heavy-metal "electron-stains" commonly used in electron microscopy. The electron-scan image is not affected by the specimen support and is affected very little by a thick evaporated coating.

Staining usually entails a danger of displacing or removing the elements which are to be studied. The staining of freeze-dried tissues or sections by *vapors* should be more widely developed and used (as described by Ingram and Hogben, 1968), since the risk is presumably much less. But the cleanest procedure is to work with unstained material if an adequate image can possibly be obtained.

Optical microscopy is generally not very effective with unstained specimens of soft tissue. Of course one can locate isolated entities such as dispersed cells, and certain tissues may be viewed quite satisfactorily—for example, epiphyseal cartilage with its cells at different stages of degeneration and mineralizing trabeculae. Birefringent structures can be well seen with the good polarization-optics available in many instruments. But cytological detail is usually not apparent. Phase and interference optics, which have not been incorporated into electron-probe instruments, would not be much use because the specimens are generally dry and much thicker than optimum.

The electron-scan image of an unstained soft tissue generally shows only boundaries where there are large changes of density. Cytological entities like nuclei and mitochondria tend to be invisible. However, changes in surface *texture* can be sensitively observed, and on this basis nuclei may sometimes stand out after a preparative procedure like freeze-fracturing and freeze-etching. For such observations a high-quality electron-scan system is needed offering the choice of backscattered-electron or secondary-electron images (see Sec. 5.E.1 for the description of these imaging modes and the imaging techniques considered in the next two paragraphs).

Histologically and cytologically the best views of unstained tissue sections may be given by the conventional transmission images produced in combination electron microscope-microanalyzers, not only because of the high spatial resolution but mainly because of high contrast. To illustrate this point, Figure 76 compares images of sections of embryonic rat incisor, one obtained by conventional electron microscopy in the combination instrument and the other by means of backscattered electrons in a scanning electron microscope. The cell-free matrix is readily distinguished from the odontoblasts in the transmission image, but cannot be distinguished in the backscattered-electron image.

Transmission electron images can be obtained also in scanning electron microscopes and microprobes, simply by putting the electron detector on the exit side of the specimen. In current commercial instruments such images are below the standard of conventional electron microscopy in spatial resolution, but they are quite similar in contrast, and they are obtainable in specimens much too thick for observation in a conventional transmission electron microscope at 100 kV.

To achieve close histological correlation, it is not only necessary to see the specimen adequately; it must be possible to place the probe precisely. The point of impact of the probe is sometimes faintly luminescent, so that a

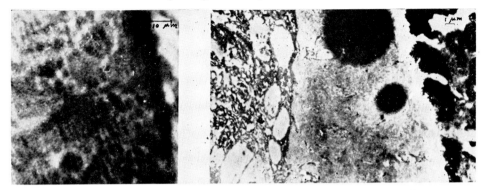

Figure 76. Micrographs of unstained sections of embryonic rat incisor. Left: Scan image, backscattered electrons, 6μm section of frozen tissue. Right: Transmission electron image, 1000 Å section. The mineralizing border is at the right of each image; it appears dark by transmission and bright by backscattering. The dark circles in the pre-dentin of the transmission image are contamination deposited during static-probe measurements. (From Hall and Höhling, 1969. Courtesy of Springer-Verlag.)

precise placement may be achieved if the relevant histological structures can be seen by optical microscopy. When the relevant structures are apparent in an electron-scan image, one might expect to pinpoint the probe with ease, since there is supposed to be perfect correspondence between the position of the spot forming the image on the display tube and the position of the probe on the specimen. When the spot on the display tube is stopped and located as desired with reference to the somewhat persistent, phosphorescent scanning image, the probe is supposed to be located perfectly at the corresponding point in the specimen itself. Here, however, caution is necessary. If the probe is driven over the specimen by magnetic scanning coils, transient inductive effects can disrupt the proportionality between the deflections of the probe and the beam of the display tube. One consequence is a distortion of the scanning image at the leading edge of the scanned field. A second consequence, even in a part of the field where the image is not distorted, can be a failure of registration—when the spot is stopped at a point in the scanning image, the probe itself is really at a point slightly removed on the specimen. Some commercial probes with magnetic scanning coils have a compensating circuit to eliminate this effect, but perfect adjustment of the circuit should not be assumed. A similar effect can occur in instruments with electrostatic scanning plates if there are transient fields due to accumulation of charge on dirty apertures or plates. The moral for the analyst is that the probe can be positioned with great convenience by reference to the electron-scan image, but one must check carefully if this method is to be relied on for precise location. The registration of scanning image and static probe can be checked readily around the edges of the specimen or with a test grid as specimen.

With combination electron microscope-microanalyzers, usually the transmission electron image is formed in the conventional way and scanning is not involved. By changing the current in a condenser lens, one goes from an irradiated field with a conventional image to a focussed microprobe. The shrinking image can be viewed continuously as the irradiated area is reduced, and there is no doubt about the location of the probe in relation to the image.

While histological correlation is not a problem in many investigations, it is sometimes a limitation standing in the way of a successful microprobe study.

CHAPTER 5

PHYSICAL ASPECTS OF THE
METHODS OF X-RAY MICROSCOPY

IN THIS CHAPTER we shall consider in some detail the physical capabilities
and limitations of the different methods of x-ray microscopy, and also
practical matters such as cost, convenience and the availability of instruments. The discussion will be confined mainly to the instrumental systems
which have actually been used in medical research and which are exemplified in Chapters 2 to 4.

Some knowledge of x-ray physics is essential for the analysis of capabilities
and limitations. For the interested reader with little or no prior knowledge,
the exposition of fundamentals in Appendix 1 should be sufficient.

5.A. PHYSICAL ASPECTS OF CONTACT MICRORADIOGRAPHY

Contact microradiography is a type of radiography designed to reveal
microstructure. The specimen is placed in direct contact with the photographic emulsion or in very close proximity to it. The requirement of fine
spatial resolution dictates several features of the method, some of which are
illustrated in the elementary arrangement of Figure 77.

1. The emulsion must be fine-grained.
2. The area of the source as "seen" from the specimen must be small. Therefore, if a conventional x-ray tube is used, the focal spot of the electrons impinging onto the anode should be small, and the orientation of the anode should provide a favorable foreshortening.
3. The distance between the emulsion and all parts of the specimen must be
small. This implies, of course, that the specimen must be thin.
4. In order to obtain useful contrast in radiographs of thin specimens, the
radiation must be relatively nonpenetrating ("soft"). Suitable wave lengths are
generally in the range 2 to 25 Å for noncalcified tissues, and 0.5 to 2 Å for
bone.
5. The window between tube and camera must pass the soft radiation, and
therefore should be less absorbent than the specimen. Commercial tubes with
thin windows can be used for many applications, but work with very soft radiation requires custom-built arrangements. (In some designs a vacuum-tight win-

179

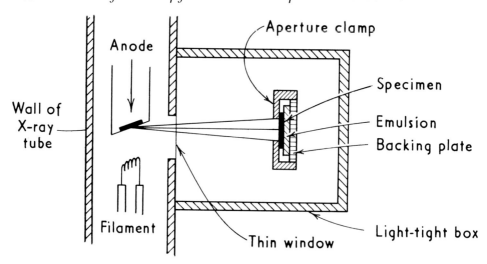

Figure 77. An arrangement for contact microradiography. The narrow rectangular focus of a conventional tube is foreshortened to an apparent square when viewed from the position of the specimen.

dow is not needed, but a window of some sort has always been placed between source and specimen to shield the emulsion from light produced at the filament.)

In this section we shall consider the limitations of conventional contact microradiography in qualitative studies, the application of the method to measurements of total mass and of amounts of individual chemical elements, some of the more common instrumental arrangements and their availability, and the range of medical material to which the method is applicable.

5.A.1. Limitations in Qualitative Medical Studies: Resolution, "Depth of Focus," Exposure Time and Operating Time, Contrast

Resolution in contact microradiography is limited chiefly by "geometrical blurring" and by the properties of the photographic emulsion.

Geometrical blurring is illustrated in Figure 78. A *point* in the specimen casts on the emulsion a shadow of diameter

$$W = \frac{c\,d}{b} \tag{5-1}$$

(The symbols are defined in Figure 78.) Thus, for example, if the source diameter as "seen" from the specimen is 0.2 cm and the distance b is 10 cm, and one wishes to restrict the width of the shadow of a point to no more than 1μm, then the distance c cannot be permitted to exceed 50μm. Equation 5–1 not only sets a limit to attainable resolution; it also explains the necessity

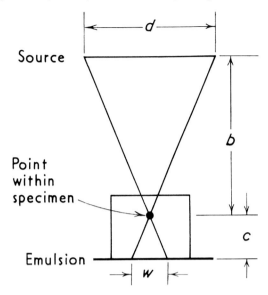

Figure 78. The diameter of the shadow of a point in contact microradiography.

of assuring close proximity between emulsion and specimen, and sets a limit to the thickness of specimen compatible with a given resolution.

A fine-grained emulsion is obviously needed for high resolution. Among the emulsions routinely used (Kodak emulsions 548 and 549, Lippmann-Gevaert), graininess limits resolution to approximately 0.2μm. While it is possible to produce less grainy recording materials, including silver-halide, dye or plastic types, these finer films have not come into use because the sensitivity decreases rapidly as the resolving power is improved (see Cosslett, 1965, p. 386; also C & N,* pp. 37–39, 365 and 366).

An approximate relationship between exposure time and maximum resolution is easily obtained. The requisite exposure time t is related to the distance a from anode to emulsion and to the speed s of the emulsion through the proportionality

$$t \propto \frac{a^2}{s} \qquad (5\text{--}2)$$

To minimize exposure time, both the geometrical blurring and the limit imposed by the emulsion should be approximately equal to the desired spatial resolution, R (the limit R cannot be achieved if either contributing limitation is larger, and exposure is unnecessarily prolonged if either is much smaller). Therefore, in Equation 5–1 we set $W = R$ and obtain

$$a \cong b = \frac{cd}{R} \qquad (5\text{--}3)$$

* We shall use the abbreviation "C&N" in our many references to the book *X-ray Microscopy*, by Cosslett and Nixon. The full reference is given in the bibliography.

Another approximate proportionality is known to link the speed and the resolving power of the emulsion: $s \propto R^2$ (C&N, p. 38). By combining these relationships we obtain

$$t \propto a^2 \frac{1}{s} \propto \frac{c^2 \, d^2}{R^2} \frac{1}{R^2} \propto \frac{1}{R^4} \qquad (5\text{--}4)$$

In connection with Equations 5–3 and 5–4, several points are noteworthy.

1. The dependence of t on R is so steep that we can speak of a virtually absolute limit of resolution in the neighborhood of $0.2 - 0.5\mu$m, for which exposure times of the order of one hour are necessary in most instuments.[°] Resolution can be improved further in a given instrument by varying emulsion and/or camera length, but the increase in exposure time is almost immediately prohibitive. And instrumental improvements yield a meager gain in terms of resolution. For example, an increase in tube output permits a directly proportional reduction in exposure time, but suppose one wishes to exploit increased output to improve resolution while keeping exposure time constant; Equation 5–4 then indicates that, with optimum choice of camera length and emulsion, a sixteenfold increase in output would be needed to improve resolution by a factor of two.

On the other hand Equation 5–4 shows that exposure time quickly ceases to be a problem when poorer resolution is acceptable.

2. The camera length a disappears from Equation 5–4. This means that in principle, radical gains in resolution are not to be expected from instrumental designs employing very long or very short cameras.

3. In work with a given instrument, the readily variable conditions are the camera length and the emulsion. The emulsion should have as high a speed as is compatible with the required resolution, and the camera length should be as short as is compatible with Equation 5–3 and the thickness of the specimen.

Resolution is also affected by the excitation of fluorescent x-rays within the specimen, and by the spread of photoelectrons generated by the x-rays within the emulsion. Usually one need not be concerned with these effects in conventional contact microradiography, but they would remain as important limitations if geometrical blurring and graininess were successfully reduced by unconventional methods.

Diffraction does not significantly limit resolution in contact microradiography. A general formula for diffractive effects in radiography, Equation 5–25, is quoted in Section 5.B.1. In the contact case the formula reduces to a limit of resolution

---

[°] Tube output and the sensitivity of the emulsion are poor when the radiation is restricted to long wave lengths, as it must be for the study of soft tissues. Exposure time is generally much less when shorter wave lengths are used, for calcified tissues or specimens of hi h average atomic number.

$$x_o \cong (c\lambda)^{\frac{1}{2}} \qquad\qquad (5\text{–}5)$$

and usually the smallness of c renders the effect completely negligible.

We have not yet mentioned one integral part of the contact method, the *study* of the developed microradiograph. As has been stressed by Cosslett (1965), a high-resolution, one-to-one image has not served its purpose until it is studied, photographed or scanned through a microscope at high magnification. The overall resolution of the method obviously cannot be better than that of the viewing system, conventionally an optical microscope with a limit in the neighborhood of $0.2\mu m$. Also, the time spent with the optical microscope must be considered in any realistic assessment of the overall speed of the method.

For adequate *contrast* in thin specimens, the radiation must be "soft" enough for absorption in the specimen to reduce appreciably the blackening of the emulsion. Can this condition generally be satisfied? As an example of the most transparent biological specimen one might want to study, we shall consider a section of soft tissue consisting essentially of protein, with weight-fractions carbon ½, nitrogen ⅙ and oxygen ⅕, plus negligible hydrogen and sulphur, cut frozen with no embedding medium, $1\mu m$ thick, and desiccated (there is generally little point in thinner specimens when the resolution of the method is several tenths of a micron at best). We suppose that the fresh tissue had a density of 1 gm/cm³ and that ¾ of its weight was water; the dried $1\mu m$ section might then weigh $\frac{1}{4} \times 10^{-4}$ g/cm².

The absorption in the specimen can be estimated from Equation A-13 (Sec. A.4). Although this equation refers to monochromatic radiation and the incident radiation usually consists largely of continuum, for present purposes it is sufficient to use the very crude rule of thumb that the radiation from a thin-window tube, run at kilovoltage V_o, will produce approximately the same contrast as monochromatic radiation of quantum energy $E = \frac{2}{3} V_o$. Suppose that the tube is run at 900 volts,[*] so that the representative quantum energy is 600 volts. The mass absorption cross sections for C, N and O (from Henke and Elgin, 1970; see also Fig. 114, Sec. A.4) are respectively 8800, 12,500 and 16,800 cm²/g. Then

$$T = e^{-\frac{1}{4}(10^{-4})(8800/2 + 12,500/6 + 16,800/5)} = 0.76$$

so that approximately 24 percent of the radiation is absorbed in the specimen. This effect would be noticed readily in an image.

Most specimens would be thicker, denser or of higher average atomic number, and higher tube voltages could be used in the study of them. We conclude that in general one can operate a suitable tube so as to render

---

[*] A higher tube voltage always produces a spectrum of higher average quantum energy and shorter average wave length. As is clear from Figures 113 and 114 (Appendix A.4), such a spectrum is usually more penetrating ("harder"), i.e. less absorbed. For our hypothetical weakly absorbing specimen, a very low tube voltage is necessary.

perceptible in the microradiograph any specimen which is otherwise suitable for study by the contact method.

However, for the study of internal structure, perceptibility of the entire specimen is not sufficient; contrast *between* parts of the specimen is needed. A quantitative formulation of this requirement is much more difficult, but several points should be noted.

1. Even if the image of a specimen of mass per unit area S is readily rendered perceptible, the contrast between two parts of a specimen *differing* by the same amount S may not be perceptible when the mass per unit area of each part is much more than S. This is because, in the second case, much harder radiation would have to be used to darken the film at all.

2. There is little intrinsic contrast between the various organic materials, which are composed almost entirely of carbon, nitrogen, oxygen and hydrogen, i.e. at most wavelengths the absorption coefficient is almost the same for all of them. Lipids and proteins are the most important example. Therefore the best way to distinguish between such groups of compounds is to compare microradiographs taken before and after selective extraction, for example extraction of lipids in xylol.

3. At common tube voltages, fresh (nondesiccated) soft tissue generally yields little internal contrast, because the mean linear absorption coefficients for water and most organic compounds are similar. But considerable internal structure may be apparent in radiographs of *dried* tissues such as conventional deparaffinized sections.

4. Contrast in soft, wet tissue can be improved by the use of very soft radiation which is mainly between the oxygen and carbon absorption edges, as can be obtained from a thin-window tube run at 500 volts. Figure 79 implies that for such radiation, the contrast of the tissue in the wet state should be almost as good as it is after desiccation.

5. If two areas differ especially in the weight-fraction of one element, monochromatic radiation at the peak absorption of that element can provide much better contrast than continuum radiation, which must consist largely of radiation far from the peak or on the wrong side of it.

6. In most of the microradiographic work to date, it has not been necessary to use refined techniques to gain adequate contrast.

The question of contrast in microradiography is discussed more extensively in C&N, and by Lindström (1955). An especially concise and full analysis has been provided by Henke (1959).

5.A.2. Measurement of Mass

Measurements of photographic density * in the microradiograph can be calibrated to give the local mass per unit area in any microscopic region

---

* Photographic density is defined as $D = \log I_o/I$, where I_o and I are respectively the intensities of visible light incident onto and transmitted through the developed emulsion, as observed in a photodensitometer.

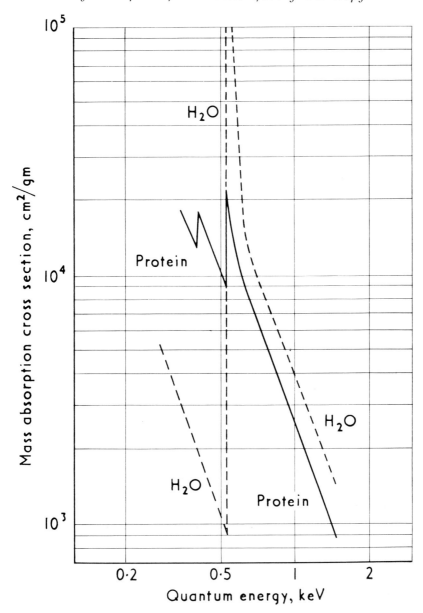

Figure 79. Mass absorption cross sections vs. quantum energy for typical protein and for water. (Based on Henke, 1959, p. 120.)

of the specimen. A calibration curve can be obtained with a step wedge, consisting of overlapping layers of plastic film of known composition and mass per unit area, placed alongside the specimen in *each* exposure (see Fig. 80).

For *monochromatic* radiation, we can apply Equation A-12 (Sec. A.4),

$$T = N/N_o = e^{-\bar{\sigma}s}$$

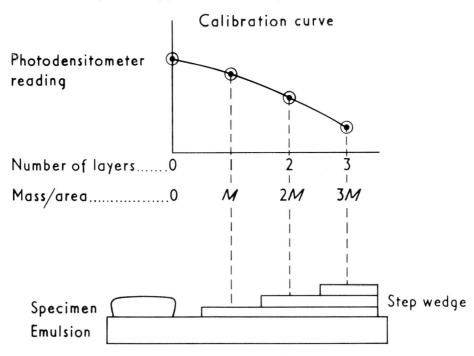

Figure 80. Measurement of mass per unit area with step-wedge foils. Specimen and foils are mounted side by side on the emulsion. Each layer of the wedge has a known mass per unit area M.

The photographic density depends on the local flux of x-ray photons reaching the emulsion N, which (for a given exposure N_o) depends only on $\bar{\sigma} S$, the product of average mass absorption coefficient and local mass per unit area. Hence, for points of equal darkening in the image of the specimen and on the calibration curve,

$$\bar{\sigma}_s \, S_s = \bar{\sigma}_c \, S_c \qquad \text{and} \qquad S_s = \frac{\bar{\sigma}_c}{\bar{\sigma}_s} \, S_c \qquad\qquad (5\text{--}6)$$

subscripts s and c referring to specimen and calibration wedge. S_c is known; the deduction of specimen mass per unit area S_s depends on knowledge of $\bar{\sigma}_c/\bar{\sigma}_s$. This ratio has been explored most extensively by Lindström (1955). He shows that for radiation between the oxygen and aluminum absorption edges, when nitrocellulose (Parlodion) is used for the calibration film, the ratio $\bar{\sigma}_c/\bar{\sigma}_s$ is quite near unity for most soft tissues, and he evaluates the deviations from unity which may occur depending on the exact composition of the tissue (protein, lipid, sulphur, phosphorus and heavy element concentrations). Within the uncertainty of the deviation, generally a few percent or less, Equation 5–6 provides a measurement of S_s, the local mass per unit area in the specimen.

Because it is laborious to apply a step wedge to every exposure, Combée and Engström (1954) have suggested a simplified procedure. Only one thickness of calibrating material is used, but different areas of emulsion are exposed for different times to get a calibration curve of photographic density vs. time.[*] It is not even necessary to provide a solid calibration film; the space between specimen and tube window, normally evacuated, may conveniently be filled with nitrogen to provide the absorption standard. Now, if a part of the specimen, exposed for time t_s, gives the same darkening as an exposure time t_c on the calibration curve, then the number of incident x-ray photons must be the same in both cases,[†] so that

$$t_s e^{-\bar{\sigma}_s S_s} = t_c e^{-\bar{\sigma}_c S_c} \tag{5-7}$$

and

$$\bar{\sigma}_s S_s = \bar{\sigma}_c S_c + \ln t_s/t_c \tag{5-8}$$

Thus, for monochromatic radiation, the step-wedge and the single-thickness methods both measure the quantity $\bar{\sigma}_s S_s$.

However monochromatic radiation is not often used, because the radiation from conventional tubes, mainly continuum, is more conveniently produced and more intense. An extension of the theory shows that such radiation can be used for the measurement of mass, so long as the x-ray absorption edges of the chief constituents of the specimen (and the calibration material) are outside the radiation spectrum. The reasoning is as follows:

For continuum radiation, the darkening of the emulsion does not depend simply on the number of transmitted photons, but we expect it to depend on a quantity which we may call the "exposure," conveniently expressed in the form tP, where t is the exposure time and P is defined as

$$P = \int_0^{V_o} r(E) \, N_o(E) \, e^{-\bar{\sigma}(E) S} \, dE \tag{5-9}$$

Here $N_o(E) \, dE$ is defined to be the number of photons per unit area per unit time incident onto the specimen with quantum energies between E and $E + dE$; $N_o(E) e^{-\bar{\sigma}(E) S} dE$ is therefore the corresponding number locally transmitted to the emulsion, and $r(E)$ is the relative sensitivity of the emulsion for radiation of quantum energy E. P is thus the sum of the properly weighted contributions of all the photons up to the maximum quantum energy, which matches the voltage V_o applied to the tube.

[*] The obvious procedure is to do the calibration exposures in succession and rely on the stability of the tube. With a motor-driven rotating sector-wedge (Henke, 1959, p. 142), one needs only a single exposure and drifts in tube output do not matter.

[†] Here we are assuming the validity of the "reciprocity law" that darkening depends on total exposure independently of exposure time. The reciprocity law is in fact valid for x-rays (see Lindström, 1955, p. 53; also C&N, p. 162).

We can write $\bar{\sigma}(E)$ in the form

$$\bar{\sigma}(E) = \bar{\sigma}_m f(E) \qquad (5\text{--}10)$$

where $\bar{\sigma}_m$ is the mean mass absorption coefficient at any arbitrary quantum energy E_m (which one would naturally imagine near the middle of the spectrum) and $f(E)$ is a function of quantum energy. Then

$$P = \int_o^{V_o} r(E) N_o(E) \, e^{-S\bar{\sigma}_m f(E)} \, dE \qquad (5\text{--}11)$$

This shows that so long as exposure time t, emulsion response $r(E)$, flux $N_o(E)$, tube voltage V_o and dependence of absorption on quantum energy $f(E)$ all remain the same, the local darkening will depend *only* on $\bar{\sigma}_m S$. In the step-wedge method, t, V_o, r (E) and $N_o(E)$ are automatically the same for specimen and calibration wedge, so that points of the same darkness in specimen and calibration curve must have equal values of $\bar{\sigma}_m S$, *provided that* $f(E)$ *is the same for specimen and wedge.*

Equivalent statements of the italicized condition are: *The mean absorption coefficients of specimen and wedge, $\bar{\sigma}_s$ and $\bar{\sigma}_c$, must depend on quantum energy in the same way,* or *the ratio $\bar{\sigma}_s/\bar{\sigma}_c$ must be independent of quantum energy.* When this condition is met, we can conclude that for points of equal darkness

$$\bar{\sigma}_{ms} S_s = \bar{\sigma}_{mc} S_c$$

and

$$S_s = \frac{\bar{\sigma}_{mc}}{\bar{\sigma}_{ms}} S_c = \frac{\bar{\sigma}_c}{\bar{\sigma}_s} S_c \qquad (5\text{--}12)$$

just as we had in Equation 5–6 for monochromatic radiation.

In practice, $f(E)$ may be the same for specimen and calibration film in two situations. First, obviously, when specimen and film have the same composition. This sufficient condition cannot be helpful when the composition of the specimen is not well established. The second favorable situation occurs when the incident spectrum does not encompass the absorption edges of the elements in the specimen and the calibration material. Equations A-10 and A-14 (Sec. A.4) then imply, for either specimen or wedge, that

$$\bar{\sigma}(E) = f_1\sigma_1(E) + f_2\sigma_2(E) + \cdots = f_1\sigma_{1m}(E_m/E)^3 + f_2\sigma_{2m}(E_m/E)^3 + \cdots$$
$$= (f_1\sigma_{1m} + f_2\sigma_{2m} + \cdots)(E_m/E)^3 = \bar{\sigma}_m(E_m/E)^3 \qquad (5\text{--}13)$$

so that the dependence of $\bar{\sigma}$ on E is the same for specimen and wedge.

For measurements on thin soft specimens, Engström and Lindström have generally used a tube with a 9μm aluminum window, run at 3 kV or less. The window removes almost all the radiation harder than aluminum K and almost all the radiation softer than oxygen K. This spectrum avoids the absorption edges of the major constituents C, N, O, P and S as well as the K edges of heavier elements; the worst intruder is the sodium K edge. (In addition, the spectrum gives good contrast and avoids giving undue emphasis to the minor, variable heavy elements.) However, the proportionality of σ and E^{-3} between edges is only an approximation and the equality of $f(E)$ for specimen and wedge should be confirmed by reference to absorption tables. Lindström (1955) has done this and has shown that with this spectrum, for typical soft tissues, the relation (5–6) is virtually as valid as it is for monochromatic radiation. Henke (1959) has shown that for specimens of protein, equality of $f(E)$ is also obtained when a tube is run at 500 volts and a nitrocellulose wedge is used.

The simplified procedures based on a single calibrating thickness (Combée and Engström, 1954; also Carlson, 1957; Wallgren and Holmstrand, 1957) are in principle not valid for continuum radiation. Consider the simplest case, where specimen and standard have the same composition and hence the same absorption coefficient. If exposure of the specimen for time t_s gives the same darkening as exposure of the standard for time t_c, then we can write

$$t_s P(S_s) = t_c P(S_c) \tag{5-14}$$

Since the calibration for a single mass S_c cannot determine the curve $P(S)$, it is not possible to deduce the specimen mass S_s from Equation 5–14.

The single-thickness method can be used as an approximation. Suppose that the same photographic density D_s is produced by exposures of specimen, calibrating material and bare emulsion for the respective times t_s, t_c and t_o (see Fig. 81). If one *assumes* that exposure depends exponentially on mass, then

$$t_o = t_s e^{-\bar{\bar{\sigma}}_s S_s} = t_c e^{-\bar{\bar{\sigma}}_c S_c} \tag{5-15}$$

(the double bar signifying an averaging over constituents and wave length), and

$$S_c \bar{\bar{\sigma}}_c = \ln t_c / t_o \tag{5-16}$$

and

$$S_s \bar{\bar{\sigma}}_s = \ln t_s / t_o \tag{5-17}$$

Equation 5–16 must be used to determine $\bar{\bar{\sigma}}_c$. Knowledge of the composition of specimen and calibrating material suggests a corresponding value for $\bar{\bar{\sigma}}_s$, and Equation 5–17 then provides a value for mass, S_s.

For continuum radiation the method is approximate because the spectrum becomes harder as the radiation penetrates the specimen, so that the exposure of the emulsion does not really depend exponentially on specimen mass. The approximation is good if the incident radiation is hard, but this implies poor contrast and large uncertainties from the errors of photodensitometry. Of course, when the transmission of the specimen and of the single calibration thickness are almost the same, the small extrapolation from the curve, though faulty in principle, will not be too damaging. But the error is difficult to estimate, and the single-thickness method with continuum radiation should be regarded as semiquantitative.

The errors and limits of detection of quantitative microradiography have been discussed extensively by C&N (pp. 149–151), Brattgard and Hyden

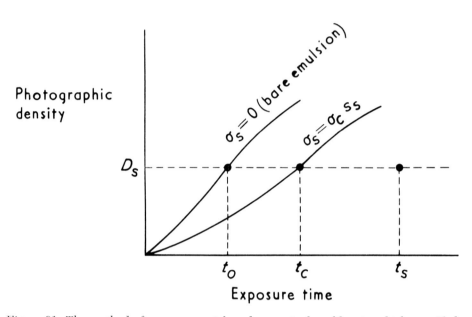

Figure 81. The method of measurement based on a single calibration thickness. If the exposure of a point in the specimen for time t_s gives density D_s, the times t_o and t_c in Equations 5–15 to 5–17 are obtained as indicated from exposure curves.

(1952), Lindström (1955), and Henke (1959). The major considerations are as follows:

Imperfections in photodensitometry and standardization lead to a probable error of approximately 3 percent in the quantity which is directly measured, σS.

The deduction of local mass per unit area S_s proceeds via Equation 5–6 and requires knowledge of the ratio of the absorption coefficients of specimen and standard. This ratio is often known more reliably than the separate coefficients, and the latter are now generally established to within a few percent.

Gross density ρ can be deduced through $\rho = S/T$, where T is the specimen thickness. However, the uncertainty in T is generally of the order of 10 percent or more in thin specimens.

The total *mass* of an object like a large cell can be obtained by the integration of local values of S over the area of the object.

It has been tacitly assumed that on a fine scale, the specimen mass is homogeneously distributed. To reveal the nature of this assumption, consider a hypothetical inhomogeneous region where all the mass is segregated into half of the area; in contradiction to Equation A-12, the overall transmission of such an area obviously could never fall below 50 percent no matter how large S might be. A study by Engström and Weissbluth (1951) indicates that the errors due to inhomogeneity are generally small in practice.

The total mass of microscopic objects can also be measured by a closely related technique, absorption microanalysis (Sec. 5.C), with which the problems of integration and inhomogeneity are much milder.

While quantitation has attracted the efforts of physicists, absolute measurements of mass do not seem to have been widely useful in medical research. Variations in mass are more often interesting. Here the limit of detectability is important. As indicated in Section 5.A.1, this limit is of the order of 2×10^{-5} g/cm². In contact microradiography there are several factors, prominently the grain size of the emulsion, which make it difficult to reduce the assayed area to less than one square micron, and the minimum detectable mass within this area would be of the order of 2×10^{-5} g/cm² $\times$ 10^{-8} cm² $= 2 \times 10^{-13}$ g.

5.A.3. Microradiographic Assay of Chemical Elements

Consider *two* radiographs of a specimen, each produced with monochromatic x-rays, the two wavelengths being close to an absorption edge of some element of interest but on opposite sides. It is apparent from Figure 113 (Sec. A.4) that the difference between the two images may be attributable mainly to the "straddled" element, the absorption coefficients of the other constituents being almost the same in both cases. Qualitative localization by inspection of the two radiographs is favored by the fact that the absorption coefficient of the straddled element is larger at the shorter wavelength, where the coefficients of the other elements are smaller.

Quantitatively, the transmissions through any one area of the specimen are

$$T_1 = e^{-(\sigma_{e1} S_e + \sigma_{r1} S_r)}$$

and

$$T_2 = e^{-(\sigma_{e2} S_e + \sigma_{r2} S_r)} \tag{5–18}$$

where the two wavelengths are designated by the subscripts 1 and 2, σ_e and σ_r are respectively the absorption coefficient for the straddled element and the average coefficient for all the other constituents, S_e is the local mass per unit area of the straddled element and S_r is the local total mass per unit area of all the other constituents. These equations may be combined to give

$$S_e = \frac{\frac{\sigma_{r1}}{\sigma_{r2}} \ln T_2 - \ln T_1}{\sigma_{e1} - \frac{\sigma_{r1}}{\sigma_{r2}} \sigma_{e2}} \qquad (5\text{-}19)$$

If the wavelengths are very close, the approximation $\sigma_{r1} = \sigma_{r2}$ may be adequate and Equation 5–19 simplifies to

$$S_e \cong \frac{\ln T_2/T_1}{\sigma_{e1} - \sigma_{e2}} \qquad (5\text{-}20)$$

More often, when selected characteristic lines are used, the wave lengths are not close enough for the approximation $\sigma_{r1} = \sigma_{r2}$, but one can use

$$\sigma_{r1}/\sigma_{r2} \cong (\lambda_1/\lambda_2)^3 \qquad (5\text{-}21)$$

(See Equations A-14 and A-1.) Equation 5–19 then takes the form

$$S_e \cong \frac{(\lambda_1/\lambda_2)^3 \ln T_2 - \ln T_1}{\sigma_{e1} - (\lambda_1/\lambda_2)^3 \sigma_{e2}} \qquad (5\text{-}22)$$

The specimen transmissions T_1 and T_2 can be obtained by photodensitometry, after one has obtained a calibration curve of density vs. exposure for the emulsion. The absorption coefficients can be taken from published values. Equation 5–19, 5–20 or 5–22 can then be used for the local assay of the straddled element.

The straddling method is based on monochromatic radiations, which may be obtained in various ways:

1. Conventional x-ray tubes with anodes selected for suitable characteristic lines. A demountable tube with interchangeable anodes is useful. Although the continuum can be reduced by filtration, at conventional operating voltages the continuum background is generally too strong for accurate quantitative analysis and such sources are used for qualitative localizations.

2. Ultrasoft x-rays monochromatized by total reflection plus filtration. Henke has pointed out that the continuum is relatively weak at very low tube voltages. At long wavelengths, one can interpose and adjust a mirror to reject wavelengths shorter than the characteristic line while totally reflecting the line itself. With filtration added, the resulting spectra are monochromatized moderately well but the radiation is not uniform over a wide area and the method has not been widely used (for full details see Henke, 1959).

3. Secondary radiators (Fig. 82). When a pure element is exposed to the radiation from a conventional tube, most of the incident radiation is absorbed, leading to excitation of the element's characteristic lines. The output, almost entirely characteristic radiation, has been used for elemental analysis (Splettstosser and Seeman, 1952; Rogers, 1952; Hall, 1958, 1961; Goldsztaub and Schmitt, 1960, 1963). Applications have been limited because the radiating area must be large to produce sufficient intensity, and because the choice of characteristic lines is limited.

4. Crystal monochromator. A diffracting crystal may be interposed between tube and specimen to provide either a characteristic line of the tube anode or a

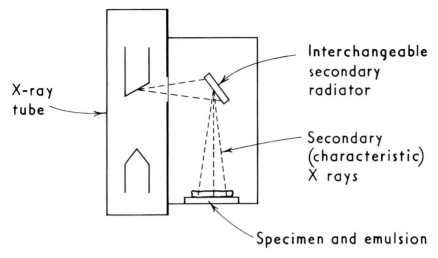

Figure 82. The use of a secondary radiator to produce a characteristic line spectrum for microradiography.

very narrow band of the continuum (Fig. 83). Adjustment of the crystal can provide two very close straddling wavelengths. This arrangement, developed extensively by Lindström (1955), is elaborate but more widely applicable than the preceding ones.

The sensitivity of the straddling method is shown in Figure 84, taken from Engström (1951). He supposed that photodensitometry can measure T_2/T_1 within 1 percent (somewhat optimistic), and he calculated the amounts of element per unit area for which the corresponding error in S_e, according to

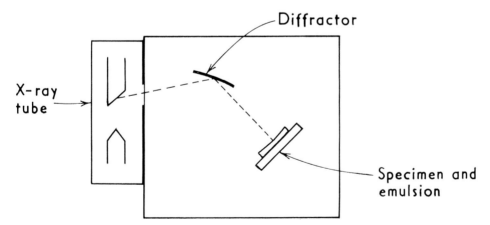

Figure 83. The use of a diffractor to produce a monochromatic spectrum for micro-radiography. Diffractor and specimen must be correctly oriented by mechanical linkages in accord with Equation A-16 and Figure 123. When the anode area is small, the diffractor should be curved to focus maximum intensity at the specimen.

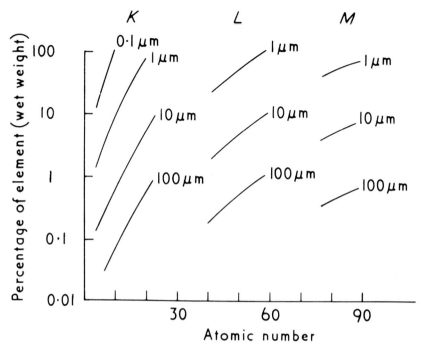

Figure 84. A calculation of the minimum detectable per cent of an element in tissues of different thicknesses, using K, L, or M absorption edges, as a function of atomic number of the element. (From Engström, 1951. Courtesy of *Acta Radiol.*)

Equation 5–20, would be 20 percent; he then converted to weight percent by assuming a tissue density of 1 gm/cm³ (so that the figure represents per-cent of *wet* weight). He concluded that for sections 10μm thick, the ele-ments measurable in their usual concentrations would be carbon, nitrogen and oxygen (plus calcium and phosphorus in bone), while phosphorus and sulphur as well would usually be measurable in sections 20μm thick.

The straddling method is fundamentally not suited for the measurement of *low* concentrations where the element absorbs only a slight fraction of the incident radiation, because one must then measure the relatively small change in a large flux. With emission methods (fluorescence and electron-probe) one need not contend with the total flux and the minimum detectable concentrations are more favorable.

Figure 84 may suggest that one should exploit L edges as far as possible by using very soft radiations. Indeed, near atomic number 30, the important quantity $(\sigma_{e1} - \sigma_{e2})$ is approximately fortyfold larger for the L edge than it is for the K edge. However, when the concentration of an element is low, most of the absorption occurs in the other constituents of the specimen; in order for the absorption in the element of interest to be appreciable, the overall transmission would have to be very low and difficult to measure accurately. Engström's analysis and Figure 84 take no account of this limitation. The absorption coefficients of the main constituents rise even more steeply than fortyfold in going from a K to an L edge, and the net effect is not a reduction in minimum measurable concentration.

A practical drawback is that the straddling method always requires two exposures, and quantitation requires the identification and photodensitometry of exactly corresponding areas in the two images. This is especially onerous with the contact arrangement, where close contact between specimen and emulsion and clean recovery for a second exposure are both vital, and where the two primary images are not magnified. For these reasons, both fundamental and practical, the straddling method has been used most often for direct qualitative visualization of relatively high elemental concentrations.

5.A.4. Equipment for Contact Microradiography

Among the x-ray microscopic methods discussed in this book, contact microradiography is the only one which can be practiced with standard x-ray equipment originally designed for other purposes. One can do qualitative microradiography and measurements of total mass per unit area by fitting a light-tight box, with provision for mounting specimen and emulsion to a commercial, thin-window, fine-focus x-ray tube. Specific apparatus and designs are discussed in Sections 2.B.4 and 2.C.5.

The chief limitation with the commercial tubes is in their windows, which are too thick to transmit the soft radiations required for the radiography of poorly absorbing thin specimens, such as thin sections of ordinary soft tissue with no absorbent additives. The requirements with respect to windows may be summarized as follows.

1. For the study of *bone*, so long as one wishes to give full weight in the radiograph to the distribution of calcium, the radiation must be predominantly on the high-absorption side of the calcium K absorption edge, i.e. less than 3 Å in wavelength. Such radiation is adequately transmitted by a 1-mm beryllium window, which is standard in commercial thin-window tubes.

If calcium is to be de-emphasized in the radiograph (perhaps in favor of phosphorus), the radiation must be predominantly more than 3 Å in wavelength. The 1-mm Be window is then too thick, but a 200μm Be window would usually be satisfactory. Such a window was routinely supplied by Philips in their simple and compact "MRG" apparatus manufactured until recently for microradiography after a design by Combée and Engström (1954), and it is not difficult for a manufacturer to incorporate a beryllium window of this thickness during production.

2. For the study of thin specimens of *soft tissue*, if no highly absorbing material is present in the specimen, beryllium windows are inadequate and

9μm aluminum windows have been used most often. Although such a window was successfully sealed to a tube by Rosengren (1959), it is usually fitted to custom-built demountable tubes, with a vacuum maintained by pumping on both sides of the window during operation.

In fact, 9μm of aluminum does not transmit very soft radiation much better than the thinnest practical beryllium window (approximately 50μm). Nevertheless the aluminum type is superior because it cuts out a band of radiation on the "hard" side of the aluminum absorption edge at 8 Å. This makes it possible to run the tube at higher voltages (perhaps 5 kV) and consequently at higher output with the wavelengths shorter than 8 Å absorbed in the window and the emergent spectrum restricted to wavelengths well absorbed by carbon, nitrogen and oxygen.

With the aluminum window, the emergent radiation is mainly in the range 8 to 10 Å. If still longer wavelengths are needed, either for extremely thin and weakly absorbing specimens or for discrimination among the elements C, N and O (see Sec. 5.A.1), one may resort to a variety of special very thin plastic windows, including Mylar, polypropylene, Formvar, and aluminum leaf dipped in Formvar. The preparation, deployment and performance of these windows have been described lucidly by Henke (1965). The very thin plastic films do not selectively cut out the short wavelengths, but they are so transparent that one can run the tube at quite low voltage (from approximately 1½ kV down to less than 1 kV) to avoid the excitation of this radiation entirely.

3. Most contact microradiography has been done with sealed x-ray tubes with conventional thin beryllium windows. In addition to the study of mineralized tissues, as indicated in Chapter 2 there is ample scope for the application of such tubes to the study of soft tissues containing highly absorbing material, which may either be present naturally (e.g. iodine in the thyroid gland), or be added by the investigator (e.g. contrast media injected into blood vessels for microangiography, or x-ray-absorbing tissue stains).

For *quantitative* microradiography, one must measure photographic densities in microareas of the radiographs. This can be done with equipment now offered by most major optical companies for quantitative microspectrophotometric cytochemistry and histochemistry. The equipment and techniques have been reviewed by Ruthmann (1970), and particular problems associated with measurements of microradiographs have been discussed by Lindström (1955), and Brattgard and Hyden (1952).

5.B. PHYSICAL ASPECTS OF PROJECTION MICRORADIOGRAPHY

The physical fundamentals of projection microradiography have been outlined in Section 1.B. As is clear from Figures 1 and 2, the main difference between the contact and projection modes of microradiography is in the spacings between source and specimen, and between specimen and photographic emulsion. In projection microradiography the specimen is much closer to the source than to the emulsion. The consequences, as explained later, are the following:

1. The specimen is considerably magnified in the primary image formed in the emulsion.

2. Because of the primary magnification, there is no need to use very fine-grained emulsions.

3. The x-ray source must be very fine (a "point" source).

4. There is tremendous "depth of focus," i.e. all parts of a thick specimen are imaged simultaneously with good spatial resolution. Of course the image, being two-dimensional, consists of superposed projections of planes throughout the specimen. Such an image of a thick specimen is confusing unless the subject is favorable (e.g. vascular networks, insects, etc.).

5. As in contact microradiography, the opacity of the specimen dictates the choice of penetrating power ("hardness") of the radiation, which may be varied by adjustment of the tube voltage. However, since the electrons in the x-ray tube do not pass all the way through the target-window, the x-rays are generated within the window and are filtered before they can emerge. The resulting dependence of the spectrum on voltage and on window thickness and composition is complicated and difficult to control accurately, especially when soft x-rays are needed.*

6. Spatial resolution depends on the diameter of the electron beam as it strikes the target. In order to focus the electrons into a fine spot on the target one generally puts a fluorescent screen in place of the emulsion, and observes the image of a test specimen (a grid) as the electron lens is adjusted (see Sec. 2.B.4.b). However, the maximum electron current which can be focussed into a fine spot unfortunately decreases steeply as the spot diameter is reduced (Sec. 5.E.2) so that, before the lens settings for maximum resolution have been established, the intensity may become too weak to give a visible image. To achieve the very highest resolution one must therefore couple the fluorescent screen to an image intensifier or resort to other elaborate techniques.

For spatial resolution and good contrast in very thin specimens, projection microradiography can perform at least as well as contact microradiography in principle, but contact microradiography may be more convenient due to points 5 and 6 above. The projection method is decidedly better for thick specimens.

5.B.1. Limitations in Qualitative Studies: Magnification, Field of View, Resolution, "Depth of Focus," Exposure Time, Contrast

The *magnification* in projection microradiography is easily worked out on the basis of Figure 85. If o is the lateral separation between two points in

---

* Projection microradiography can also be carried out by focussing the electrons into a fine spot on the surface of a conventional x-ray tube anode. But it is difficult to get the specimen close to the x-ray source with this method.

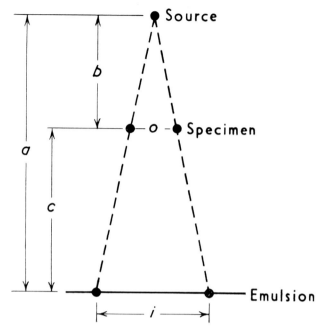

Figure 85. The dimensions which determine magnification in projection microradiography.

the specimen at a distance b from the source, i is the distance between the corresponding points in the plane of the photographic emulsion and the x-ray source is very small, then by definition the magnification m is $m = i/o$ and according to simple geometry

$$m = \frac{i}{o} = \frac{a}{b} = \frac{b+c}{b} \qquad (5\text{--}23)$$

For thin specimens the distance b from source to specimen can easily be made as small as $50\mu m$, in which case even the short camera length $a = 5$ cm would give a magnification of 5 cm/$50\mu m = 1000$ x. For thick specimens if the magnification is to be uniform throughout the object, b must be several times the specimen thickness; high magnifications are then not practical, but are usually not desired anyway.

The *field of view* is limited by the geometrical distortion which sets in if very divergent x-rays are used. According to Nixon (1961), the useful cone of x-rays fills an angle of about 60°. The diameter f of the field of view is then

$$f \cong 2b \tan(60°/2) \cong b \qquad (5\text{--}24)$$

i.e. the field of view approximately equals the distance from window to specimen. Thus high primary magnification is always accompanied by a small field of view.

Spatial resolution is limited by two effects, a geometric one (penumbra) and diffraction. According to Cosslett and Nixon (1960, p. 54), the limit R_d imposed by diffraction is

$$R_d = (bc\lambda/a)^{1/2} \qquad (5\text{--}25)$$

where λ is the x-ray wavelength and b, c and a are defined in Figure 85. In projection, usually $c \cong a$ and (5–25) reduces to

$$R_d \cong (b\lambda)^{1/2} \qquad (5\text{--}26)$$

The finest resolution is most likely to be desired with thin specimens. For a thin specimen of soft tissue the mean wavelength λ might have to be, say, 10 Å. According to (5–26), for a resolution of 0.1μm the specimen would then have to be no more than 10μm from the window, a spacing which is attainable at the cost of some effort. Equation 5–26 also indicates that the specimen can be much further from the window when the resolution does not have to be so good and when the average wavelength can be shorter (e.g. for mineralized or thicker specimens).

The "geometrical limit" to spatial resolution is estimated from the assumption that the radiation consists of rays travelling in straight lines (an approximation which is rectified when diffraction is taken into account as above). The geometrical blurring is just as formulated in Equation 5–1 and Figure 78 of Section 5.A.1, i.e. the blurring in the image w is related to source diameter d and the specimen spacings b and c by

$$w = d\,\frac{c}{b} \qquad (5\text{--}27)$$

However, the corresponding geometrical limit of resolution within the specimen, R_g, is obtained by dividing w by the magnification, $m = (b+c)/b$:

$$R_g = \frac{w}{m} = d\,\frac{c}{b}\frac{b}{b+c} = d\,\frac{c}{b+c} \qquad (5\text{--}28)$$

When $c \gg b$, Equation 5–28 reduces to

$$R_g \cong d \qquad (5\text{--}29)$$

Thus the geometrical limit of resolution is equal to the diameter of the x-ray source.

The diameter of the x-ray source depends on two factors, the diameter of the electron beam as it enters the target, and the spread of the electrons due to scattering within the target. The maximum electron current which can be focussed into a spot of diameter d is known to vary as $d^{8/3}$ and when the spot diameter is reduced below 0.1μm the available current is usually too small to produce x-ray intensities adequate for photographic images. Even at 0.1μm, and even if especially bright electron guns are used in the tube

(Cordier *et al.*, 1965), the intensity is too low to produce visible x-ray images on a fluorescent screen and in order to focus the electrons, one must either couple the screen to an image intensifier (Anderton, 1966a), or focus the electrons in a preliminary procedure which brings them out of the tube (Cordier *et al.*, 1965; Nixon, 1960) or which involves extra manoeuvres inside the tube (Ong, 1959). These procedures are burdensome in practice but one is freed from them only when an image is directly visible on the fluorescent screen. This generally occurs only for focal spots approximately 1μm or more in diameter.

To avoid excessive electron spread within the target, the tube voltage is usually 10 kV or less and dense targets of high electron stopping-power are favored, gold for example. The spread of 10-kV electrons in gold is less than 0.1μm.

In practice, projection microradiography readily provides a spatial resolution of 1μm, while there are considerable complications in going from 1μm to the present limit of about 0.1μm.

The *"depth of focus"* in projection microradiography is very large because both limits of lateral spatial resolution, R_g and R_d, can be kept small throughout a thick specimen. Equation 5–28 shows that the geometric limit R_g is equal to or less than the size of the electron spot for *any* point in the object space. Equation 5–26 shows that the diffraction limit R_d is poorest for the object points furthest from the source, but even there the limit R_d is usually unimportant for thick specimens in practice. Thus an entire thick specimen can be "in focus" simultaneously. This fact is not entirely a blessing, since it implies that there can be no effect analogous to "optical sectioning" in light microscopy. The superposition in the emulsion of the absorption patterns from all levels of an *unsuitable* thick specimen can be hopelessly confusing. The great depth of focus is accompanied by a lack of resolution in depth, which can be alleviated to some extent when stereo pairs are obtained (Sec. 2.B.4.b).

Exposure time becomes a limiting factor at the highest spatial resolutions. With the apparatus of Cordier *et al.* (1965), incorporating an especially bright electron source, spatial resolutions of 0.1μm to 0.2μm were achieved at operating voltages of 7 to 10 kV with exposures of about fifteen minutes. Much shorter exposures, of the order of one minute, are common in the usual work at more modest resolution (Chap. 2).

In first approximation, exposure time is not affected by the distance from source to emulsion. Specifically, if distances b and c (Fig. 85) and the coarseness (graininess or spatial resolution) of the emulsion are all multiplied by the same factor, magnification and the geometrical limit of resolution R_g are unchanged, while the lowered intensity at the emulsion is compensated by its higher speed. Such a scaling-up of the camera improves the field of view at a given magnification, but there are two limitations: if a fluorescent screen is used for focussing, it must still be close to the source in order to receive sufficient intensity and, more seriously, the diffraction limit to resolution, R_d, becomes worse as b is increased, so that b must be minimized for work at the highest resolution.

Cosslett (1959) has compared the speeds of contact and projection microradiography, also evaluating an intermediate method proposed by Le Poole and Ong (1956, 1957) in which the specimen is placed midway between source and emulsion. Cosslett shows that the exposure time to *record* a field of given size is least with the contact method, but projection microradiography is faster overall if the contact microradiographs must subsequently be surveyed with a high-power optical microscope. (In addition to his explicit conclusions we must note that a shorter recording time may be a great asset if specimen or instrumentation is unstable.) He shows further that if the two methods are used with the same short *camera length,* the recording times are similar but the field is much smaller and the ultimate limit of resolution is better in the projection mode.

Contrast depends on the wavelength spectrum of the x-rays. One can readily produce spectra centered around 2 to 4 Å, produced with tube voltages of 5 to 10 kV, but self-filtration in the target-window makes it difficult to get much softer radiation. Thus projection microradiography cannot be applied to very thin, poorly absorbing specimens as can contact microradiography with special soft x-ray sources like Henke's (1959, 1970).

The relationship between resolution, speed and contrast is extremely complex in projection microradiography. Detailed studies have been presented by Anderton (1966b) and Dyson (1956). Practice can be guided by the valid generalizations that the projection method gives prompter visualization of micro fields and better resolution in thick specimens, while the contact method gives a faster record of large fields and, with special tubes, can give better contrast in very weakly absorbing specimens.

5.B.2. Quantitative Measurements

In Sections 5.A.2 and 5.A.3 we have discussed the measurement of total mass and of amounts of chemical elements in micro areas by contact microradiography. The projection method does not lend itself to similar elemental measurements because it cannot readily provide the requisite radiation "straddling" the absorption edge of an element of interest.[*] Measurements of total mass in micro areas may be carried out just as described in Section 5.A.2. With the contact method, one cannot expect to measure photographic densities in spots smaller than $1 \mu m$ in the microradiograph, whereas with the projection method one may expect to assay spots corresponding to diam-

---

[*] But elemental measurements can be made in a projection geometry by the related technique of microabsorption, as described in Section 5.C.3.

eters of about 0.2μm in the specimen. Thus the minimal *area* measurable by projection is about twenty times smaller than with the contact method. The superior figure for mass measurement by the projection method in Table II of Section 1.C reflects this factor, it having been assumed that both methods extend to approximately the same minimum mass per unit area. Actually the superiority of the projection mode in terms of total local mass is only nominal, because it is difficult in the projection geometry to produce the very soft radiation needed to assay very low mass per unit area. In fact, although photographic densitometry is greatly simplified in the projection mode because the assayed spot in the radiograph is enlarged by the primary magnification, quantitative measurements in this mode are infrequent because they are usually uninteresting in the specimens to which it is best suited.

When local mass is measured by the projection method, one must keep in mind that the path length of highly divergent x-rays through the specimen varies substantially. Measurements should therefore be restricted to nearly axial radiation. For an x-ray making a small angle $a°$ with the central axis, the length of absorbing path is increased by a factor $(1 + a/57)$ with respect to the absorption path of the central ray, and the mass would be overestimated by the same factor if the effect were ignored.

5.B.3. Apparatus for Projection Microradiography

An x-ray tube specifically suited to projection microradiography is essential. The technology of such tubes is unfamiliar to most x-ray physicists but is very closely connected with the hardware of electron microscopy, so that the apparatus is not really novel and should be quite reliable. While many sets have been custom-built in or for physics laboratories, some have been made commercially in the United States by the General Electric Company (Schenectady, N. Y.), and have been produced and supplied in England by Hilger and Watts (London). The cost varies greatly with basic capabilities and accessories, but ten to fifteen thousand dollars is likely.

5.C. PHYSICAL ASPECTS OF ABSORPTION MICROANALYSIS

The physical principles of absorption microanalysis have been outlined in Section 1.B and Figures 3 and 4. The technique produces no image, but measurements of total mass or amounts of individual elements are made in selected micro areas of a thin specimen.

5.C.1. Measurement of Total Mass, Contact Mode

The system depicted in Figure 3 has been developed chiefly by Rosengren (1959). We shall refer to this arrangement as a "contact" mode of measurement because of its obviously close relationship to contact microradiography.

Both procedures can be based on the same ordinary x-ray tube. However, for microanalysis it is not really necessary for specimen and detector to be very close. In order for the assayed area to be well defined, it is clearly the aperture which must be close to the specimen.

The measurement of mass is based on the following formulation: Let C_o be the count recorded by the detector (a gas counter or solid-state counter, as described in Section A.6) in a fixed time under fixed operating conditions, in the absence of any specimen. If the area of the aperture is A, then the average count dC_o due to radiation passing through any small ("infinitesimal") bit of area dA is

$$dC_o = C_o \frac{dA}{A} \tag{5–30}$$

After the specimen is inserted, if the transmission through the corresponding bit of area in the specimen is T, the recorded count under the same conditions will be

$$C = \int_A T C_o \frac{dA}{A} = \frac{C_o}{A} \int_A T \, dA \tag{5–31}$$

The transmission T through the bit dA of the specimen is

$$T = e^{-\bar{\sigma}S} \tag{5–32}$$

where $\bar{\sigma}$ and S are the average mass absorption coefficient and local mass per unit area in the bit dA (as explained in the Appendix A.4). Now, *if relatively penetrating radiation is used*, so that $\bar{\sigma}S \ll 1$, with sufficient accuracy one may use the approximation

$$T = e^{-\bar{\sigma}S} \simeq 1 - \bar{\sigma}S \tag{5–33}$$

Substitution of (5–33) into (5–31) gives

$$C = \frac{C_o}{A} \int_A (1 - \bar{\sigma}S) \, dA = C_o - \frac{C_o}{A} \int_A \bar{\sigma}S \, dA \tag{5–34}$$

If, in addition, the absorption coefficient $\bar{\sigma}$ is uniform throughout the assayed spot (wherever mass is present), Equation 5–34 reduces to

$$C_o - C = \frac{C_o}{A} \int_A \bar{\sigma}S \, dA = \frac{C_o}{A} \bar{\sigma} \int_A S \, dA = C_o \frac{\bar{\sigma}M}{A} \tag{5–35}$$

where $M (= \int S \, dA)$ is the *total* mass of specimen within the entire assayed region. Thus the reduction in count due to absorption in the specimen is proportional to the total irradiated mass.

For quantitation, a calibration curve may be obtained with a step wedge of the kind described in Section 5.A.2, with known composition (hence known $\bar{\sigma}$) and known mass per unit area (i.e. known M/A). The curve may be drawn with $\bar{\sigma}M/A$ as ordinate vs. C as abscissa, so that a subsequent count C_s/A with a specimen measures the quantity $\bar\sigma M/A$ in the specimen. If calibration wedge and specimen are both organic, usually $\bar\sigma$ is the same for both (as discussed in Section 5.A.2), in which case the quantity directly determined from the specimen count and the calibration curve is M_s/A, where M_s is the total irradiated mass of the specimen and A is the area of the aperture.

We must stress that it has *not* been assumed that density or thickness are uniform within the assayed spot. The result (5–35) is valid irrespective of inhomogeneities, so long as the condition $\bar\sigma S \ll 1$ is satisfied throughout the irradiated region.[*]

Two situations are common in practice:

1. The specimen "fills" the aperture. Typically the assayed spot might be part of a tissue section. Here, usually, total mass is uninteresting and the directly measured mass per unit area, M/A, is the desired quantity.

2. The aperture is larger than the specimen and totally encloses it. An isolated cell would be typical. Here the total mass M of the specimen would probably be wanted, and is to be obtained by multiplying the measured quantity M/A by the area A of the aperture. If one does want to know average mass per unit area in the cell, S_s, it can be obtained by dividing the total mass of the cell by the area of the cell, A_s, so that the specimen mass per unit area S_s is obtained via

$$S_s = (M/A)A/A_s \qquad\qquad (5\text{–}36)$$

Equation 5–35 suggests that, within its range of validity, the calibration curve of count *vs.* mass/area should be simply a straight line. In fact this is not precisely true because the absorption coefficient $\bar\sigma$ varies slightly with mass/area S, due to hardening of the spectrum as the radiation passes through the specimen. However, the effect is very small since the radiation has been adjusted to be only slightly absorbed. Consequently $\bar\sigma$ is virtually independent of S and the calibration curve is virtually a straight line, which means that the curve is usually established well enough when one records the count C_o and just one other count C_c with one calibration thickness of mass/area S_c. In this case (still assuming that the absorption coefficient $\bar\sigma$ is the same in specimen and calibration material), Equation 5–35 reduces to

$$\frac{C_o - C_s}{C_o - C_c} = \frac{M_s/A}{S_c} \qquad \text{or} \qquad M_s/A = S_c\frac{C_o - C_s}{C_o - C_c} \qquad (5\text{–}37)$$

where M_s and C_s are respectively the total irradiated mass of specimen and the specimen count. Once again, the total mass of an isolated object enclosed within the aperture can be obtained by multiplying M_s/A from Equation 5–37 by the area A of the aperture.

---

[*] It is easy to see physically why the "inhomogeneity effect" disappears. The mathematical analysis boils down to this: when the absorption is low along all the radiation paths through the specimen, the radiation flux is virtually the same everywhere; consequently each bit of mass in the specimen absorbs the same amount of radiation no matter where it is located.

The pulse counter is more sensitive than photographic film in the sense that smaller reductions in radiation can be measured. In calculating the minimum mass measurable by microradiography (Sec. 5.A.1), we concluded that the absorption in about 0.2×10^{-4} gm/cm^2 of organic material would be measurable in photographic emulsion if very soft radiation was used. The same radiation would be approximately 5 percent absorbed in a specimen layer of approximately 0.7×10^{-5} gm/cm^2; this absorption is too low for photographic assay but satisfies the condition preceding Equation 5–33, i.e. $\bar{\sigma}S = \frac{1}{20} \ll 1$, and is perhaps near the practical limit for assay by a pulse detector. The minimum area which can be assayed depends primarily on the aperture, which with modern techniques might be 1μm^2, the same as the minimum area estimated for microradiography. This leads to a minimum measurable mass of approximately 0.7×10^{-5} gm/cm$^2 \times 10^{-8}$ cm$^2 = 0.7 \times 10^{-13}$ gm.

The essential feature of the technique is the accurate measurement of small changes in the x-ray flux. This is possible because the total number of counts in each measurement can be quite large, with correspondingly small statistical fluctuation. The possibility of working at high transmission leads to Equations 5–34 and 5–35, with the consequence that the measurement is not affected by the degree of homogeneity within the assayed area. This avoids one of the major question marks affecting microradiographic assay (Sec. 5.A.2).

It is not easy to realize the highest sensitivity because extreme demands are made on the stability of the x-ray tube and the detector. In practice, the method is sometimes preferable to microradiographic assay because it is a simpler, faster and more direct way of measuring total mass or mean mass per unit area over an entire micro field.

5.C.2. Measurement of Total Mass, Projection Mode

The systems now under consideration differ from microradiography in that the photographic emulsion is replaced by a pulse counter and the area of measurement in the specimen is defined by an aperture. The arrangement of Figure 4 (Sec. 1.B), which does not seem to have been tried, is the projection version of such a system for the measurement of mass, based on the special type of x-ray tube used for projection microradiography. One would first view the projection image on a fluorescent screen, then position the aperture as desired by reference to the viewed image, and then substitute the pulse detector in place of the screen. Quantitation and calibration would be just as outlined in Section 5.C.1, except that in all of the equations, the quantity A must now be understood to be the area of the aperture A_{ap} projected back to the plane of the specimen, so that

$$A = A_{ap}\left(\frac{b}{b+c}\right)^2 \qquad (5\text{–}38)$$

In this scheme there would be no difficulty in producing apertures for the assay of very small specimen areas, since the area of the aperture itself can be much larger. One might expect to assay areas down to the resolution limit of the projection equipment, approximately 0.2μm. The favorable value for sensitivity, entry 4 in Table II (Sec. 1.C) stems from the excellent minimum area. But, as in the case of projection microradiography, the minimum value could not easily be realized because the x-ray tube does not readily supply the extremely soft x-ray spectrum which would be needed for the assay of specimens of very low mass per unit area.

5.C.3. Assay of Individual Chemical Elements

Chemical elemental analysis can be carried out with the arrangement of Figure 86, a projection system developed and tested by Long (1958). The principle is just as described in Section 5.A.3; again pairs of measurements must be made at wavelengths straddling the absorption edge of the element of interest. With the present scheme the specimen is exposed to the

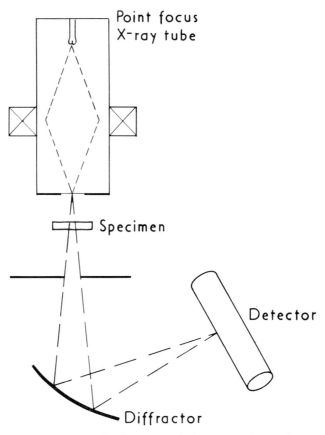

Figure 86. An arrangement for "projection" absorption chemical microanalysis.

continuum of wave lengths generated in the target-window of the tube, but the diffracting crystal is first set to diffract only one, then reset to diffract only the other chosen wave length into the detector, so that the end result ideally is the same as if the source were monochromatic.* Quantitative analysis is based on Equation 5–19. The minimum measureable amount of element depends strongly on atomic number, but Long's work shows that it may be as low as 10^{-12} gm.

The analogous contact mode of analysis (Fig. 87) has been used by Lindström (1955). This system is indeed the same as Lindström's apparatus for microradiographic assay (Fig. 83, Sec. 5.A.3) except that the photographic emulsion is replaced by an aperture and a pulse counter, and very

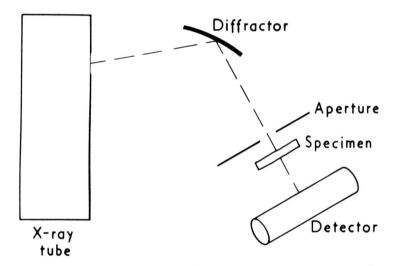

Figure 87. An arrangement for "contact" absorption chemical microanalysis.

close contact between specimen and detector is no longer necessary. Here the radiation is monochromatized by the diffracting crystal before reaching the specimen; the principle of quantitation remains exactly the same (Equation 5–19). Again the sensitivity depends strongly on atomic number, and the minimum measureable amount has been estimated by Lindström at about 10^{-12} gm in favorable cases.

The estimates of sensitivity in Table II (Sec. 1.C) are rather speculative here as some of the relevant theoretical analysis and experimental tests do not seem to have been done. As to minimum measureable amount of element, Long's theoretical analysis (1958) for the projection system is complete, but for the contact system Lindström (1955) based his estimates on

---

* Unless a characteristic spectral line of the window material is luckily suitable, the "wavelength" selected by the crystal is really a very narrow band of wave lengths in the continuum.

sufficient changes in specimen transmission without regard to limitations due to low intensity at the detector or restrictions in measurement time. As to minimum measurable weight-fractions, neither Long nor Lindström dealt theoretically with the limitation resulting from the fact that almost all of the absorption occurs in other constituents when the concentration of the assayed element is low (cf. Sec. 5.A.3). Experimentally, while Lindström performed quite a few microradiographic elemental assays, microanalysis with a pulse detector does not seem to have been applied to elemental analysis extensively in either the projection or contact mode. The entries in Table II have therefore been based mainly on the performance achieved in practice by Lindström's *microradiographic* system, and on the fact that the pulse-detector systems should do at least as well.

Alternative methods have undercut the motivation for the full development of the absorption microanalysis of chemical elements. For qualitative studies microradiography has the major advantage of providing images. When no image is required, lower elemental weight-fractions can be measured by fluorescence microanalysis (Sec. 5.D); and smaller elemental amounts and lower weight-fractions can be measured, with the provision of images as well, by electron-probe microanalysis (Sec. 5.E).

5.C.4. Equipment for Absorption Microanalysis

Rosengren's system for the measurement of mass (Fig. 3) can be set up readily with an x-ray tube suitable for contact microradiography; it is only necessary to fit apertures and a commercially obtainable x-ray pulse detector with its associated electrical circuits. Equipment for projection microradiography is similarly adaptable along the lines of Figure 4. However, elemental analysis according to the arrangements of Figures 86 and 87 requires the incorporation of diffraction-crystal monochromators, an extensive and specialized project whose advisability must be evaluated in view of the alternatives, i.e. x-ray fluorescence and electron-probe analysis.

5.D. PHYSICAL ASPECTS OF X-RAY FLUORESCENCE MICROANALYSIS

The fundamentals of x-ray fluorescence microanalysis have been described in Section 1.B. (Fig. 5). We have already noted there that one must choose between two instrumental arrangements: either one may use a diffracting crystal to restrict detection to a single characteristic X radiation as in Figure 6 (see also Appendix A.7), or one may omit the crystal as in Figure 5, and rely on pulse-height analysis of the pulses produced in the counter for the identification of the characteristic radiations and the measurement of their intensities (Appendix A.7). When relatively large specimen volumes are to be analyzed, as in conventional x-ray fluorescence analysis, the intensity of the fluorescent x-rays is ample and the best performance is obtained with diffractors, since they provide the best wavelength discrimination. With extremely small specimens the x-ray intensity is so low that one must omit the diffractor and bring the counter close to the speci- men in order to get adequate counting rates. The performance of diffrac- tive and nondiffractive systems for a variety of specimens has been com- pared extensively by Zeitz (1969), who has shown that a well-designed diffractive system is superior down to quite small analyzed volumes con- taining of the order of $10\mu g$ of tissue. However, *microanalysis*, i.e. the ele- mental analysis of microscopic volumes, is concerned with amounts which are smaller still by one or more orders of magnitude. While Zeitz (1962) has explored the possibilities for analyzing even such small amounts diffrac- tively, the medical studies in Chapter 3 have been based exclusively on nondiffractive analysis, to which we shall restrict our attention here.

5.D.1. The Minimum Measurable Amount of an Element

In measurements of potassium in nerve cells, Röckert and Rosengren (1966) reported minimum measureable amounts in the range of 10^{-10}— 10^{-11} gm of potassium. In order to estimate minima for other elements in other specimens and to assess the potential of instrumental modification, we must consider the many factors on which this minimum depends.

a. *Dependence on Atomic Number.* The window-target produces both characteristic and continuum radiation (Appendix, Fig. 111), and both may contribute significantly to the generation of the observed x-ray fluores- cence in the specimen. If the target can be well matched to the element of interest in the specimen [*] most of the signal will be generated by the tar- get's characteristic radiation. Because relatively few elements are physi- cally suited to be target-windows, the match of target to assayed element is usually not close, and most of the signal may be generated by that part of the *continuum* radiation which is close to the absorption peak of the as- sayed element.

To estimate the dependence on atomic number of excitations by the continuum, we can start with an expression (Kulenkampff, 1922; Amrehn, 1956; Amrehn and Kulenkampff, 1955) for the number $N(V)$ dV of con- tinuum quanta of energy between V and $(V + dV)$ generated when one electron of energy E_o strikes the target window:

$$N(V) \ dV \propto Z_t \frac{E_o - V}{V} \ dV \qquad (5\text{--}39)$$

[*] When no diffractor is used, the characteristic x-rays of the target must be sufficiently dif- ferent from the recorded x-ray lines of the specimen for pulse-height analysis to distinguish effectively between them; otherwise the background from the scattered target line would be intolerable. But too great a separation implies inefficient excitation of fluorescence. Ideally the quantum energy of the target line should be greater than the quantum energy of the observed line by a factor of 1.2 to 1.5, depending on the pulse-height resolution of the detector.

where Z_t is the atomic number of the target material. For each continuum quantum of energy V incident on a thin specimen, the number n_B of characteristic quanta of one spectral line emitted by the assayed element is

$$n_B = w_s \sigma_s S_s \tag{5-40}$$

where w_s is the fluorescence yield of the element, σ_s is the relevant mass absorption cross section in cm^2/gm, and S_s gm/cm^2 of the element are present. The *atomic* absorption cross section σ_a, in $cm^2/atom$, has the form

$$\sigma_a \propto Z_s^4 V^{-3} \tag{5-41}$$

and the conversion factor from σ_a to σ_s, i.e. atoms/gm, is approximately proportional to $1/Z_s$, so σ_s has the form

$$\sigma_s \propto Z_s^3 V^{-3} \tag{5-42}$$

where Z_s is the atomic number of the assayed element. By combining these equations, we obtain an expression for the number of quanta of the assayed element generated in the specimen per electron incident on the target:

$$n_B \propto w_s S_s Z_s^3 \int_{E_{xs}}^{E_o} V^{-3} Z_t \frac{E_o - V}{V} dV \tag{5-43}$$

where E_{xs} is the absorption-edge energy of the assayed element. We note that

$$E_{xs} \propto Z_s^2 \tag{5-44}$$

Then, after integration, Equation 5–43 reduces to

$$n_B \propto Z_t w_s S_s Z_s^{-1} [\tfrac{1}{3} U_s - \tfrac{1}{2} + \tfrac{1}{6} U_s^{-2}] \tag{5-45}$$

where $U_s = E_o/E_{xs}$.

Before interpreting this expression, we shall derive a corresponding result for fluorescence excited by *characteristic* radiation from the target. According to Green and Cosslett (1961), each electron incident on the target generates a number N_c of characteristic quanta in the target,

$$N_c \propto w_t \frac{R}{A_t C} [U_t \ln U_t - U_t + 1] \tag{5-46}$$

where w_t is the fluorescence yield for the characteristic radiation in question, R is a factor allowing for loss of incident electrons due to backscattering from the target, A_t is the atomic number of the target material, C is a "constant" related to the rate of electron energy loss in the target and varying only slowly with atomic number, $U_t = E_o/E_{xt}$ and E_{xt} is the absorption-edge energy corresponding to the target's characteristic radiation. We can again use the forms (5–40) and (5–42) to estimate the conversion of this radiation to x-ray fluorescence in the specimen. The result for the number n_c of fluorescence quanta generated from the assayed element in the specimen by incident characteristic x-rays is

$$n_c \propto \frac{R}{A_t C} w_t w_s S_s Z_s{}^3 V_t{}^{-3}[U_t \ln U_t - U_t + 1] \qquad (5\text{-}47)$$

where V_t is the quantum energy of the target's characteristic radiation. For the sake of estimating how the yield may vary with the atomic number of the assayed element we shall assume that for any Z_s, one works with a target whose atomic number Z_t is a fixed multiple of Z_s.* Then

$$V_t \propto Z_t{}^2 \propto Z_s{}^2 \qquad (5\text{-}48)$$

and

$$A_t \propto Z_t \propto Z_s \qquad (5\text{-}49)$$

Insertion of (5–48) and (5–49) into (5–47) gives

$$n_c \propto \frac{R}{C} w_t w_s S_s Z_s{}^{-4}[U_t \ln U_t - U_t + 1] \qquad (5\text{-}50)$$

The Equations 5–45 and 5–50 represent the signals generated by continuum and characteristic radiation *per electron* incident on the window-target. To convert to counts per second, we must multiply by the electron current i which must be limited to avoid melting the window. This restriction implies a dependence of the permissible current on atomic number. The power dissipated in the target is $P = E_o i$. The maximum tolerable power depends on the target material and will not vary systematically with the atomic number Z_s of the assayed element, but if we assume that E_o is adjusted with Z_s to maintain constant overvoltage factors U_t and U_s, then the tolerable current i will vary on the average as

$$i \propto E_o{}^{-1} \propto E_x{}^{-1} \propto Z_s{}^{-2} \qquad (5\text{-}51)$$

Multiplying (5–45) and (5–50) by this factor, we obtain for the counting rates n'_B and n'_c generated by incident continuum and characteristic radiation

$$n'_B \propto Z_t w_s Z_s{}^{-3}[\tfrac{1}{3} U_s - \tfrac{1}{2} + \tfrac{1}{6} U_s{}^{-2}] \qquad (5\text{-}52)$$

and

$$n'_c \propto \frac{R}{C} w_t w_s Z_s{}^{-6}[U_t \ln U_t - U_t + 1] \qquad (5\text{-}53)$$

(For Equation 5–53, it has been assumed that the target has been matched to the assayed element. Equation 5–52 does not involve this assumption, but

---

* Actually, if the detector is a gas proportional counter, the detector pulse-spread for a characteristic radiation increases in proportion with atomic number, and one should keep Z_t/Z_s constant in order to get the best match while avoiding interference from the target's characteristic quanta scattered by the specimen. If a solid-state detector is used, the pulse spread is almost independent of atomic number so that one can utilize smaller ratios Z_t/Z_s as Z_s increases (if a target can be made of the element with ideal Z_t). This is somewhat favorable for the analysis of elements of higher atomic number.

the yield from continuum is obviously maximized by choosing a high Z_t—for example, a gold target.)

The equations both show that as a function of atomic number, the available counting rates depend mainly on the quantity wZ^{-3}.* This factor happens also to determine the minimum amounts measureable in thin specimens by electron probe analysis, and it is discussed more fully in Section 5.E.2. As shown there in Table VI the best sensitivity, in terms of counting rate per unit of mass of assayed element, is to be expected in the neighborhood of $Z_s = 20$ (calcium) but within a wide range of atomic numbers the sensitivity should not be much different.

b. *Dependence on Assayed Area.* The intensity of radiation incident onto the specimen, and hence the counting rate produced per unit mass of the assayed element, are proportional to the electron current in the tube. The permissible current, limited by the danger of melting the window, is approximately proportional to the radius r of the spot of focussed electrons (Cosslett and Nixon, pp. 217 and 219). This is a very approximate form for a dependence which has not been accurately worked out for our conditions. This radius, in turn, must be similar to the radius of the aperture which defines the irradiated specimen area and thus determines the spatial resolution of the method.

Hence the minimum measureable amount of an element will be (very approximately) inversely proportional to the radius of the defining aperture adjacent to the specimen, this radius being the determining factor for the spatial resolution of the analysis.

While the minimal measureable amount depends on atomic number and spatial resolution for physical reasons little affected by instrumental design, it depends strongly as well on three instrumental features: the distance from target to specimen, the fraction of the fluorescent radiation received and registered by the detector, and the available electron current. In the apparatus used by Röckert and Rosengren (1966), for which a limit of approximately 10^{-10} gm of element was quoted, the distance from target to specimen was 1 mm; there is little doubt (Long and Röckert, 1963) that this distance can be decreased to about 200μm or less, improving the minimal amount by a factor of at least $(1000/200)^2 = 25$. The efficacy of detection cannot be easily improved, since a closer positioning of the detector requires better control of scattering in the aperture, but the possible improvement by a factor of 5, anticipated by Long and Röckert, seems conservative. Finally, Röckert and Rosengren's operating current of 50μA was limited in fact by the available power supply, while very much higher currents (diffi-

---

* Of the other factors in (5–52) and (5–53), R and C are relatively insensitive to atomic number, and it can be shown that the best results can be obtained if U_s and U_t are maintained at values in the neighborhood of 3 to 5.

cult to estimate accurately) could certainly be produced, and tolerated by the window of the tube, for their assays where spatial resolution was of the order of 100μm. Thus it is anticipated that the absolute sensitivity inherent in the arrangement of Figure 5 (Sec. 1.B) should be better than realized in the prototype apparatus of Long, Röckert and Rosengren by a factor of 100 or more, i.e. the minimum elemental amount should be less than 10^{-12} gm. This performance remains to be achieved in practice.

5.D.2. The Minimum Measureable Weight-Fraction of an Element

For the measurement of low elemental weight-fractions, emission methods such as x-ray fluorescence analysis are inherently superior to absorption methods such as microradiographic assay (Section 5.A.3). With an absorption method one must always measure incident and transmitted fluxes I_o and I and the important quantity is the absorbed fraction $(I_o - I)I_o^{-1}$. When the elemental weight-fraction is low, the experimental errors in measuring I_o and I are too large for a valid determination of the small difference $(I_o - I)$ produced by the assayed element. In the case of emission methods, on the other hand, the transmitted flux is irrelevant. In fluorescence analysis, the signal consists of characteristic x-rays emitted from the assayed element, and the minimum measureable weight-fraction is determined by interference from two kinds of background x-radiation: first, continuum radiation from the anode, scattered by the specimen, and secondly characteristic radiations of similar wave length generated in other elements in the specimen.

Let us consider first the limit set by scattered continuum. Since the scattering occurs with no change or with insignificant change in quantum energy, and since the detection system is set to respond only to a small range of quantum energies around the energy V_x of the selected characteristic line, we have to assess only the scattering of photons incident with quantum energies near V_x. For elements in the atomic-number range 8 to 38, near the absorption edge the cross section for the generation of fluorescent K-radiation $(w\sigma)$ is somewhat more than 100 cm²/gm, while the scattering cross section in soft tissue is less than 1 cm²/gm. From the ratio of these cross sections, and from the fact that fluorescence is excited by all incident radiation above the absorption edge while the counter registers scattered radiation only if it is near the characteristic line, we may anticipate that for an element with atomic number $8 \leq Z \leq 38$ present in low concentration in soft tissue, the characteristic signal will be equal to the continuum background for weight-fractions of the order of 0.1 percent. In fact Hall (1963) found, using a gas-flow proportional counter to detect the K radiation from zinc in an organic matrix, that the zinc-K signal and the scattered background were equal at a zinc weight-fraction of just 0.1 percent. This result suggests a minimum measurable weight-fraction in the range 0.1 to 0.01 percent.*

Neither extensive data nor detailed theory seem to be available to generalize this datum to other elements or detectors, but a simple estimation will be useful. From Equation 5–45 we see that the fluorescence signal per electron in the x-ray tube n_B is proportional to $w\ Z^{-1}$. The intensity of scattered background, n_{sc}, can be expressed in a form similar to Equation 5–43:

$$n_{sc} \propto \int_0^{E_o} \sigma_{sc}(V)\ R(V)\frac{E_o - V}{V}\ dV \qquad (5\text{–}54)$$

where $\sigma_{sc}(V)$ is the cross section for scattering photons of energy V into the detector and $R(V)$ is the probability that they are registered, the detector having been set up with a pulse-height window centered on the selected characteristic line. Since the window of the pulse-height analyzer accepts pulses only within a narrow range around the characteristic quantum energy V_x, the function $R(V)$ can be regarded for present purposes as effectively unity within some range $V_x \pm \Delta$ and zero elsewhere, so that the integral reduces to

$$n_{sc} \propto 2\ \Delta\ \sigma_{sc}(V_x)\frac{E_o - V_x}{V_x} - 2\ \Delta\ \sigma_{sc}(V_x)(U' - 1) \qquad (5\text{–}55)$$

with $U' = E_o/V_x$. Now a gas counter converts a monochromatic radiation V_x into a spread of pulse heights with a statistical deviation $\delta \propto V_x^{1/2} \propto Z$, so that the counting efficiency for the characteristic line may be kept independent of Z by setting the window 2Δ proportional to Z. As to the other factors in Equation 5–55, the scattering cross section σ_{sc} depends on quantum energy and hence on atomic number in a complicated way. However, Figure 88, calculated from Compton and Allison (1935), pp. 781 and 782, shows that at least over the most interesting range of atomic numbers, $15 \le Z \le 30$, the relationship is well approximated by $\sigma_{sc} \propto V^{-0.8} \propto Z^{-1.6}$.†
The remaining factor in Equation 5–55 $U' - 1$, will be very nearly independent of Z if the over-voltage $U_s = E_o/E_{xs}$ (Equation 5–45) is chosen by the operator to be independent of Z. Hence under operating conditions for which the signal $n_B \propto w\ Z^{-1}$, Equation 5–55 implies

$$n_{sc} \propto Z \cdot Z^{-1.6} = Z^{-0.6}\ \text{and}\ n_B/n_{sc} \propto wZ^{-1}/Z^{-0.6} = wZ^{-0.4}$$

Since $w\ Z^{-3}$ is approximately constant (cf. Table VI, Sec. 5.E.2), we conclude finally that the ratio of signal to background varies with the atomic number of the assayed element approximately according to

* Weight-fraction for the specimen "as is," of course: dry weight-fraction if the specimen is dried.

† The other estimates of scattering in this section are based on the same tabulations of Compton and Allison.

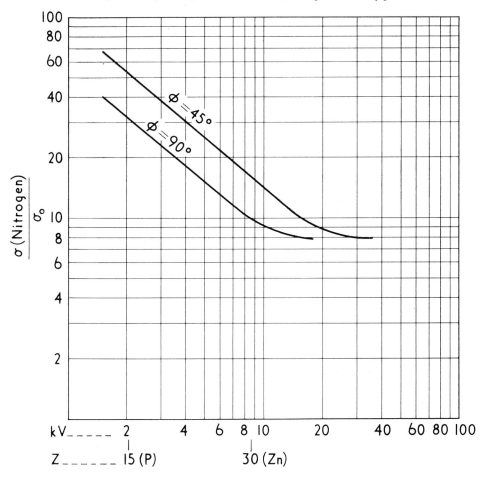

Figure 88. Scattering cross sections in nitrogen for x-radiations in a quantum-energy range corresponding roughly to the K radiations from elements of atomic number 15–30. (The curves for the other chief scatterers in soft tissue, carbon and oxygen, are similar.)

$$n_B/n_{sc} \propto Z^{3-0.4} = Z^{2.6} \quad \text{(for } 15 \leqq Z \leqq 30\text{)} \qquad (5\text{–}56)$$

This implies, for example, that the signal-to-background ratio at a given weight-fraction should be only about one-sixth as good for phosphorus $(Z = 15)$ as it is for zinc $(Z = 30)$. The poorer signal-to-background ratio at the lower atomic numbers is manifest in the measurements of potassium in nerve cells (Long and Röckert, 1963). Fortunately, the interesting elements below atomic number 20 are naturally present in such high concentrations in tissues that they can be assayed without a very good instrumental performance so far as signal/background ratio is concerned.

To recapitulate, when the detector is a gas proportional counter coupled to a pulse-height analyzer, the fluorescence signal and the scattered background are approximately equal at a weight-fraction of 0.1 percent for assays of zinc $(Z = 30)$ in an organic matrix, and the signal/background ratio at a given weight-fraction depends on atomic number Z approximately as $Z^{2.6}$. But in deducing minimum measurable weight-fractions, one must take the following points into account.

1. A filter can be placed between target and specimen (Fig. 89), to make the spectrum of incident radiation "harder" and remove almost all of the incident radiation near the quantum energy of the selected characteristic line. In the case of zinc in an organic matrix, the insertion of a titanium filter, 1.2 10^{-2} cm thick, reduced the zinc weight-fraction corresponding to equal signal and background from 0.1 percent to 0.005 percent (Hall, 1963). In this way, zinc weight-fractions down to 0.001 percent (10 ppm) were successfully measured.

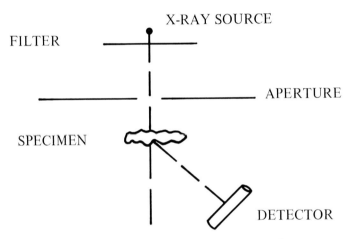

Figure 89. Arrangement with filter between x-ray source and specimen, to improve signal-to-background ratio.

2. By removing incident radiation near the energy of the characteristic line, the filter also seriously reduces the intensity of the generated fluorescence signal, so that the improvement in the minimum measureable weight-fraction is at the cost of a poorer value for the minimum measureable amount of element. For example, the titanium filter just mentioned, while improving signal/background by a factor of 20, reduced the fluorescence intensity by a factor of approximately 5. In order to judge the feasibility of a particular analysis one must consider separately the minimum weight-fractions and the minimum absolute amounts which can be assayed in a given instrumental configuration; both criteria must be satisfied for the successful analysis of a specimen.

3. Lithium-drifted silicon solid-state detectors will provide a considerable improvement, as has been confirmed in several commercial x-ray fluorescence analyzers, although their use in x-ray fluorescence *microanalysis* is

just beginning (Nicholson, 1971). At present (July, 1971), the pulse spread produced by monochromatic radiation in the best commercial solid-state detectors is approximately 160 eV full-width at half-height (cf. Appendix A.7) regardless of quantum energy in the range of K radiations from elements in the atomic number range 15 to 30. For gas counters the spread is proportional to atomic number Z and ranges from approximately 600 eV at $Z = 15$ to approximately 1200 eV at $Z = 30$. Consequently solid-state detectors will improve signal-to-background ratios by a factor near 4 at $Z = 15$ and near 8 at $Z = 30$.

4. Our estimations have been pegged to a measurment of zinc fluorescence and scattered background on an organic matrix (Hall, 1963). In this study the radiation was observed at a scattering angle ϕ of 90° between the directions of the incident and the scattered rays. However, the intensity of scattering varies with angle and $\phi = 45°$ is a more common angle of observation in fluorescence microanalysis. Figure 88 shows that over the range $15 \leq Z \leq 30$, the ratio σ_{sc}/σ_o at 45° is larger than at 90° by a factor of approximately 1.6. The quantity σ_o, the scattering cross section of a single free electron, has the angular dependence $\sigma_o \propto (1 + \cos^2 \phi)$ and thus is larger at 45° than at 90° by a factor of 1.5. Consequently signal-to-background ratios at $\phi = 45°$ are poorer than at 90° by a factor of 1.5 (1.6) = $2\frac{1}{2}$.

5. Background scattering per unit mass increases with the mean atomic number of the specimen. Therefore the worst background is to be expected from bone or teeth. A calculation from the scattering tables of Compton and Allison indicates that for radiations corresponding to the K lines of elements $15 \leq Z \leq 30$, scattering per unit mass in apatite is approximately twice as intense as in soft tissue.

6. So far we have considered scattering from the specimen only. Other sources of scattered background are gas, the walls and the other surfaces in the specimen chamber, and especially the aperture between tube and specimen. Air in the specimen chamber is sometimes tolerable, but gas scattering can be much reduced with a helium atmosphere, and can be eliminated entirely by evacuation of the specimen chamber. In the instrument of Hall (1963), background from all extraneous sources was successfully controlled so that the observed background could be attributed entirely to the specimen and its very thin supporting film. However Hall's arrangement was not designed for high spatial resolution and does not include a fine aperture. It remains to be established how successfully one can control scatter from the aperture or system of apertures.

Given the qualifications above, fluorescence microanalysis is suited to measure elemental weight-fractions of the order of 0.01 percent in biological tissues, with one further serious limitation to which we now turn: interference between the selected radiation and other characteristic lines.

When a gas proportional detector and pulse-height analyzer are "tuned" to the K_a radiation from an element of atomic number Z, it is apparent from Figure 121 (Sec. A.7) that the system will respond as well to K_a radiations from elements $Z \pm 1$ and even, to a considerable extent, $Z \pm 2$. Thus, in the studies of potassium ($Z = 19$) described in Section 3.B.2, it was recognized that the analytical system itself did not discriminate fully against chlorine ($Z = 17$) and calcium ($Z = 20$). If the detector is a solid-state rather than a gas type, the discrimination is more effective, as suggested by Figure 122, but still may not be complete against the K_a lines of elements $Z \pm 1$. Furthermore, there may be serious interferences between the assayed K_a line and the K_β radiation of a neighboring element. However, well-established techniques are now available for discriminating between the radiations of neighboring elements without recourse to diffracting crystals. We shall discuss three methods: the use of a single filter, the use of paired filters (usually "balanced"), and pulse-height "network" analysis.

1. *When there is only a single seriously interfering element,* the arrangement of Figure 90 (Hall, 1963) may be used. Of the two proportional counters, one registers both the desired and the interfering radiation, while the single filter selectively prevents the higher-energy radiations from reaching the second counter. For K_a radiations from elements of atomic

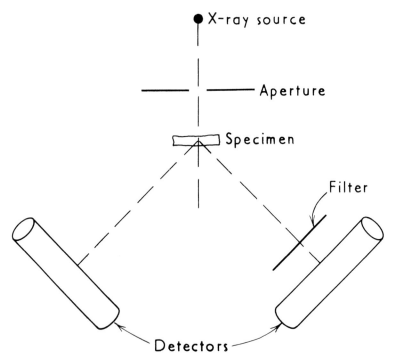

Figure 90. Single-filter arrangement for analyzing radiations from elements of adjacent atomic number.

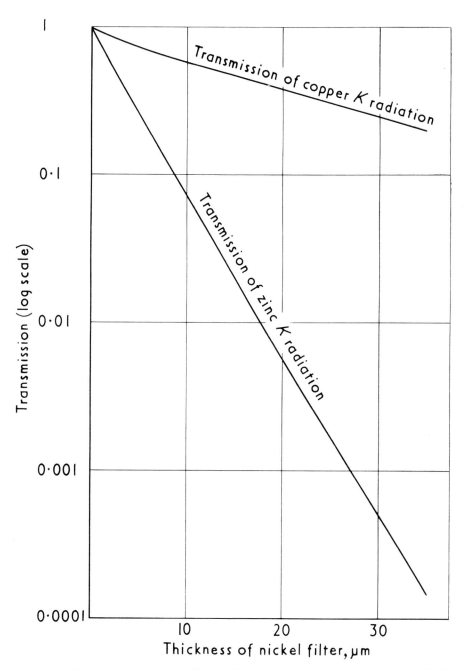

Figure 91. Effectiveness of a nickel filter in discriminating against zinc radiation in favor of copper radiation. (The curves would be straight lines if only K_α radiations were involved. The depicted curves are not exactly straight lines as the K_β radiations have also been taken into account.)

number Z and Z + 1, the appropriate filter contains element Z − 1. Figure 91 shows, for example, how a nickel filter (Z − 28) can pass most of the copper K_a radiation (Z = 29) while removing almost all of the zinc K_a radiation (Z = 30).

Before considering how the single-filter method can be made quantitative, we shall describe the other two nondiffractive methods, since a single formalism applies to all three. Because the radiation Z + 1 can be quite effectively removed, element Z can be assayed in the presence of much larger amounts of Z + 1. However, there is no filter which discriminates strongly against radiation Z in favor of radiation Z + 1, so that the method is not equally effective for the detection of element Z + 1 in the presence of much larger amounts of Z. For example, in an organic matrix, twenty parts per million of copper was assayed in the presence of 2000 ppm of zinc, but the assay of zinc was unreliable when the ratio of concentrations copper/zinc was more than 10 (Hall, 1963).

2. If radiation from element Z is to be assayed in the presence of interference from neighboring radiations of *both* shorter and longer wave lengths, the arrangement of Figure 92 may be used. One filter passes radiations Z − 1 and Z while discriminating against neighboring higher atomic numbers; the other filter passes Z − 1 but discriminates against Z as well as the neighboring higher atomic numbers. The neatest form of this two-filter method is the "balanced filter" technique (Fig. 93) introduced by Ross (1928). "Balance" means that the filter thicknesses are adjusted so that the transmissions are equal for an interfering line on one side of the absorption edges.* If the "jump ratio"—the ratio of absorption cross sections at the two sides of the absorption edge—is the same for the two filters, then the transmissions will also be equal for an interfering line at the other side of the edges. Furthermore, insofar as the wavelength dependence of the absorption cross section is quite similar for different absorbers, varying approximately as λ^3 in wavelength intervals which do not include absorption edges, balanced filters will have equal transmission for *all* radiations except for the lines or continuum falling between their absorption edges. Thus with ideal balanced filters, all of the interfering radiations can be made to produce equal counts in the two counters, and the difference between the counts is simply proportional to the intensity of the radiation which is to be assayed.

Fine points of the Ross method, especially complications due to con-

* The requirement is not simply that an interfering line shall produce equal count rates in the two counters; the filter transmissions themselves must be equal. Therefore, to avoid the effects of unequal solid angles or counter efficiencies, the balance of a pair of filters must be tested by alternate runs with a single counter.

Obviously, the fluorescence analysis itself can also be carried out with only a single detector, if successive runs are made with alternated filters, as is done in some commercial x-ray fluorescence analyzers.

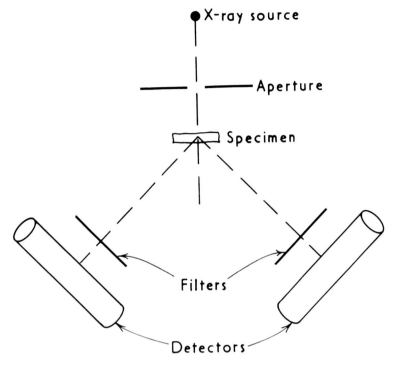

Figure 92. Two-filter arrangement for analyzing a radiation which is between two inter-fering radiations.

tinuum radiation, have recently been analysed thoroughly by Castaing and Pichoir (1966), who use *gases* like oxygen and nitrogen as readily balanced filters for certain assays at long wave lengths. Two serious difficulties must be noted here. First, the quantitative validity of the balanced-filter prin-ciple depends on the equality of the jump ratios in the two filters. If the jump ratios are unequal and the filters are balanced with respect to radia-tion $Z - 1$, they will be unbalanced for radiation $Z + 1$, and *vice-versa*, as indicated in Figure 94. In this case, it is impossible for the two counters to respond equally to all interfering characteristic lines. In fact, the jump ratios are not always very nearly equal in otherwise suitable pairs of filters. A second serious difficulty is the practical one of producing pairs with just the correct relative thickness. Gas filters can be balanced easily by adjusting the pressure, but generally one must use solid filters, for which the existing methods of balancing (Rhodes, 1968) are laborious at best.

However, with the calibration technique described below, quantitative analysis is possible with the same filters and configuration of detectors even if the filters are not balanced. We shall see that the unbalanced scheme is precise in principle if only three characteristic lines are present, for exam-ple $Z - 1$, Z and $Z + 1$, and three pure-element standards are used for

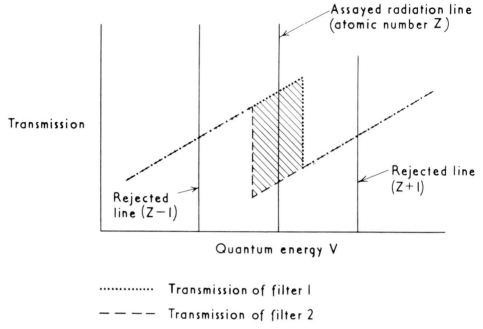

Figure 93. The balanced-filter method of x-ray analysis. The graphs show the transmissions *T* of two balanced filters as functions of the quantum energy *V* of the incident radiation. (The depicted quantities are actually—*log log* $(1/T)$ vs. *log V*. These quantities are related in the same way as *T* vs. *V* but with the advantage that the graphs are straight lines.)

calibration. If additionally, one or both of the radiations $Z \pm 2$ are present, and still only three standards are used, the correction provided by unbalanced filters is imperfect but the radiations $Z \pm 2$ or more remote characteristic lines can be rejected more or less effectively by simultaneous pulse-height analysis.

3. It is also possible to analyze quantitatively by means of pulse-height discrimination alone, even when the pulse-height distributions of the constituent elements overlap badly. Obviously this technique will be most effective when the overlap is not so severe, for example when solid-state detectors are applied to K-radiations from elements $22 \leq Z \leq 30$ (Fig. 122, Sec. A.7). The overlap of distributions from neighboring atomic numbers is worse both for solid-state detectors with $Z < 20$, and for gas proportional counters at all wave lengths (Fig. 121). (The degree of overlap of K_a radiations from neighboring elements is independent of atomic number in the case of the gas counters.) Nevertheless, even neighboring elements may be analysed with a gas counter, as shown by Dolby (1963), by means of the pulse-height "network" method which we shall now describe.

If the analyst must cope with radiation from *n* elements in the specimen,

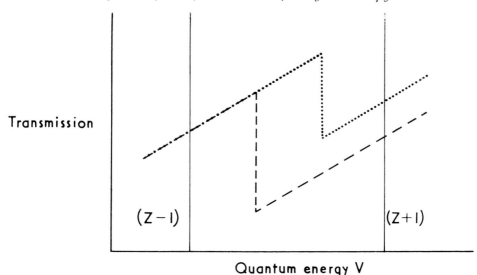

Figure 94. Breakdown of the balanced filter method when the filter jump ratios are unequal. When the filters are balanced to reject radiation $(Z - 1)$ as shown, radiation $(Z + 1)$ will not be fully rejected, and *vice versa*.

he sets n pulse-height "channels" within which pulses are counted. Usually each channel will be centered around the average pulse height produced by the most prominent x-ray line of the corresponding element (cf. Fig. 95). Then the counts C_r produced in each channel r are

$$C_r = K_{r1}M_1 + K_{r2}M_2 + \cdots + K_{rn}M_n \qquad (5\text{--}57)$$

where M_1 is the mass of element 1, etc. The "count" C_r may be either an integral (scaler) count or the reading of a count-rate meter. The set of n

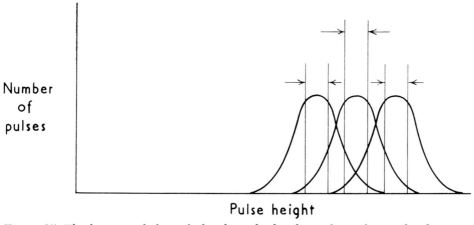

Figure 95. The location of channels for the pulse-height analysis of several radiations. The three curves depict pulse-height distributions produced by three neighboring radiations. For analysis by the "network" method three channels would be used, located as indicated.

equations 5–57 expresses the count in each channel as the sum of contributions from the constituent elements, and each coefficient K_{rs} is simply the average count introduced into channel r per unit mass of element s.

The coefficients K_{rs} can all be determined by calibration with standards under conditions of uniform excitation, each standard consisting of a known mass of one element.* One run with the specimen then gives n quantities C_r, and the n unknown elemental amounts M_s can be obtained from the n simultaneous equations 5–57.

The equations may be solved automatically by an analog electrical network once calibration has set the coefficients K_{rs} into the network in the form of potentiometer settings (Dolby, 1963; Mitra and Hall, 1971). The network "unscrambles" the interfering radiations on a statistical basis and produces n output signals, each of which is the measure of the amount of one constituent alone.

The network method may be regarded as a synthesis of the specimen's pulse-height spectrum from the properly weighted pulse-height spectra of the individual constituents. The synthesis is crude since each spectrum is represented merely by the counts observed within a few broad channels. A much more refined synthesis can be provided by multichannel analyzers which faithfully represent each pulse-height spectrum by division into 100 or more channels. When the number of channels exceeds the number of constituents, the number of Equations 5–57 is larger than the number of unknowns and the equations are "overdetermined." The analysis may then be optimized by finding the solution with a least-squares deviation, readily determined if the data are fed into a computer. Multichannel analysis is quite useful if one does not know the identity of the important constituents in advance. However the electronic equipment is very elaborate and expensive, and the results must wait on the services of a remote computer. When the important constituents are known and the intrinsic resolution of the detector is not very good anyway, it has been shown (Dolby, 1961) that virtually the same sensitivity and accuracy are provided with the few broad channels of the much simpler network method, which provides the results of a run directly and immediately.

The same formalism, Equations 5–57, is equally applicable to the one-filter and the two-filter methods. One needs merely to enlarge the concept

---

* In the *electron probe* analysis of *thick* specimens each constituent contributes to the count in each channel simply in proportion to its weight-fraction. Then Equations 5–57 are still valid if the coefficients K_{rs} are taken to be the counts produced in pure thick standards and the quantities M_s are taken to be the *weight-fractions*. Thus in this case pure thick standards may be used, and thin standards of known mass are not needed.

One can also avoid the need for standards of known mass in the fluorescence measurement of weight-fractions in thin specimens, by means of the extension of the network equations described in Section 5.D.3.

of "channel" to refer to *any* kind of discrimination applied to the radiation, e.g. restriction of the count either to a range of pulse heights, or to radiation which has traversed a certain filter (or to a combination of both restrictions: registration of a certain band of pulse heights produced in a filtered counter). In this enlarged concept, the coefficient K_{rs} is the efficiency of contribution of the element s to the count in the r th channel, whatever is the condition defining the channel. The one-filter method is seen then to be simply a form of network-analysis for specimens containing only two important fluorescing elements. Similarly one can perform network analysis of three fluorescing elements with two *unbalanced* filters combined with pulse-height bands to give a total of three channels. For the trio of elements $Z-1$, Z, $Z+1$ the statistical error depends on choice of filtration and pulse-bands and is different for each element; the analysis for Z is optimum when the filters are near balance, but the network scheme removes the need for precise balance.* Finally the balanced-filter method can be seen as a special case of the network method in which the equations 5–57 reduce to

$$C_1 = K_{1,z-}M_{z-} + K_{1,z}M_z + K_{1,z+}M_{z+} \qquad (5\text{--}58)$$
$$C_2 = K_{2,z-}M_{z-} + K_{2,z}M_z + K_{z,z+}M_{z+}$$

and balance implies $K_{1,z-} = K_{2,z-}$; $K_{1,z+} = K_{2,z+}$. Hence subtraction gives simply

$$(C_1 - C_2) = (K_{1,z} - K_{2,z})M_z \qquad (5\text{--}59)$$

and the difference-count $(C_1 - C_2)$ is seen to be the measure of M_z alone, as stated above.

In sum, nondiffractive fluorescence analysis can be seriously complicated by interferences between elements, but techniques do exist for the assay of two, or even three or more elements of neighboring atomic number. It would be too complicated to consider here in detail the evaluation of the effectiveness of these techniques, and the determination of the most effective combination of filters and pulse-height channels for any given case. In general, one can assay an element in the presence of larger amounts of neighboring elements ranging up to excess concentrations of ten to one-hundredfold by means of the network techniques, whereas in their absence a clean nondiffractive separation may be impossible even at equal concentrations.

---

* To analyze for Z in the presence of $Z-1$ and $Z+1$, one can use two filters and define three channels as follows: two channels centered on the average pulse height of radiation Z, one associated with each filter, and a third channel centered on the average pulse of $Z-1$, associated with filter number 2 (Fig. 92). Such a third channel is arranged especially to help assay for Z in the presence of excess amounts of $Z-1$, since the filters themselves are least effective in this case.

5.D.3. Quantitation

We shall consider three types of quantitative assay: measurement of the absolute amount of an element, of the total mass of the specimen, and of the weight-fraction of an element.

Measurement of the Absolute Amount of an Element

Our discussion is restricted to thin specimens, in which the fluorescence signal from an element is linearly proportional to the amount of the element in the analyzed field. Calibration therefore requires simply a run with a known amount of the assayed element.

A known amount of element may be laid onto a thin supporting film by depositing a drop of measured volume from a solution of known concentration. The solvent must be either volatile or noninterfering. For assays of potassium, Long and Röckert (1963) introduced a convenient procedure which avoids the need for accurate measurement of the volume of a small drop (see Sec. 3.B). Their standards were evaporated drops from a dilute solution of potassium hydroxide of known concentration. The amounts of potassium in individual standards were determined by adding I^{131} to the solution and comparing the activity of each standard with that of an aliquot of the bulk, by means of a scintillation counter (the radioactivity was too weak to affect the fluorescence data).

Of course the conditions of excitation should be the same for standard and specimen. Since the x-ray tube output may be somewhat unsteady, when counts are recorded from a scaler it is necessary to monitor the excitation integrated over each run. A suitable normalization can be provided by the count in an additional detector placed in the path of the transmitted beam. Long and Röckert have managed without the extra counter by keeping a small amount of iron in the analyzed volume (mounted on a separate film) and recording the iron fluorescence in each run in a separate pulse-height band. The characteristic spectral lines of iron and of their assayed element (potassium) were separated well enough for interference to be negligible.

While the fluorescence signal itself is proportional to elemental amount in thin specimens, it is usually necessary to subtract background from the total count to arrive at the genuine fluorescence signal. Background from continuum may be estimated by measurements at pulse heights near the fluorescence peak, and background from interfering elements may be handled as described above.

Measurement of Total Specimen Mass

One can measure total specimen mass within the irradiated volume by counting scattered radiation in a pulse-height band well away from the spectral peaks of the fluorescing elements. A good intense signal can usually be obtained by centering the pulse band on the main spectral line of the anode of the x-ray tube. For thin specimens the counting rate will be linearly proportional to the total irradiated mass and to the average scattering cross section of the specimen. Hence calibration can be achieved against a standard of known mass (for example a piece of uniform plastic film), and the calibration will be absolute and valid insofar as the ratio of the mean scattering cross sections of specimen and standard are known.

In general, scattering cross sections per unit mass vary considerably with atomic number as indicated by Table III. Therefore the scattered in-

TABLE III

RELATIVE SCATTERING CROSS SECTIONS *PER UNIT MASS* AS A FUNCTION OF ATOMIC NUMBER

Element	*Atomic Number*	*Relative Cross Section at 6.4 keV*	*Relative Cross Section at 9.7 keV*
H	1	1.0	1.0
C	6	1.02	0.75
N	7	1.52	1.01
O	8	2.00	1.30
P	15	3.42	2.58
Ca	20	5.05	3.52

The cross sections are calculated from the tabulations of Compton and Allison (1935, pp 781 and 782), for a scattering angle of 45°.

tensity cannot be a useful measure of total mass unless the weight-fractions of the major constituents of the specimen are fairly well known. However this condition is satisfied for most biological soft tissues. Table IV (from

TABLE IV

RELATIVE SCATTERING PER UNIT MASS IN SUCROSE, SOFT TISSUE AND DNA (ADAPTED FROM ZEITZ, 1969, p. 49)

Specimen	*Relative Scattered Intensity*		
	at 5.4 keV	*at 17.4 keV*	*at 22.2 keV*
Sucrose	1.00	1.00	1.00
Soft Tissue	1.05	1.03	1.02
DNA	1.05	0.97	0.94

The intensities in each column are normalized relative to the intensity from sucrose.

Zeitz, 1969) shows that even carbohydrate and protein differ by less than 10 percent in scattering cross section per unit mass, and the variation among proteins, nucleic acids and most soft tissues is considerably less. In fact, as a function of atomic number, the scattering cross section per unit mass does not vary nearly as steeply as the absorption cross section per unit mass. Consequently, in comparison with the well-established microradiographic and microabsorption techniques for the measurement of total mass

(Sec. 5.A.2 and Sec. 5.C.1), the scattering technique is less susceptible to error arising from variations in composition.

In all of our estimations of scattering we have assumed that the total scattered intensity is the sum of the intensities from the individual atoms in the specimen. This is true for amorphous specimens but not true for single crystals where interference effects are important. While biological tissues are certainly crystalline to some extent, the crystals are generally small and not well oriented, and the intensity scattered by the entire specimen into the large solid angle subtended by the counter is usually the same as it would be if the specimen were amorphous.

Measurement of Weight-fractions

Since elemental mass can be measured by means of a standard containing a known amount of element, and total mass can be measured by means of a standard of known mass, clearly it is possible to measure a weight-fraction, which by definition is elemental mass/total mass. But there is a different approach, based on the network technique, which obviates the need for any standard of known mass. A two-channel network is set up, leading to a pair of simultaneous equations of the form (5–57), with one pulse-height channel centered on the characteristic radiation of the assayed element and the second pulse-height channel centered on scattered radiation, responding chiefly to total mass. We noted above that a standard of known mass was wanted for the measurement of an elemental *amount* by fluorescence analysis, but we shall see now that standards of known mass are not necessary to measure a weight-fraction, i.e. a *ratio* of elemental to specimen mass.

For the assay of a single element in an organic matrix, Equations 5–57 can be written

$$C_e = K_{ee}M_e + K_{em}M_m$$
$$C_m = K_{me}M_e + K_{mm}M_m \qquad (5\text{--}60)$$

where C_e and C_m are the total counts registered in the "element" and "mass" channels, M_e is the mass of the assayed element in the specimen, and M_m is the "residual" specimen mass (i.e. total specimen mass equals $M_m + M_e$). The quantities K are instrumental constants which are to be sufficiently determined by calibration with standards.

From Equation 5–60 one obtains, by simple algebra,

$$\phi \equiv \frac{M_e}{M_m} = \frac{r - r_m}{r_e - r} \frac{K_{mm}}{K_{me}} \qquad (5\text{--}61)$$

where r is the ratio of counts C_e/C_m obtained with the specimen; $r_m = K_{em}/K_{mm}$ is the count ratio observed with a thin calibration standard of pure organic material; and $r_e = K_{ee}/K_{me}$ is the ratio with a thin calibration standard of pure element. A further calibration standard is introduced, containing element mass M_{ec} and additional organic mass M_{mc} in a known ratio ϕ_c. For this standard, Equation 5–61 gives

$$\phi_c \equiv \frac{M_{ec}}{M_{mc}} = \frac{r_c - r_m}{r_e - r_c} \frac{K_{mm}}{K_{me}} \tag{5–62}$$

where r_c is the count ratio C_e/C_m observed with the standard. Division of (5–61) by (5–62) gives the formula for the assay:

$$\phi = \frac{M_e}{M_m} = \phi_c \frac{\dfrac{r - r_m}{r_e - r}}{\dfrac{r_c - r_m}{r_e - r_c}} \tag{5–63}$$

Equation 5–63 can be used for the measurement of weight-fractions. Standards are still required, as described, but it is not necessary to know the mass of any of them; in the mixed standard, it is necessary to know only the *ratio* M_{ec}/M_{mc}.

In connection with this method, the following points should be noted.

1. It is a great advantage not to have to know the mass of any standard. Standards of known mass are at best not easy to prepare. And in micro-fluorescence work with quite fine apertures, even if the actual mass of a standard is known, the requisite quantity—the mass *within the excited volume*—may be very difficult to establish accurately.

One does need a standard containing the assayed element in known weight-fraction in an organic matrix. Such a standard is usually easily prepared: for example by evaporation from a drop of solution containing known concentrations of a salt of the element plus an organic solute such as sucrose (Hall, 1963; Zeitz, 1969).

2. It is implicitly assumed in Equations 5–61 and 5–62 that K_{mm} and K_{em} are constant for specimens and standard, i.e. that there is a constant counting efficiency per unit of residual mass. As we have seen, this amounts to assuming a matrix composition sufficiently invariant to give a steady average scattering cross section, an assumption which is usually satisfactory for soft tissues, and usually unsatisfactory for hard tissues. (There is no need to assume an equal scattering cross section for the assayed element, no matter how large its weight-fraction, since the assayed element is given its own coefficients in Equations 5–61 and 5–62.)

3. The measured quantity ϕ is not quite the same as the weight-fraction f.

$$\phi \equiv \frac{M_e}{M_m} \quad \text{and} \quad f \equiv \frac{M_e}{M_m + M_e} \tag{5–64}$$

It is readily seen that

$$\frac{1}{f} = \frac{1}{\phi} + 1 \tag{5-65}$$

so that the weight-fraction is easily obtained from ϕ. When the weight-fraction is low ($M_e \ll M_m$), it is virtually equal to ϕ.

4. When the weight-fraction is low ($M_e \ll M_m$) in both specimen and calibration mixture, $r \ll r_e$ and $r_c \ll r_e$ and Equation 5-63 reduces to the simpler form

$$\frac{M_e}{M_m} = \phi_c \frac{r - r_m}{r_c - r_m} \tag{5-66}$$

The pure-element standard is then unnecessary.

5. The same approach can be used for the assay of several elements in an organic matrix. For example, for the assay of two elements 1 and 2, a three-channel network system is set up, the channels being designated below as a, b and m. If the element spectra do not overlap badly, one counter with three pulse-height channels is sufficient. For neighboring elements two counters with a filter (Fig. 90) may be desirable. Then one "element" channel is associated with each counter and restricted to a pulse-height band corresponding to the characteristic spectrum, while the third channel "for mass" is connected to either of the counters and restricted to larger pulse heights. Whether one or two counters are used, the mathematical formalism is identical. Equations 5-57 take the form

$$\begin{aligned}
C_a &= K_{a1}M_1 + K_{a2}M_2 + K_{am}M_m \\
C_b &= K_{b1}M_1 + K_{b2}M_2 + K_{bm}M_m \\
C_m &= K_{m1}M_1 + K_{m2}M_2 + K_{mm}M_m
\end{aligned} \tag{5-67}$$

The general solution is straightforward but complicated. However, for the important case where the weight-fractions of the assayed elements are low, the terms $K_{m1}M_1$ and $K_{m2}M_2$ in the last equation are negligible. Then Equation 5-67 reduces to

$$r_a \equiv \frac{C_a}{C_m} = \frac{K_{a1}}{K_{mm}}\phi_1 + \frac{K_{a2}}{K_{mm}}\phi_2 + \frac{K_{am}}{K_{mm}}$$

and

$$r_b \equiv \frac{C_b}{C_m} = \frac{K_{b1}}{K_{mm}}\phi_1 + \frac{K_{b2}}{K_{mm}}\phi_2 + \frac{K_{bm}}{K_{mm}} \tag{5-68}$$

To use Equations 5-68, three thin calibration standards are required, one with pure organic matrix, one with element "1" in known weight-fraction in organic matrix, and one with element "2" in known weight-fraction in organic matrix. The observations are all made in the three channels simultaneously, and we make (or recall) the definitions

r_a = count ratio C_a/C_m for the specimen (likewise for subscript b)

r_{a1} = count ratio C_a/C_m for standard containing element 1 (likewise for subscripts b and 2)

r_{am} = count ratio C_a/C_m for pure organic standard (likewise for subscript b)

$\Delta_a = r_a - r_{am}$ (likewise for subscript b)

$\Delta_{a1} = r_{a1} - r_{am}$ (likewise for subscripts b and 2).

f_{1c} = weight-fraction of element 1 in its standard (likewise for subscript 2),

and ϕ_{1c} is defined by $\dfrac{1}{\phi_{1c}} = \dfrac{1}{f_{1c}} - 1$.

We note that $K_{am}/K_{mm} = r_{am}$ and $K_{bm}/K_{mm} = r_{bm}$. Application of Equations 5–68 to the standards gives the result

$$\Delta_{a1} = \frac{K_{a1}}{K_{mm}} \phi_{1c} \qquad \text{(likewise for subscripts } b \text{ and } 2) \qquad (5\text{–}69)$$

Hence the calibration data determine all the ratios of the form K_{a1}/K_{mm}. The Equations 5–68 may then be solved in terms of the measured ratios, giving the result

$$\phi_1 = \phi_{1c} \frac{\Delta_a \Delta_{b2} - \Delta_b \Delta_{a2}}{\Delta_{a1} \Delta_{b2} - \Delta_{b1} \Delta_{a2}}$$

and

$$\phi_2 = \phi_{2c} \frac{\Delta_b \Delta_{a1} - \Delta_a \Delta_{b1}}{\Delta_{b2} \Delta_{a1} - \Delta_{a2} \Delta_{b1}} \qquad (5\text{–}70)$$

Equations 5–70 give the desired weight-fractions entirely in terms of measured quantities and the known weight-fractions of the calibration standards. These equations have been used for the fluorescence assay of zinc and copper in thin organic specimens (Hall, 1963).

For the assay of three or more elements, the explicit solution of the simultaneous equations is tedious. It should be possible to solve the equations and get the weight-fractions automatically by means of a modified version of the analog electrical network method described above (Sec. 5D2), but this does not seem to have been done yet.

Limitation on Specimen Thickness

In the entire discussion of quantitation we have assumed that the intensity of a characteristic radiation emerging from the specimen is directly proportional to the amount of the corresponding element. This is valid only for thin specimens. As thickness increases the signal fails to grow in full proportion because of attenuation of the incident exciting radiation and absorption of the emerging fluorescent radiation within the specimen.

Zeitz (1969) has worked out in detail the magnitude of these effects and the limits of thickness for which they are negligible. Here we shall give a simplified estimation.

For nondiffractive analysis, to avoid interference between the assayed and scattered radiation, the exciting radiation must on the average have quantum energies considerably higher than the assayed radiation, and is hence considerably more penetrating (less absorbed) in organic matrices. Furthermore, the incident radiation is normal to the specimen while the observed radiation emerges obliquely. We have therefore estimated the average absorption of the fluorescent radiation while assuming that the absorption of the incident radiation is negligible.

If a flat specimen has thickness T cm and radiation is observed at an angle of emergence θ (Fig. 96), the average attenuation factor will be

$$\text{transmission} \simeq e^{-\frac{T}{2} \sec \theta \ (\rho_1\sigma_1 + \rho_2\sigma_2 + \ldots)} \tag{5-71}$$

where the quantities ρ are the densities of the constituent elements in gm/cc and the quantities σ are their absorption cross sections in cm^2/gm. (The quantity $T/2$ is used because the fluorescence is generated, on average, in the middle of the specimen.) Table V, based on Equation 5–71, gives the thickness in microns for which the average absorption in a proteinaceous

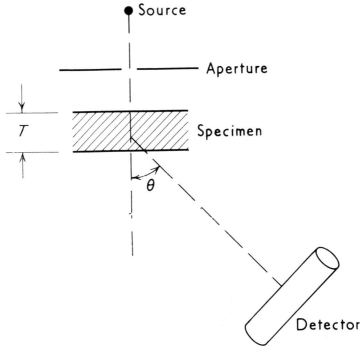

Figure 96. Geometry for the estimation of the mean absorption path of the fluorescence radiation within the specimen.

specimen will be 10 percent, for an emergence angle $\theta = 45°$. The specimen is assumed to be originally one-third protein and two-thirds water by weight; the density of protein is taken to be 1.4 gm/cc and the composition of the protein is taken to be H 7%, C 52.5%, N 16.5%, O 22.5% and S 1.5%. The absorption cross sections of Heinrich (1966) have been used. In the Table, the column "wet" refers to the specimen with its water present, and the column "dry" refers to the same specimen changed only by removal of the water (for example, a section cut frozen and simply dried. The result is the same whether or not the specimen collapses on drying, since the products ρT in Equation 5–71 are unaffected by collapse.)

TABLE V

THICKNESS (μm) FOR 10% ABSORPTION
(CALCULATED FOR EMERGENCE ANGLE OF 45°)

Radiation		Thickness	
Element	*keV*	*Undried Tissue Protein*	*Dried Tissue Protein **
Si	1.74	1.7	7.6
P	2.01	2.6	11.5
S	2.31	3.8	16.9
Cl	2.62	5.3	20.8
K	3.31	10.3	40.2
Ca	3.69	13.9	54.
Fe	6.40	66.	256.
Zn	8.64	153.	595.

* Thickness prior to drying and possible collapse (collapse does not affect absorption).

Two facts appear strikingly in the Table: the maximum thickness for which absorption is negligible is much less in wet than in dry specimens, and this thickness varies drastically with the atomic number of the assayed element (which determines the penetrating power of the assayed radiation). For zinc K_a radiation, the simple theory of quantitation for thin specimens can still be applied up to thicknesses like .5 mm, while the corresponding thickness for the assay of silicon is less than 10μm.

With hard tissues, absorption is so high that the quantitative theory for thin specimens is generally inapplicable.

5.D.4. Spatial Resolution

The spatial resolution of the fluorescence technique can be no finer than the diameter of the system of apertures fitted between the x-ray tube and the specimen. For the studies described in Chapter 3, apertures and spatial resolution of the order of 100μm were satisfactory.

Two factors must limit the attainable resolution. Firstly, as the analyzed volume is reduced, the amount of the assayed element in this volume may fall below the detectable minimum. Not only will the amount of element generally be less in smaller volumes, but the minimum detectable amount rises as the aperture is made finer (Sec. 5.D.1). Secondly, scattering and edge effects become more difficult to control as the aperture is reduced, so that the irradiated specimen area may not be proportionately diminished. The attainable resolution remains to be established in practice. At present it is clear that the spatial resolution can be considerably better than 100μm, while resolution better than 10μm seems unlikely.

5.D.5. Availability of Apparatus

The commercially available x-ray fluorescence instruments are not suitable for the microanalysis of biological material. One standard type of equipment uses a conventional x-ray tube for excitation, and diffracting-crystal spectrometers to analyze the fluorescence radiation. In such instruments the exciting radiation cannot be used efficiently for microanalysis because the anode is too far from the specimen and the anode spot is too large, while the x-ray collecting power of the spectrometers is too low to work with the low intensities available from microvolumes of tissue. A newer commercial type of equipment uses a radioactive source for excitation and solid-state pulse counters or balanced filters for nondiffractive x-ray analysis, providing a relatively cheap and simple apparatus very effective for factory or field work. In this system the x-ray source could be brought quite close to the specimen and the x-ray collecting power is high, but the intensity available from radioactive sources is insufficient for our purpose. The one x-ray fluorescence arrangement which has been used successfully for biological microanalysis has been that of Figure 5, and this design has been realized only in a few research laboratories.

5.D.6. Range of Application

A glance at Table II (Sec. 1.C) shows that in quantitative terms, especially with respect to spatial resolution and minimum detectable amounts, x-ray fluorescence is not nearly as impressive as electron microprobe analysis. This may explain why electron microprobe analysis is more widely used and why there are no commercial fluorescence microanalyzers. However, with fluorescence analysis the specimens are not necessarily in a vacuum, and studies of the wet state are possible. There is much less danger of beam damage, and even dynamical experiments *in vivo* should be possible. The sensitivity, while limited, is reasonably matched to the total amounts of important elements in individual biological cells.

The range of applications is rather restricted for the instrument which was used for the fluorescence analyses described in Chapter 3. But as indicated above, the ultimate performance of the technique was far from being achieved in this instrument. The advent of improved methods of nondiffractive x-ray spectroscopy, especially with solid-state detectors, and a tighter overall design will definitely lead to much greater sensitivity (cf.

Nicholson, 1971). Therefore, among all the techniques discussed in this book, the applicability of x-ray fluorescence analysis has been the least realized in practice, and one must consider the somewhat speculative estimations above in order to evaluate the potential of the method.

5.E. PHYSICAL ASPECTS OF ELECTRON-PROBE X-RAY MICROANALYSIS

In the electron-probe x-ray microanalyzer, local concentrations of chemical elements are determined within microscopic regions of a specimen by

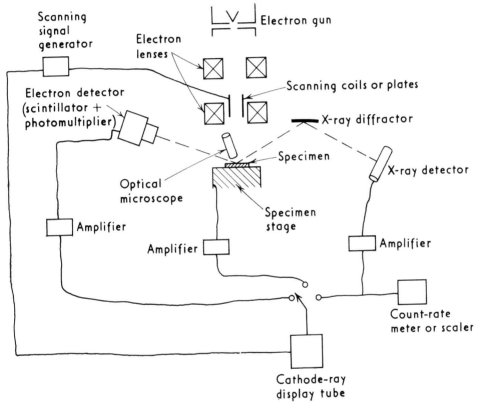

Figure 97. Components of a standard electron-probe x-ray microanalyzer.

the spectroscopic analysis of the x-rays generated by a finely focussed beam of electrons. The most important instrumental components, indicated in Figure 97, are the following:

1. An electron gun, accelerating electrodes and magnetic lenses, similar to those in an electron microscope but used to produce the probe.
2. Scanning coils or deflecting plates to position the probe on the specimen or to perform scans.

3. A detector for backscattered or absorbed electrons, to produce scanning images as in the scanning electron microscope.

4. A mechanical stage to hold the specimen and to bring particular areas within reach of the probe as desired.

5. An optical microscope to help in placing the probe precisely where it is desired on the specimen.

6. One or more x-ray spectroscopes to analyze the generated x-rays.

In this section we shall discuss the different modes of operation of electron-probe instruments, the limits of spatial resolution and of sensitivity, methods of quantitative analysis, and the speed, convenience and cost of electron-probe instruments. The problems of specimen preparation, probe-induced damage and histological correlation, and the range of medical problems susceptible to electron probe study, have already been discussed in Chapter 4.

5.E.1. Modes of Operation of Electron-probe Instruments

With most probe instruments, one can scan the probe over a small rectangular area of the specimen by electrical deflection of the beam. The largest area which can be scanned in this way is usually a square in the range of 0.2 to 2 mm on a side. Several types of scanning image may be formed by regulating the brightness on a synchronously scanning cathode-ray display tube, modulated by any one of several signals which vary as the probe moves from point to point over the specimen.

The signal used most widely in scanning images is the intensity of electrons scattered back from the specimen, generally detected by a scintillator which in turn activates a photomultiplier. The image formed in this way is called the *backscattered-electron* image.

A modified image is obtained when the electron detector is moved out of the direct path of the backscattered electrons and a small voltage is applied to collect *low-energy* electrons. The image is then based on electrons originally within the specimen and knocked out by the probe, the so-called *secondary electrons*.

In comparison with backscattered-electron images, secondary-electron images may be more revealing because they are more sensitive to topographical features and chemical variations at the surface of the specimen; for this reason the secondary-electron image is used the most in the scanning electron microscope.

One may also insulate the specimen and use as the modulating signal the current which flows from specimen to ground potential; this gives a *specimen-current* image. It is complementary to the other two types of electron image since the specimen current is equal to the (constant) probe current minus the sum of the backscattered and secondary-electron currents. Since the specimen-current signal is relatively large, the quality of the corresponding image may benefit from a favorable signal-to-noise ratio.

There are many possible refinements of scanning-electron images, based on special procedures such as the application of electrical biases to the specimen. These refinements have been especially useful in the study of solid-state electronic devices, magnetic specimens, etc., but have not yet appeared to be relevant for biology.

In terms of chemical elements, scanning-electron images are non-specific * and are merely a means of visualization of the specimen. But if one sets the x-ray spectroscope for a particular element, and regulates the brightness of the display tube by the *x-ray signal* as the probe is scanned over an area, one can get an image of the distribution of that element.†
(Because the x-ray signal is relatively weak, such images may have to be built up over long times, sometimes ten minutes or more, recording from the display tube by camera while the beam is scanned.)

It is also possible to produce an x-ray image showing the distribution of total dry mass in a thin specimen (Sec. 5.E.3).

While the x-ray image can be chemically specific, it is generally not quantitative; even if one bothered to do densitometry on the photographic record, there would be no simple relationship between photographic effect and x-ray intensity.‡ One way to get a quantitative record of the spatial distribution of an element is to record the x-ray intensity, for example on a moving-chart recorder, as the probe is made to scan along a line on the specimen surface. Figure 98 typifies such a linear scan. It is even more convenient to record the information in the form of Figure 99 made by photographing a triple exposure from the display tube, successively displaying the electron-scan image, the line along which the probe was scanned, and the graph of x-ray intensity along the line (the last obtained by applying the x-ray count-rate signal to the deflecting plates of the display tube). Such a picture puts the quantitative information immediately and precisely in register with the structure of the specimen as seen in the electron image. Obviously the linear scan is at its best when the desired information is essentially "one-dimensional." When local concentrations depend only on distance from some straight-line boundary in a specimen,

---

* As shown by Heinrich (1964), one can base a method of local quantitative analysis of binary specimens on the relationship between atomic number and backscattered-electron intensity, but the method is not applicable to most biological probe studies.

† If the area is large and the x-ray spectrometer is of high quality and sharply focused, the spectrometer may be out of focus at the edges of the scan giving a misleading image. In fact the defocussing effect occurs in one direction but not at right angles, so that this difficulty can be overcome by mechanically scanning the specimen instead of the probe in one direction. Some commercial probes are equipped to do this.

‡ By arranging electronically to activate the cathode-ray tube only when the x-ray intensity falls within a certain range, one can map contours of equal concentration. This technique has not been used much in biological work, where the x-ray counting rates are usually too low for it to be effective.

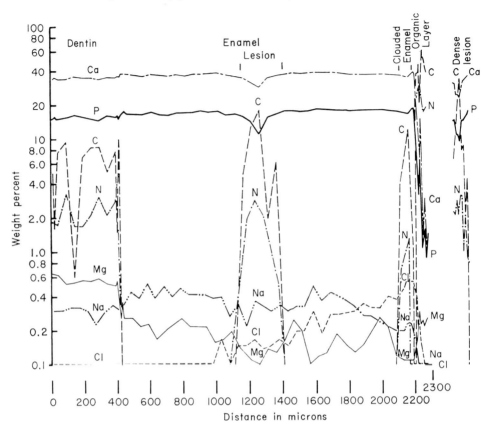

Figure 98. X-ray signals for several elements recorded along a line of scan extending from the outside of a tooth inwards to the enamel-dentin junction. (From Andersen, 1967a. Courtesy of Interscience Publishers.)

the linear scan may contain all the information that the instrument can provide.

When accurate quantitative analysis is needed, especially when elemental concentrations and x-ray intensities are low, it is generally necessary to record the data while the probe is stationary. If the constituent elements of the specimen must first be identified, one commonly works with a static probe and records x-ray intensity as the diffracting crystal is rotated, to detect and identify the different characteristic x-ray emissions in succession. The electron-scanning images, x-ray scanning images or optical microscope are then used simply to help pick out the most suitable spots for static-probe analysis.*

While conventional microprobe instruments are mainly limited to the modes of operation described above, recently combination electron-micro-

* A display tube with a persistent phosphor is always provided, so that one can place the static probe precisely with reference to the scanning image before it fades away.

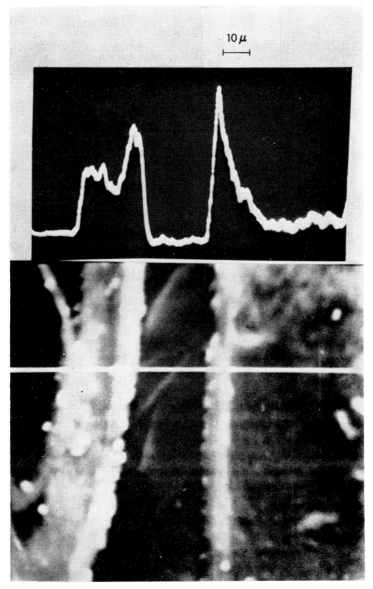

10 μ

Figure 99. A triple exposure showing a scanning electron image of a section of skin, the location of a linear scan made within the section, and the graph of the sulphur x-ray intensity along the linear scan. (One could of course put all three exposures on a single frame, without "winding on" the camera, giving perhaps even closer correlation at the cost of some confusion in the picture.)

scope x-ray microanalyzers ("EM-MA"S) have been coming into use (Duncumb, 1966). In these devices (Fig. 100), the electron beam produces a transmission image as in the conventional electron microscope, and may then be focussed into a probe and positioned precisely with reference to

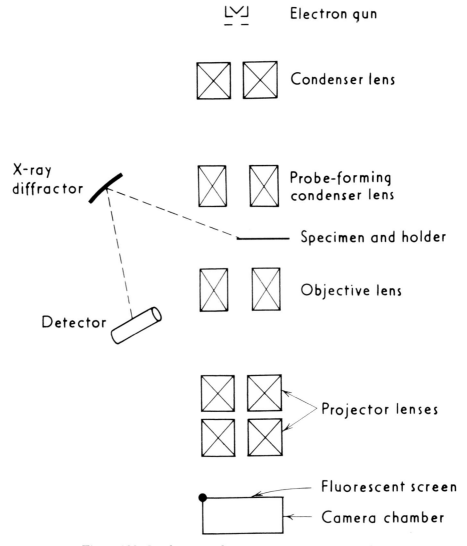

Figure 100. Combination electron microscope-microprobe.

the EM image. The x-rays generated in the specimen are analyzed as in conventional microprobes by highly efficient spectroscopes. Transmission electron microscopy opens prospects of improved visualization of the structures and fields which one wants to analyze, and the EM-MA instruments have already been used extensively in biology.

Quite recently attachments have become available to perform microprobe analysis in the commercial scanning electron microscope ("SEM"). The electron images in a microprobe are formed in the same way as in the scanning electron microscope. Thus the conventional microprobe on the one hand, and the scanning electron microscope with an x-ray spectrometer attachment on the other, may be regarded as basically the same instrument, only in the "microprobe" the x-ray spectrometers are made as good as possible at some cost to the quality of the electron imaging, while in the "SEM with x-ray attachments" the electron images are favored and the x-ray spectroscopes may not be as efficient. With another new attachment to the commercial SEM, one can perform *transmission* scanning electron microscopy of thin specimens, up to more than 1μm thick (Swift and Brown, 1970b; Swift *et al.*, 1969). In this mode the beam is scanned over the specimen and the signal is taken from a small electron detector placed on the *exit* side of of the specimen. The spatial resolution of the images is similar to that in other modes of scanning electron microscopy, not better than 100 Å in the commercial instruments, but the image *contrast* is as excellent as in the conventional transmission electron microscope. Thus there is the attractive possibility of performing x-ray microanalysis while viewing transmission electron images in the SEM.

5.E.2 The Minimum Measurable Amount of an Element

In all quantitative discussions of microprobe work, one must distinguish clearly between "thin" and "thick" specimens. In referring to a "thin specimen" we imply that most of the incident electrons pass completely through the specimen with the loss of only a small fraction of their energy. By "thick specimens" we imply that the incident electrons cannot reach the far surface at all.[*]

For thin specimens one can readily estimate minimum measurable amounts of elements. If the probe is focussed onto an area containing S_e gm/cm^2 of the assayed element, then each incident electron will on the average generate a number of characteristic x-rays

$$N_x = (wQ_x)S_e \qquad (5\text{--}72)$$

Here the subscript x denotes the particular x-ray line for which the x-ray spectrometer is set; Q_x is the mass cross section in cm^2/gm for ionization of the relevant electronic orbit; and w, the "fluorescence yield," is the probability that an ionization results in the emission of a quantum of the selected characteristic line. (The definition of a cross section and the logic behind Equation 5–72 are explained in the Appendix, Sec. A. 4.) The x-ray counting rate R must then be

$$R \text{ (counts per second)} = gwQ_xS_e \, 0.6 \, 10^{13}i \qquad (5\text{--}73)$$

where g is the fraction of the generated characteristic x-rays counted by the spectrometer, and we have converted to probe current i in microamperes by means of the factor $1\mu A = 0.6 \, 10^{13} \, electrons/second.$

---

[*] We shall not discuss specimens of intermediate thickness, where quantitative analysis is rather complicated, and indeed virtually hopeless if the specimen is inhomogeneous as well, as is most often the case for biological materials.

The maximum current which can be focussed into a spot of diameter d is known to be proportional to $d^{8/3}$, so we can write

$$i = i_o d^{8/3} \tag{5-74}$$

where d is the diameter of the probe in micrometers and i_o is the maximum current which can be focussed into a 1μm spot. The total amount of element, M_e, in the spot of diameter d microns is

$$M_e(gm) = S_e(gm/cm^2)\frac{\pi}{4}d^2 10^{-8}(cm^2) \tag{5-75}$$

The combination of the last three equations gives

$$M_e(gm) = 1.3 \ 10^{-21} \frac{R}{gwQ_x i_o d^{2/3}} \tag{5-76}$$

Equation 5–76 will enable us to estimate how sensitivities vary over a wide range of conditions, but for a given instrument it is best to base the estimations on an actual measurement. For example, with one highly sensitive modern instrument, with 30-kV electrons impinging on a uniform thin film of copper (atomic number 29), the counting rate was approximately five 10^8 counts per second per gm/cm² of film per microampere of probe current. At 30 kV a current of 1μA could be focussed into a minimum diameter of approximately 1μm.* Hence, under these conditions, *for a 1μm probe*, at best

$$R = \frac{cps}{\mu A \ gm/cm.} \times \mu A \times gm/cm^2$$

$$= 5 \ 10^8 \times 1 \times \frac{M_e}{\pi/4 \times 10^{-8}} \tag{5-77}$$

so that

$$M_e(gm) = 1.6 \ 10^{-17} \ R \tag{5-78}$$

In practice one needs a counting rate at least in the range $R = 3 - 10$ counts per second. Inserting this range for R into (5–78), we see that for the specified conditions, M_e, the minimum amount of element measurable, is in the neighborhood of 10^{-16} gm.

The minimum measurable amount in other situations can now be deduced from Equation 5–76. With respect first to the dependence of this minimum on the atomic number Z of the assayed element, in (5–76) the factors R and d are irrelevant. For purposes of estimation, the spectrometer efficiency g may be taken independent of Z.† The maximum sensitivity is

---

* Biological specimens in general will not tolerate a 1μA probe, but currents up to this amount may be withstood by thin specimens mounted on very thin supports and coated with a heavy evaporated layer of conducting material (see Sec. 4.B.1.b).

† Actually g varies a great deal, depending mainly on the design of the spectrometer and the

approached when the column voltage E_o is approximately three times the energy at the ionization edge, V_x, which in turn is proportional to Z^2. At $E_o = 3 V_x$, the ionization cross section Q_x has a maximum proportional to Z^{-5}, while the maximum current i_o is proportional to E_o and hence to V_x and to Z^2.[*] When these relationships are put into Equation 5–76, we get a dependence of the minimum amount M_e on atomic number,

$$M_e \; \alpha \; \frac{1}{Q_x i_o w} \; \alpha \; \frac{1}{Z^{-5} Z^2 w} = \frac{Z^3}{w} \qquad (5\text{–}79)$$

The fluorescence yield w does not depend on Z as simply as Q or i_o. For K_α radiations w is represented adequately (Campbell, 1963) by

$$w_{k\alpha} = (1 + aZ^{-4})^{-1} \left(1 - \frac{Z}{250}\right) \qquad (5\text{–}80)$$

where a depends slightly on atomic number but is always close to 10^6. When this formula is inserted into Equation 5–79, with the values of a quoted by Campbell from Burhop (1952), Table VI is obtained for the relative-sensitivity function Z^3/w.

TABLE VI

DEPENDENCE OF MINIMUM MEASUREABLE MASS ON ATOMIC NUMBER: EFFECT OF FLUORESCENCE YIELD, IONIZATION CROSS SECTION AND PROBE CURRENT

Atomic Number Z	Fluorescence Yield w	Relative Minimum Mass $10^{-3} \, Z^3/w$
6	0.0014	154
8	.0044	118
10	.011	91
12	.015	115
14	.028	99
16	.050	82
18	.074	79
20	.106	75
24	.19	73
28	.29	76
32	.39	84
38	.53	103
44	.61	140

In view of the great variation of the individual contributing factors with atomic number (especially the dependence of the ionization cross-section on the fifth power!), the remarkable aspect of the result is the relatively small change in overall sensitivity over the entire range of atomic numbers for which the K radiations are commonly used ($6 \leq Z \leq 40$). This permits

---

choice of diffracting crystal, and a given crystal cannot be used for all atomic numbers. However there is little systematic variation of g with Z, except that the efficiency is relatively poor in the range $6 \leq Z \leq 8$ when multilayers of stearate are used as diffractors instead of true crystals.

[*] At $E_o = 3V_x$, the atomic ionization cross section in cm²/atom is proportional to $(E_o \, V_x)^{-1}$ (Green and Cosslett, 1961), hence to V_x^{-2} and to Z^{-4}. An additional factor Z^{-1} enters in multiplying by atoms/gm to convert to the mass ionization cross section Q_x in cm²/gm.

us to take the minimum amount quoted above for copper, approximately 10^{-16} gm, as a valid guide for thin specimens and probes of 1-μm diameter throughout this range of atomic numbers.

We should note the following additional points about the minimum amount of element detectable in *thin* specimens.

1. Although the quantitative analysis above refers to K radiations, it can be extended to the L radiations of elements of higher atomic number. Specifically, it can be shown that the limits of detectability are not much different for the assay of an element based on its K_a radiation, and the assay of another element based on its L_{a1} radiation, if the wavelengths of the two radiations are similar (cf. Reed, 1968).

2. For many specimens, e.g. typical 5μm tissue sections, column voltages E_o of 30 kV or higher may be desirable to satisfy the conditions for the quantitative theory presented below for the analysis of thin specimens (Sec. 5.E.7.b). For elements of low atomic number, E_o may then be much greater than the value 3 V_x suggested earlier. However, this does not imply a loss of sensitivity. In fact, the sensitivity factor wQi_o rises rapidly as the column voltage is raised to approximately 3 V_x, and then continues to rise slightly if E_o is increased still further (see Hall, 1968, for a fuller analysis).

3. The minimum of 10^{-16} gm has been estimated for a probe diameter of 1μm. We see from Equation 5–76 that if the probe diameter d is reduced in the interest of better spatial resolution, the minimum detectable amount must suffer by a factor $d^{-2/3}$.

It is noteworthy that d refers to the size of the probe, not to the size of a "hot spot" of the assayed element which may be smaller than the probe. While Equation 5–76 was derived under the assumption that the assayed element was distributed uniformly within the probe, it is evident that in thin specimens the counting rate would be the same for the same amount of element distributed nonuniformly, since the lowered yield per electron in any part of the probe would be balanced by higher yields in other parts.[*] An interesting consequence of Equation 5–76 is that it may be possible to bring a "hot spot" within the detectable range by using a larger probe, even if the additional area contains none of the assayed element. The reason for this somewhat paradoxical fact is that the available maximum current increases so rapidly with probe diameter that more current can be put into the hot spot itself when the probe is enlarged.

4. The minimum measurable amount as derived here sets a necessary condition, but not a sufficient condition for an analysis to be feasible. It is also required simultaneously that the *weight-fraction* of the assayed element must exceed the minimum discussed below (Sec. 5.E.4). With a large

---

[*] We are assuming here, not quite accurately, that the flux of electrons is uniform over the area of the probe.

probe and a relatively thick specimen, the minimum amount may be exceeded but analysis may be ruled out by a subminimal weight-fraction, or conversely analysis may be ruled out for small probes in thin specimens due to insufficient amount within the probe even where the weight-fraction is relatively high.

5. Our estimate has been based on the performance of a probe equipped with a typical modern diffracting-crystal spectrometer. The efficiency of x-ray detection (i.e. the factor g) can be greatly increased by omitting the diffractor and analyzing the x-rays solely by means of the different average pulse height produced by each characteristic radiation in the detector. This method has been used very little in biological microprobe work because the discrimination of the detector against background radiations is generally inadequate, but amounts of element well below 10^{-16} grams may be measurable in this way when the weight-fraction is high in very small areas. This possibility is discussed further under the heading of spatial resolution (Sec. 5.E.5.c).

In *thick* specimens, the analytical possibilities are complicated by the scattering and diffusion of the incident electrons. No matter how finely a static probe is focussed, electrons of initial energy E_o will spread out within the specimen, generating any particular characteristic radiation "x" all the while until atomic collisions degrade them to V_x, the ionization energy (see Fig. 101). Since, for example, in soft tissue the range of a 30-kV electron may be some tens of microns, the productive volume can be distressingly large. Therefore, if spatial resolutions near one micron must be realized, one usually limits the range of the electrons by resorting to a low-voltage probe

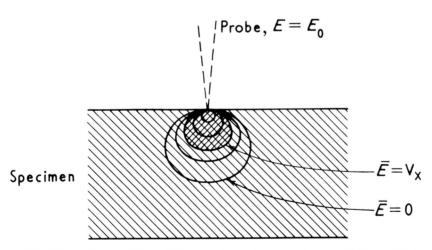

Figure 101. The volume within which a static probe generates a radiation with ionization threshhold energy V_x. On each contour the probe electrons are degraded to a particular average energy $\bar{E}$. The radiation with threshhold V_x will be generated almost entirely within the doubly hatched region of the figure.

(restricting the analysis to the long wavelength spectral lines, K, L and M, which such a probe is capable of generating).

Even when spatial resolution is not so important, the minimal measurable amount cannot be achieved *meaningfully* when the productive volume is large. When the requisite weight-fraction of the element is present, the total amount in such a volume invariably exceeds the amount which can be detected in smaller volumes with probes of lower voltage.

The relationships between spatial resolution and sensitivity in thick specimens lead to a complicated pattern of capabilities, which has been worked out and presented in detail by Andersen (1967a,b). While the minimum detectable amount of an element depends strongly on atomic number, the choice of K or L radiation, weight-fraction, and composition of the matrix, he shows that minima in the range $10^{-15} - 10^{-16}$ gm are common.

For optimal operating conditions in thick specimens, one should refer to Andersen's work. However, in any comparison of his values of sensitivity with the values presented in this chapter for thin specimens, two points must be kept in mind:

> 1. Andersen's minimum amount at any column voltage is simply the product of the minimum weight-fraction and the total mass of the analyzed region (the latter being density × productive volume). Therefore the presence of the required minimum of either absolute amount or weight-fraction guarantees the satisfaction of the other minimum under the conditions specified. In the case of thin specimens the two limits are naturally obtained independently, and one must consider both in judging if an analysis is feasible.
>
> 2. Andersen's minimal weight-fractions were determined for fixed running times and specimen currents, respectively twenty seconds and $0.05\mu A$ for biological materials. This is reasonable for thick specimens, where it is difficult to predict or assess damage from more intense probes. In thin specimens higher currents can generally be used, favorably affecting the sensitivities, and the maximum available current should be taken into account as was done above.

5.E.3. Measurement of the Distribution of Total Dry Mass

The bremsstrahlung [*] produced by an electron probe in a thin specimen can provide a measure of the local dry mass per unit area.

When an electron of energy E passes through a very thin film of thickness ds, containing N_r atoms/cm^3 of atomic number Z_r, the average number of bremsstrahlung quanta generated with energies between V and $(V + dV)$ can be expressed in the form

$$n \; dV \; ds = k \frac{dV}{V} \left(\sum_r N_r Z_r^2 B_r \right) ds \qquad (5\text{-}81)$$

so long as $E > V$. Here the sum is taken over the constituent elements, k is a constant, and the function B_r is unity if one accepts the equation proposed

[*] The generation of bremsstrahlung is described in the Appendix, Section A.3.

by Kramers (1923, quoted in Compton and Allison, p. 105), or is proportional to $(\ln E/\bar{V}_r)$ if one accepts an alternative equation (Marshall and
Hall, 1968).

For organic specimens where Z_r is predominantly 6, 7 or 8, with negligible
error one may regard B_r in either case as independent of Z_r and write that
atomic weight $A_r = 2\ Z_r$. After straightforward algebraic manipulations,
Equation 5–81 then implies that the number of bremsstrahlung quanta n_w
generated in a band of energies between V and $(V + \Delta V)$ is

$$n_w = KS\left(\sum_r f_r Z_r\right) \tag{5–82}$$

Here S is the local mass per unit area, f_r is the weight-fraction of the constituent with atomic number Z_r, and K is a constant which incorporates a
number of factors which remain fixed during a measurement.

The quantity n_w is monitored by a counter placed to receive x-rays directly from the specimen, and coupled to electronic circuits which register
only pulses corresponding to the selected band of quantum energies (Fig.
102). To be faithful to Equation 5–81, this band must be chosen to exclude
characteristic radiations and must be below the average energy of the
electrons emerging from the specimen.

The quantity $\sum_r f_r Z_r$ in Equation 5–82 is simply the mean atomic number
$\bar{Z}$. In general it may be evaluated explicitly if the weight-fractions f_r are
determined from measurements of the characteristic radiations as described

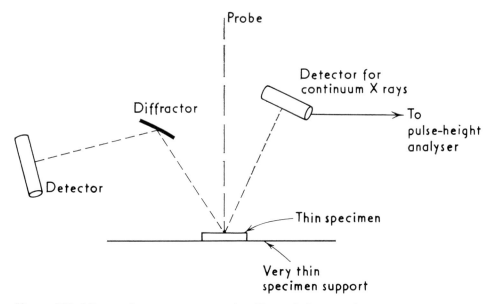

Figure 102. Microprobe arrangement with additional detector for continuum measurements.

in Section 5.E.7.b. However, at this point we shall only consider predominantly organic specimens, where $\bar{Z}$ is very nearly constant (being almost the same in proteins and carbohydrates, although approximately 15% less in fat). According to Equation 5–82 the bremsstrahlung counting rate gives a direct and immediate measure of local dry mass/area S in such thin specimens.

The following features of the method should be noted:

1. For relative measurements, no calibration is needed. If counting rates are observed under identical operating conditions for two spots in one specimen or in different specimens, the relative mass per unit area under the probe is simply the ratio of the counting rates.

2. The method can be calibrated for absolute measurements by means of a single standard, a thin uniform organic film of known composition and mass per unit area. In contrast to the step wedge required in microradiographic measurements (Sec. 5.A.2), here only a single film is needed because the counting rate is linearly proportional to mass per unit area within the operating range of the method.

3. Linearity breaks down for specimens which are too thick because electrons are degraded below the observed band of quantum energies, and because the path length in the specimen becomes more than proportional to specimen thickness due to electron scatter. For 30-kV electrons in organic materials, linearity certainly prevails up to a mass per unit area of at least $600\mu g/cm^2$ (Marshall and Hall, 1968). For a section of soft tissue which is two-thirds water and one-third protein, cut frozen and then dried, this corresponds to a section thickness of $16\mu m$. Of course the section thickness corresponding to $600\mu g/cm^2$ depends on the method of preparation, and is less for less watery tissues or embedded tissues. The density of a preparation of soft tissue will scarcely ever be greater than 1.4 gm/cm^3, which is the density of protein and more than the density of most embedding materials, so one can expect linearity in *any* preparation of soft tissue up to thicknesses of at least $600/1.4\mu cm = 4.3\mu m$.

4. The sensitivity of the method is so high that even for small probe currents ($0.01\mu A$, say) the limits of detectability are set not by counting rate but by extraneous backgrounds. The specimen support must be very thin, as the background generated in the support must be subtracted from the total count. In conventional microprobes, supporting films of nylon have been used, approximately $20\mu g$ to $40\mu g/cm^2$. The nonuniformity of these films, approximately $\pm$ 10 percent, produces an absolute uncertainty in every measurement of a few $\mu g/cm^2$. In combination transmission electron microscope-x-ray microprobes, ample bremsstrahlung counting rates are generated in ultrathin sections of 200 to 1,000 Å, but there is a high background of x-rays produced by scattered electrons hitting EM grids, pole pieces, etc., and it is not yet clear to what extent these backgrounds can be reduced or taken into account.*

* Aside from the limiting extraneous effects, the very high intrinsic sensitivity of the method is readily calculable. In one instrument, with 35-kV electrons the bremsstrahlung counting rate with a large aperture at the counter has been approx. 1.6×10^6 counts per second per nA of current per gm/cm^2 of soft tissue. If one supposes (Sec. 5.E.2) that the available current is 1,000 $d^{8/3}$ nA, where d is the diameter of the probe in microns, and that a minimum count rate of 10 cps is required, this datum implies a minimum measurable mass per unit area S of

5. Just as the distribution of an element can be mapped by scanning with the probe and modulating a synchronously scanned display tube with the characteristic x-ray signal, one can get a picture of the distribution of total dry mass in a thin specimen by modulating with the bremsstrahlung signal.

6. The scanning facility can be used to measure the entire mass or the average mass per unit area in whole fields. The total bremsstrahlung count from a long *scanning* run is proportional to the *average* mass per unit area in the entire scanned field. If the scanned field completely encloses some thin object (like a section or cell) and contains nothing else besides the supporting film, then the total count, after due subtraction of the background from the support, is proportional to the total mass of the entire object.

7. While inhomogeneity in a specimen introduces an uncertainty into the microradiographic measurement of mass per unit area (Sec. 5.A.2), it does not affect the bremmstrahlung assay of thin specimens. (In general, this effect occurs with methods based on absorption rather than emission.)

8. We have noted (Equation 5–82) that the bremsstrahlung method really measures the quantity $S\sum_r f_r Z_r$. It is interesting to compare this with the quantity actually measured by the microradiographic method. From Equations 5–6, 5–12 and A-10 it is seen that the microradiographic method determines the quantity $S\sum_r f_r \sigma_r$, where σ_r is the mass absorption cross section of element r for x-rays near the middle of the spectrum of the incident radiation.[*] But as a function of atomic number, σ_r may be written in the form $\sigma_r = \sigma_o Z_r^3$, so that the quantity measured by the microadiographic method is essentially $S\sigma_o\sum_r f_r Z_r^3$.

The quantity which one nominally wants to measure is simply S, the mass per unit area. As noted earlier, Lindström has shown that in soft tissues, the variation of the microradiographic co-factor $\sum_r f_r Z_r^3$ is generally small enough to make very little difference. With the bremsstrahlung method, the variation of the co-factor $\sum_r f_r Z_r$ must be even slighter.

5.E.4. The Minimum Measurable Weight-fraction of an Element

When a specimen containing only one element is placed under a probe and the counting rate is recorded while the diffracting crystal is slowly rotated through the position of a characteristic peak, a curve like Figure 103 is obtained. The peak height P_o is the sum of the contribution S_o from

$$S = \frac{10}{(1.6)\ 10^6\ (10^3 d^{8/3})} \cong 6\ 10^{-9} d^{-8/3}\ gm/cm^2$$

and a minimum measurable mass M of

$$M = \text{area} \times S = \frac{\pi}{4}\ 10^{-8} d^2 S \cong 5\ 10^{-17} d^{-2/3} gm \qquad (5\text{–}83)$$

Thus the minimum observable mass M becomes less as the probe is enlarged. For a large probe the assayed mass would always be much greater than the minimal amount, so that the limit set by Equation 5–83 is relevant only for quite small probes. For a probe diameter $d = 0.1\mu m$, Equation 5–83 gives a minimal mass $M \cong 2\ 10^{-16}$ gm, while the preceding equation gives a minimal mass per unit area S corresponding to a section of soft tissue only a few hundred Å in thickness.

[*] Since $S\sum_r f_r \sigma_r$ is determined relative to a calibration wedge and it is essential that $\sigma(speci-men) / \sigma (wedge)$ should be independent of wavelength, it is not necessary to specify the x-ray wavelength more precisely.

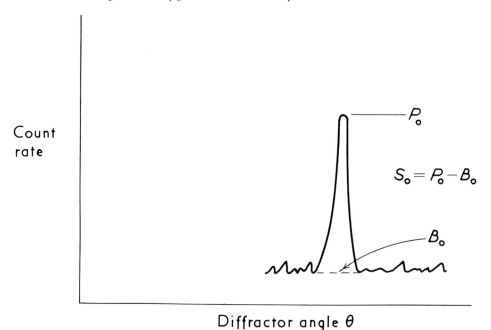

Figure 103. Peak P_o, background B_o and signal S_o when a single element is under the probe.

the characteristic radiation, and the background B_o which is due almost entirely to continuum radiation of very similar wave length.

If the same element is present in low weight-fraction f in a matrix of
nearly the same atomic number, the curve will look like Figure 104. The
signal S will be reduced to $S = f\, S_o$, while the background B will be virtually
unchanged. When the weight-fraction is so low that the signal cannot be

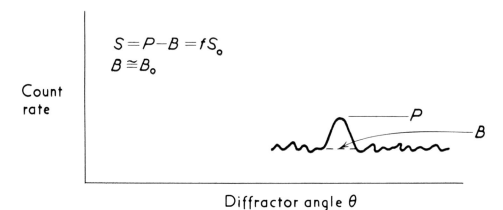

Figure 104. Peak P, background B and signal S when an element is present in a matrix
of nearly equal atomic number.

reliably distinguished from statistical fluctuations in the background, the measurement is impossible. In practice this limit usually occurs when S is of the order of one-fourth of B. Thus

$$S_{min} \cong f_{min} S_o \cong \frac{1}{4} B_o$$

so

$$f_{min} \cong \frac{1}{4} \frac{B_o}{S_o} \tag{5-84}$$

The quantity B_o/S_o depends primarily on the quality of the diffracting crystal and the x-ray spectrometer, and in modern instruments is usually in the range 1/500 to 1/1,000. According to Equation 5–84, f_{min} is then in the range 2.5 to 5×10^{-4}. Of course the criterion (5–84) is very crude, but actual measurements of a precisely defined f_{min} under a wide variety of conditions (Andersen, 1967a,b) almost always fall within the range 1 to 10×10^{-4} (i.e. 100 to 1,000 ppm).

In biological work, the assayed element is often present in low concentration in a matrix of much smaller atomic number. In this case, because the efficiency of continuum production is proportional to atomic number, the background B will be

$$B \cong B_o \frac{Z_m}{Z} \tag{5-85}$$

where Z refers to the assayed element and Z_m is the mean atomic number of the matrix. Equation 5–85 applies to both thin and thick specimens. When $Z_m \ll Z$, the reduced background of continuum radiation leads to a substantial improvement in the minimum measurable weight-fraction.

We have so far considered only the continuum background generated in the specimen itself. For thin specimens, the specimen support should be even thinner and of low atomic number; otherwise the background generated in the support can swamp the signal from a low concentration of an element.

5.E.5. Spatial Resolution

The factors which determine the attainable spatial resolution are quite different for thick, thin and ultrathin specimens.

5.E.5.a. *Thick Specimens*

After entering a specimen, the probe electrons generally lose their energy in small steps in a series of atomic collisions. The spatial resolution for x-ray microanalysis is determined chiefly by the length of "effective path," i.e. by the distance which the electrons may travel before their energy drops below the ionization energy associated with the observed x-rays and these x-rays can no longer be generated * (see Fig. 105). In order to estimate the length of effective path we may use the Bethe formula † for election energy loss per unit of path length in material of atomic number Z:

$$\frac{dE'}{ds'} = \frac{2\pi e^4}{E'} \, NZ \, ln \, \frac{1.17E''}{11.5Z} \tag{5-86}$$

Here N is the number of atoms/cc, e is the electrical charge of the electron, E' is the electron energy expressed in ergs, E'' is the electron energy ex-

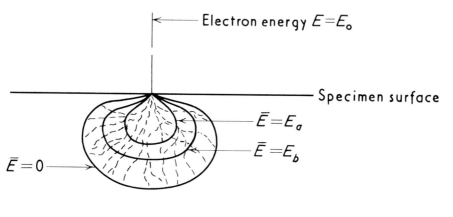

Figure 105. The different excited volumes for radiations with different ionization threshholds. The inner contour defines the surface where the probe electrons are degraded on the average to an energy E_a; a radiation with threshhold energy E_a will be excited mainly within this contour. A radiation with lower threshhold E_b will be generated within a larger volume as shown.

pressed in electron volts and s' is the path length in centimeters. For composite material this formula reduces with some approximations to

$$\frac{dE}{ds} = \frac{4\rho}{E} \, ln \, \frac{100E}{\bar{Z}} \tag{5-87}$$

where $\bar{Z}$ is the average atomic number, ρ is the density of the specimen in gm/cc, E is the electron energy in keV, and s is path length in microns. For

---

* We must distinguish between the spatial resolution for x-ray microanalysis and the spatial resolution of the scanning-electron images. The contrast in scanning-electron images of thick specimens depends mainly on the intensities of electron backscattering and secondary-electron emission at the point of impact of the probe. Hence the resolution attainable in the scanning-electron image depends primarily on the probe diameter. If the intention is merely to obtain scanning-electron images, the probe diameter can usually be reduced to approx. 0.1μm, or even much less in certain instruments.

† Bethe, H. A.: *Handbuch der Physik*, 24:491–523, 1933. The constants 1.17 and 11.5 are modifications recommended by Nelms (1956).

soft tissue, $100/Z$ is approximately 16 and Equation 5–87 can be integrated to give an effective range ρs in units of $\mu m \times gm/cc$:

$$I(E_o, E_x) = s = \int_{E_x}^{E_o} \frac{E\ dE}{4ln16E} \tag{5-88}$$

where E_o is the energy of the probe electrons at incidence and E_x is the ionization-edge energy associated with the observed x-rays.

The integral $I(E_o, E_x)$ of Equation 5–88 has been worked out numerically for $E_o \leq 40$ keV, $E_x \cong 0$, and is displayed in Figure 106. This graph enables one to estimate the effective range in soft tissue for *any* practical probe voltage E_o and ionization energy E_x. For example, suppose the probe is run at 15 kV, the ionization energy E_x is 9 keV (the copper K ionization edge), and the density of the specimen is $\rho = \frac{1}{2}$ gm/cc (a possible value for a section cut frozen and dehydrated). Then

$$I(15,9) = I(15,0) - I(9,0) = 5.8 - 2.4 = 3.4\mu m\ gm/cm^3$$

The effective path length for $\rho = \frac{1}{2}$ is then $3.4/\frac{1}{2} \cong 7\mu m$. (Thus we see that a probe voltage of 15 kV will give a rather poor spatial resolution when copper K x-rays are detected in a thick specimen of soft tissue.) Or, to consider a more favorable example, suppose that $\rho \cong 1\frac{1}{2}$ (embedded tissue), $E_x = 4$ keV (the calcium K edge) and one runs the probe at $E_o = 8$ keV. Then

$$I(8,4) = I(8,0) - I(4,0) = 1.9 - 0.7 = 1.2\mu m\ gm/cm^3$$

and the length of effective path is $s \cong 1.2/1.5 \cong 0.8\mu m$.

In order to use Figure 106 for the estimation of effective path length, one must know the ionization energy associated with the observed x-ray line. Some ionization energies are also indicated in the figure.

To conclude our very brief treatment of spatial resolution in thick biological specimens, we should note:

1. The estimation just outlined is very crude in two respects. First, because of scattering the electron paths are not straight lines, which reduces the total travel of the electrons from their point of entry into the specimen. Secondly, some electrons will scatter at approximately 90° soon after entering the specimen, and the area of x-ray excitation clearly has a diameter which is *twice* the range of these electrons. These two effects tend to cancel, and the results obtained with our crude procedure are close to the results given by more elaborate procedures.

2. In order to have a spatial resolution of $1\mu m$ or better in thick specimens of soft tissue, the difference $(E_o - E_x)$ should be no more than a few keV. On the other hand, as already noted, for efficient excitation and good sensitivity, one wants $E_o/E_x \geq 2$ (preferably $E_o/E_x \geq 3$). The only way to satisfy both conditions is to work with relatively soft radiations, $E_x \leq 4$ keV, using the L or M radiations of elements of atomic number $Z \geq 21$ (cf. Andersen, 1967a,b).

3. In embedded specimens the density ρ may be well known. In unembedded

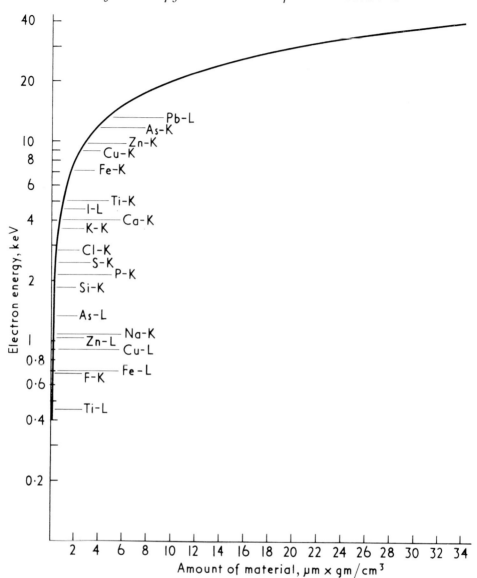

Figure 106. Graph for estimating residual electron energy as a function of penetration. Excitation threshhold energies of some commonly used x-ray lines are also indicated.

tissues like frozen sections, ρ may be only vaguely known, with a corresponding uncertainty in the length of effective path.

  4. In comparison with soft tissues, spatial resolution in hard tissues (tooth and bone) is improved in proportion to the higher value of the density, which is usually in the range 2–3 gm/cc (the exact value depending on the degree of mineralization). The factor $ln\ 100\ E/\overline{Z}$ in equation 5–87 is unfavorably affected but makes little difference.

  5. While effective path length is the dominating factor in determining spatial resolution in thick biological specimens, in order to establish the limits of per-

formance precisely one must also take into account the size of the probe. The spatial resolution of course will not be finer than the size of the focussed probe, unless extraordinary techniques are used. Detailed quantitative analyses suggesting optimum operating conditions, in particular optimum probe voltages, have been presented by Duncumb (1959).

6. In practice, the spatial resolution attainable for x-ray analysis in thick specimens has usually been in the range $0.3\,\mu$m to $2\,\mu$m.

5.E.5.b. *Thin Specimens*

In this section we shall consider chiefly specimens of soft tissue, $1\,\mu$m to $10\,\mu$m in thickness, subjected to electron probes with energies above 30 keV. Since most of the incident electrons will pass completely through such specimens with the loss of only a small fraction of their energy, the spatial resolution along the direction of the electrons cannot be finer than the thickness of the specimen. This is not too unsatisfactory, since the resolution may still match what can be seen in the specimen by optical microscopy or in the scanning-electron image. The main question is how badly the beam spreads laterally in transit.

The focussed electrons do not impinge on the specimen as a parallel beam; their directions lie within a cone whose apex angle is determined by an aperture placed within the probe-forming electron lens (Fig. 107).

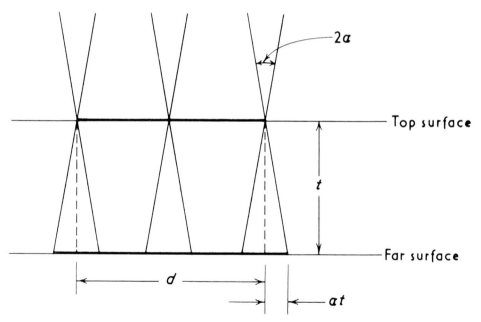

Figure 107. The spread of the beam within the specimen due to divergence at incidence. When the probe is focussed into a circle of diameter d on the top surface of a thin specimen of thickness t, a cone of rays of semiangle a arrives at each point within the circle. In the absence of scatter within the specimen, the beam would arrive at the far surface with diameter $(d + 2\,a\,t)$.

Hence, if the beam is focussed at the upper surface of the specimen, it would have to be broader in the plane of emergence even if the electrons suffered no deflection within the specimen. However, in practice the cone is quite narrow, usually with a half-angle[a] less than 4°, and the consequent spread in the course of a few microns is generally negligible.

The spread is due mainly to atomic collisions which deflect the probe electrons. In a specimen $\geqq 1\mu m$ in thickness, the spread results from the compounding of many individual collisions, and while good methods of estimation exist, the calculation of the spread for any given set of conditions is quite laborious. None of the published calculations have been intended for the present conditions of thickness and density but the results provided by Bishop (1965), while centered on higher values of mass per

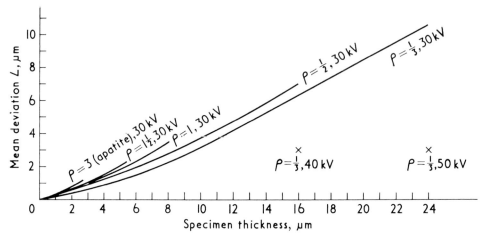

Figure 108. The lateral spread of an electron beam passing through a thin specimen.

unit area, extend to values just low enough to serve as the basis for the conclusions presented here.

Bishop determined the mean square lateral spread of electrons as they penetrate a thick target, as a function of residual electron energy. By estimating the average residual electron energy on emergence from a thin specimen through the energy-loss Equation 5–87, we have converted his data to lateral spread as a function of specimen thickness and density, with the results shown in Figure 108. The noteworthy aspects of the results are the following:

>   1. Most of Figure 108 relates to displacements and thicknesses too large to interest us. This was necessary for purposes of estimation because unfortunately the data provided by Bishop can be used only marginally at the smaller lengths.
>   2. Figure 108 certainly *overestimates* the actual spread, since Bishop's results are for thick specimens and include the effect of *backscattering* from depths greater than the thickness of our specimens.

3. From Figure 108, one can choose operating conditions to keep the spread within desired limits. For instance, in specimens of density $\rho = 1\frac{1}{2}$ *gm/cc*, with 30 kV electrons, if the root mean square deviation L is to be kept below 1μm, one should keep the specimen thickness t less than 3μm. Unfortunately, Bishop did not provide enough data to encompass a similar estimate for low densities like $\rho = \frac{1}{3}$ but useful conclusions can still be reached, based on the fact that the spread increases with thickness more rapidly than in direct proportion. Thus for example, since L is approximately 4μm when 30-keV electrons are incident on a 12μm layer of density $\frac{1}{3}$ gm/cc, we can conclude that if the thickness is reduced to 3μm, L will be reduced to less than $4 \times \frac{3}{12} = 1\mu$m. The same conclusion can be reached from the facts that L decreases with ρ at fixed t and that L (30 keV, $t = 3\mu$m, $\rho = 1\frac{1}{2}$) is approximately 1μm.

Figure 108 shows that the spread in thin specimens can be greatly reduced by increasing the probe voltage. From the one point available at 40 keV we may conclude that at this voltage the mean deviation is less than 1μm for sections up to 5μm in thickness. And with 50-keV probes one can go up to a specimen thickness of 8μm in material of density $\frac{1}{3}$ before L exceeds 1μm.

4. There is a special problem in estimating spatial resolution in frozen-dried sections of fresh-frozen tissues. (As discussed in Sec. 4.B.1.d, artefacts are minimized by this method of preparation.) Spatial resolutions can be estimated directly from Figure 108 if the dry-weight fraction and the section thickness are known * and *it is assumed that the specimen does not collapse after sectioning*. In fact, a considerable but unknown degree of collapse probably occurs, making the analyzed specimen considerably thinner and denser. This effect may improve the spatial resolution a great deal. As an extreme hypothetical example: according to Figure 108, at 30 keV an uncollapsed 12μm section with density $\frac{1}{3}$ gm/cc would produce a lateral mean spread L of 4.2μm; if the section collapsed to density 1 gm/cc and thickness 4μm, L would be reduced to 1.4μm.

5. The analyst can often choose to work either with high-energy electrons, with respect to which his specimen may be "thin," or with low-energy electrons, with respect to which the same specimen may be "thick." Except for quite thin specimens ($\leq 2\mu$m), the short-range low-voltage electrons give better spatial resolution along the direction of the electron beam. The lateral resolutions may be estimated and compared from Figures 106 and 108. Most often the low-voltage electrons give the better lateral resolution as well although this is not always so, especially for quite thin specimens ($\leq 3\mu$m), low densities ($\rho \leq \frac{1}{2}$), or if the ionization energy must be inconveniently high (for example, the assay of calcium with a K ionization edge at 4 keV).

Thus for qualitative studies with maximum spatial resolution, low-voltage electrons are most often preferable. However, in work with the long-wave-length x-rays generated by low-voltage electrons there are

* When a specimen consists of water with density $\rho_w = 1.0$ gm/cc and protein with density $\rho_p = 1.4$ gm/cc, if the water is removed *with no other change in dimensions*, the final density ρ_{dry} is related to the original dry-weight fraction f_p by

$$\rho_{dry}(gm/cc) = \frac{\rho_p\rho_w f_p}{\rho_p - f_p(\rho_p - \rho_w)} = \frac{1.4 f_p}{1.4 - 0.4 f_p} = \frac{7 f_p}{7 - 2 f_p} \tag{5-89}$$

The initial dry-weight fraction and the final density are then numerically almost equal in the range $\frac{1}{3} \leq f_p \leq \frac{1}{2}$, which includes most soft tissues.

greater instrumental problems and serious complications in quantitation due to severe x-ray absorption effects. In addition, with high-voltage electrons one can measure not only elemental weight-fractions but also total mass per unit area and hence elemental amounts per unit volume as well (see Secs. 5.E.7.b and 5.E.3).

5.E.5.c. Ultrathin Specimens

By "ultrathin" we mean specimens suitable for conventional transmission electron microscopy, i.e. ≤ 1000 Å in thickness. In combination electron microscope-microanalyzers ("EM-MA"s), such specimens can be viewed with spatial resolution similar to that in ordinary electron microscopes, so that one can take advantage of the best possible spatial resolution provided by the microanalytical procedure.

The microanalytical spatial resolution in this case is not limited by the spread of the probe, which is negligible within the ultrathin specimens. The resolution is essentially equal to the size of the probe, and the limit is set by the sharp decrease in available probe current and hence in x-ray signal (see Sec. 5.E.2) as the probe size is reduced.

There is a formula (Hall, 1968) for the limits of performance in these conditions:

$$f_w = \frac{f}{1+f}, \text{ with } f = \frac{1}{tE^2MSg} \frac{1+(1+8tE^2MB)^{\frac{1}{2}}}{2} \qquad (5\text{-}90)$$

Here f_w is the local weight-fraction of an element to be analyzed, t is the running time of a static-probe analysis, E is the probable statistical error for the run, M is the local specimen organic mass per unit area, S is the instrumental sensitivity (characteristic counts/second/μg of element/cm^2), and B is the background coefficient (background counts/sec/μg of organic mass/cm^2). The probe diameter d enters through the relationship $S = constant \times d^{8/3}$, the constant typically being approximately 700 in current instruments when d is expressed in μm. If the element is located within an area smaller than the probe, g is defined as $g = \dfrac{\text{"hot" area}}{\text{probe area.}}$

Based on the performance of current instruments, the following limits were estimated (Hall, 1971) for 20-second runs and relative probable error $E = 0.20$ (i.e., 20%):

Regarding spatial resolution in ultrathin specimens, the following points should be noted:

    1. As Table VII indicates (entries 1 and 16), analyses can be performed down to quite low weight-fractions with an analytical spatial resolution of 1μm. This has been confirmed in practice (see Sec. 4.A.3). When diffractive spectrometers are used (as is usual), analyses with probe diameters and spatial res-

olution of $0.1\mu m$ can be achieved only if the local elemental concentration is quite high (entries 6 and 21).

2. The performance of entries 9–15 and 24–28 is estimated for nondiffractive spectrometers relying on solid-state detectors or gas-filled proportional counters and pulse-height analysis techniques which are not yet fully developed. Thus this performance has not yet been achieved, but will quite probably be available soon.

TABLE VII

ESTIMATED MINIMUM MEASUREABLE ELEMENTAL AMOUNTS M_x AND WEIGHT-FRACTIONS f_w[a,b]

Entry No.	Detector	Hot-spot Diameter (μm)	Probe Diameter (μm)	Mass per Unit Area $M(\mu g/cm^2)$	Weight-Fraction f_w	Elemental Amount M_x ($10^{-16}gm$)
1	↓ diff. [c]	1.	1.	↓ 5	0.0008	0.3
2		0.32	1.		0.0078	0.3
3		0.32	0.32		0.0074	0.3
4		0.1	1.		0.075	0.3
5		0.1	0.32		0.073	0.3
6		0.1	0.1		0.142	0.7
7		0.032	0.1		0.64	0.7
8		0.032	0.032		0.76	1.2
9	solid [d]	0.1	0.32		0.012	0.05
10	solid	0.1	0.1		0.0067	0.03
11	gas [e]	0.1	0.1		0.0089	0.04
12	solid	0.032	0.032		0.043	0.02
13	gas	0.032	0.032		0.051	0.02
14	solid	0.01	0.01		0.44	0.03
15	gas	0.01	0.01		0.45	0.03
16	↓ diff.	1.	1.	↓ 12	0.0005	0.5
17		0.32	1.		0.0046	0.5
18		0.32	0.32		0.0036	0.4
19		0.1	1.		0.045	0.5
20		0.1	0.32		0.036	0.4
21		0.1	0.1		0.065	0.7
22		0.032	0.1		0.41	0.7
23		0.032	0.032		0.56	1.3
24	solid	0.1	0.32		0.0079	0.08
25	solid	0.1	0.1		0.0041	0.04
26	gas	0.1	0.1		0.0056	0.06
27	solid	0.032	0.032		0.024	0.02
28	gas	0.032	0.032		0.029	0.03

[a] The amounts and weight-fractions have been calculated for a statistical probable error of 20 percent in a running time of 20 sec. The values are functions of hot-spot diameter, probe diameter, and specimen thickness.
[b] This table is similar to the one published by Hall (1968), but is based on more modern performance figures.
[c] Fully focussing diffractive spectrometer.
[d] Solid-state detector used nondiffractively.
[e] Gas counter used nondiffractively.

3. Still further radical improvements are to be expected as certain already existing components are incorporated into EM-MA instruments, namely brighter electron sources (Crewe, 1968), electron lenses of much lower aberration and the newest solid-state nondiffractive x-ray spectrometers (Russ et al., 1971).

4. In sum, while radical improvements in spatial resolution are expected soon for ultrathin specimens, at present the best available resolution for x-ray analysis is in the range $0.1\mu m$ to $1\mu m$, depending on the elemental weight-fraction which is to be observed.

5.E.6. Depth of Focus

Because the electrons in a probe are nearly parallel, the cross section of the electron beam can remain very small through a remarkable depth. This is why the scanning electron microscope can provide electron images of nonflat surfaces with a depth of focus of hundreds of microns or more. However, in microprobe analysis one generally studies objects with a relatively flat surface, so that the depth of observation depends primarily on electron range within the object. Features below the surface will appear in the electron image only if the probe electrons have sufficient range to reach them, scatter back, and return to the surface. When hard tissues are studied, and when low-energy electrons are used on soft tissues, the range is only a few microns or less. If high-energy electrons are used for the study of histological sections of soft tissue, features at all levels in a 10-μm section may be manifest in the electron image, although with some cost to lateral spatial resolution as we have seen, and with possible confusion from the superposition of planes in the image.

5.E.7. Quantitation

With careful use, electron-probe X-ray microanalysis can provide quantitative local elemental assays in a wide variety of specimens, with probable errors of only 1 to 2 percent. But with biological specimens the method can only rarely be this accurate.

5.E.7.a. Thick Specimens

It can be shown that the intensity of a given characteristic radiation I_x generated by an electron probe in a thick specimen is, to a fair degree of approximation, directly proportional to the weight-fraction f_x of the corresponding element. Consequently, the experimental procedure for quantitative analysis is extraordinarily simple. One merely compares the intensity of the selected characteristic radiation from the chosen spot in the specimen with the intensity of the same radiation I_{xo} from a standard of known composition, of course keeping probe voltage and current constant, and

$$f_x \cong \frac{I_x}{I_{xo}} f_{xo} \tag{5-91}$$

If the standard consists of the analyzed element in the pure state, $f_{xo} = 1$ and

$$f_x \cong \frac{I_x}{I_{xo}} \tag{5-92}$$

While the analytical act itself is no more complicated than just described, for several reasons one may need equations more accurate and more complicated than (5–91) and (5–92) in order to deduce the desired weight-fractions from the observed intensities. The complications are the following:

1. X-ray absorption. The simple proportionality between x-ray intensity and weight-fraction refers to the x-rays *directly generated* by the probe, but a substantial fraction of the generated x-radiation may be absorbed before escaping from the specimen, and this fraction may be quite different in specimen and standard.

There are ways to correct for absorption. However, if the absorption effect is large the correction can be accurately made only if the specimen has a smooth flat surface and the composition is homogeneous along the paths of the incident electrons and the emerging radiation.

2. Fluorescence. In addition to x-radiation generated directly by the probe, characteristic x-rays may be produced in secondary processes of fluorescence, either by bremsstrahlung or by other characteristic x-rays. The fluorescence effect is usually negligible or correctable in biological specimens (a detailed treatment is given by Hall, 1971).

3. Backscattering. The generated intensity is affected by the fact that many of the incident electrons are scattered back out of the specimen before giving up all of their energy, and the backscattered fraction may be quite different in specimen and standard.

4. Penetration effects. Only a very small fraction of the probe electrons produce useful characteristic x-rays before being brought to rest in the specimen. The yield depends on the competition between productive ionizations and nonproductive energy losses, so that the intensity I_x will be proportional to weight-fraction only if the rate of energy loss per unit mass/area, $dE/\rho \, dx$, is the same in specimen and standard. The rate of energy loss $dE/\rho \, dx$ depends on the ratio atomic number/atomic weight Z/A and also on effects associated with chemical binding and physical state of aggregation (i.e. powder or compact block, etc.). The effect associated with Z/A can be handled accurately; the effect of chemical binding on penetration is presumably small; the effect of physical state of aggregation is not yet well appreciated and may be large (Ichinokawa, 1969).

5. Chemical shifts. Throughout most of the spectrum of characteristic x-rays, the effects of chemical binding are too slight to interfere with quantitative analysis, but for quantum energies less than approx. 2 keV (wave lengths $\geqq 6$ Å), chemical binding may appreciably affect both wave length and the relative intensities of different spectral lines. For quantitative studies based on radiations of long wave length, it is therefore important to try to use standards in which the assayed element is chemically bound in the same way as in the specimen.

The chemical shifts at long wave length can be measured with high-quality x-ray spectrometers, and the chemical state of an assayed element can sometimes be determined from such a measurement. A great deal of work is being done at present on the development of this technique as well as on the analytical difficulties imposed by the effect. The interested reader can refer to a large set of papers edited by Mallett *et al.* (1966), or to a recent set edited by Mallett and Newkirk (1970).

Under favorable conditions all of the above complications are negligible or else manageable in terms of the elaborate modern theory of quantitation.

But one must avoid mechanical application of the theory when the prerequisites for its validity are not satisfied. In particular, inhomogeneities in biological specimens may make accurate corrections for absorption, fluorescence and backscattering impossible when these effects are large (see Fig. 74, Sec. 4.B.1.e). The reader who wants to analyze thick specimens quantitatively in spite of this warning is referred to Birks (1963) for a relatively simple but early and not so precise exposition of theory, and to Philibert (1969) or Hall (1971) for a fuller exposition.

5.E.7.b. Thin Specimens

In thin specimens, the intensity of a characteristic radiation cannot suffice as a measure of weight-fraction, since the intensity obviously depends not only on weight-fraction but also on the local thickness or mass per unit area of the specimen. In the quantitative method developed for use with thin biological specimens by Marshall and Hall (1968), one always observes the *ratio* of the intensity of a characteristic radiation I_x and the simultaneously observed continuum intensity I_w. (For the instrumental arrangement see Fig. 102). Qualitatively speaking, with thin specimens, the characteristic intensity is a measure of elemental mass/area and, as discussed in Section 5.E.3, the continuum intensity can serve as a measure of total mass/area, so that suitable calibration enables one to determine elemental weight-fractions f_x according to the simple equation

$$f_x = \frac{\text{elemental mass}}{\text{total mass}} = \frac{\text{elemental mass/area}}{\text{total mass/area}} \tag{5-93}$$

It is convenient to regard a biological specimen as consisting of a matrix of known composition (elements denoted below by subscript m), plus additional elements in unknown amount (denoted below by subscript u). For example, in most soft tissues one can assume a proteinaceous matrix with weight-fractions 52% C, 22% O, 17% N, 7% H, and 2% S + P. The quantitative equation derived by Marshall and Hall (1968) for the weight-fraction f_x of element x can then be written in the form

$$f_x = \frac{Q_x G_x A_x}{\sum_u Q_u G_u A_u + H(1 - \sum_u Q_u G_u Z_u^2 B_u)} \tag{5-94}$$

with

$$\frac{1}{H} = \sum_m \frac{f_m'}{A_m} Z_m^2 B_m \quad \text{and} \quad f_k' = \frac{f_k}{\sum_m f_k}$$

Here $Q_x = (I_x/I_w)$ specimen divided by (I_x/I_w) standard,

$$G_x = \frac{N_x}{\Sigma N Z^2 B} \text{ in the standard for element } x$$

N_x = atoms/cc of element x in the standard for element x

A_x = atomic weight of element x

Z_x = atomic number of element x

B_x is defined in Section 5.E.3. (Two definitions of B_x are given there, but the results are virtually the same whichever is used.)

In order to use this equation the following points must be noted:

1. The continuum radiation is counted within a band of quantum energies set by electronically accepting only a corresponding band of pulse heights from the proportional counter. The restrictions on the choice of band are merely that characteristic radiations must be excluded, and that the highest quantum energy must be well under the mean energy of the probe electrons as they *emerge* from the specimen. In practice the band can usually be comfortably set well above the most energetic characteristic radiations and well below the energy of the incident probe electrons.

2. The quantity f_m' is the weight-fraction of element m *in the postulated matrix*. It depends solely on the postulated composition of the matrix and has nothing to do with measurements on the specimen. For instance, for the matrix mentioned above, $f_c' = 0.52$, $f_0' = 0.22$, etc.

3. All of the quantities in formula 5–94 are independent of the measurements, being known entirely from prior information, with the exception of the quantities Q, which are obtained directly from the measurements of characteristic and continuum intensities from specimen and standard(s).

4. Thin standards of known composition are required, but there is no need to know the thickness of any standard or of the specimen (Hall and Werba, 1971). A standard is needed for each element to be assayed, but one standard may suffice for more than one element. For example, for the assay of Ca and P in mineralizing tissues a single thin standard of apatite $(3Ca_3(PO_4)_2 \, Ca(OH)_2)$ is sufficient.

The quantities G_x refer only to the standards. To work out the values of G it is not necessary to know the actual numbers of atoms/cc N, since only the ratios of numbers of atoms in the standard are important. Hence the relative numbers of atoms, which are given directly by the chemical formula of each standard, may be inserted for the quantities N. The sum in the definition of G is taken over all the constituent elements of the standard.

5. In order to measure the *ratio* of two elemental weight-fractions f_a/f_b one may use the simpler equation, easily derived from (5–94),

$$\frac{f_a}{f_b} = \frac{Q_a G_a A_a}{Q_b G_b A_b} \qquad (5\text{–}95)$$

6. It is not necessary to postulate the presence of any matrix of known composition. If no matrix is postulated the factor $1 - \sum_u QGZ^2B$ becomes zero and Equation 5–94 reduces to

$$f_x = \frac{Q_x G_x A_x}{\sum_u Q_u G_u A_u} \qquad (5\text{–}96)$$

Whether or not a matrix is postulated, for the assay of a given element x, it is necessary to observe the characteristic radiations of other elements only when they contribute significantly in Equation 5–94 through the terms with subscript *u*. Thus frequently in soft tissues where the composition of the matrix may be assumed, it is not necessary to observe any characteristic radiation besides that of the assayed element. When the concentrations are low for all elements outside the postulated matrix, including the assayed element, the terms with subscript *u* in Equation 5–94 are all negligible and the formula reduces to

$$f_x = Q_x G_x A_x \sum_m f_m' \frac{Z_m^2}{A_m} B_m \tag{5-97}$$

It was *not* assumed in the derivation of the general formula that the specimen is predominantly organic. The formula 5–94 is equally valid whether the postulated matrix constitutes a major or a minor part (or no part) of the specimen, and it applies to highly or moderately mineralized as well as to soft tissues.

7. Since characteristic and continuum radiation are produced in parallel as electrons proceed through a specimen, the ratio characteristic count/white count is unaffected by variations in electron backscattering or penetration. Also, absorption and fluorescence effects in thin specimens are much less than in thick ones. Thus the analysis of thin biological specimens via Equation 5–94 avoids the main difficulties encountered in the quantitative analysis of thick specimens.

8. For probe energies of 30 keV the permissible upper limit of specimen mass/area is approximately $600\mu g/cm^2$, except that absorption corrections may already be important in thinner specimens if low-energy characteristic radiations are detected or if the tissue is mineralized.

9. Equation 5–94 is based on Equation 5–81 (Sec. 5.E.3), which cannot lead to very accurate results because of the uncertainty about the correct form of B and because k is not constant as assumed but actually varies somewhat with atomic number. The associated analytical errors may be in the range 5–10% (Hall and Werba, 1971). With ultrathin sections it is also difficult to correct accurately for the relatively large background of continuum generated in pieces of metal located near the specimen in the instrument. Thus the analytical method for thin specimens based on normalization to the continuum at present cannot attain the accuracy achieved by thick-specimen analysis under exceptionally favorable conditions. On the other hand, in most tissues the former method bypasses the problems associated with absorption, fluorescence, backscatter and penetration, which can produce completely misleading results when the conventional theory of quantitation is incautiously applied to thick specimens.

5.E.7.c. Specimens of Intermediate Thickness

There is no good method at present for the quantitative microprobe analysis of biological specimens intermediate between "thick" and "thin."

5.E.8. Availability of Instrumentation

Among x-ray microanalytical instruments, electron microprobes are by far the most expensive, costing more than $70,000. Consequently very few probes are available entirely for biological work. In addition, while specimen preparation is relatively simple and the actual running time for one static probe analysis is short, one or two hours are usually spent in preliminary adjustments for each type of specimen, and of course in terms of volume or area the amount of material actually analyzed is miniscule. For these mundane reasons, the microprobe is not suited for diagnostic or survey work and the available instrument time should be spent on carefully selected research studies.

Microprobes are made commercially by Applied Research Laboratories, Materials Analysis Corporation and North American Philips ("Norelco") in the U.S.A.; by the Cambridge Instrument Company and A.E.I. in England; by Cameca in France; by JEOL and Hitachi in Japan; and by Siemens in Germany. Several of these companies also make scanning electron microscopes with available x-ray spectroscopic attachments. Combination electron microscope-microanalyzers are made commercially by A.E.I., Cameca, and Siemens. In addition several manufacturers of electron microscopes, including Philips (Netherlands) and Siemens, have offered x-ray spectroscopic attachments to pre-existing conventional electron microscopes, but the sensitivity of these attachments may not be good enough for the weak x-ray signals generated in most ultrathin biological specimens.

The reader may find a more extensive discussion of the methods of quantitative measurement in biological specimens with the electron microprobe, and of the available types of microprobe instrumentation, in a recent review paper (Hall, 1971).

APPENDIX

SOME FUNDAMENTALS OF X-RAY PHYSICS

A.1. RELATIONSHIP BETWEEN WAVE LENGTH AND QUANTUM ENERGY

X-RAYS ARE A PART of the spectrum of electromagnetic radiation (Fig. 109). Some of the phenomena of this radiation, notably diffraction and interference, are readily understood in terms of the properties of waves; some aspects of the interactions of radiation with individual atoms are readily understood in terms of radiated particles, called "photons" or "light-

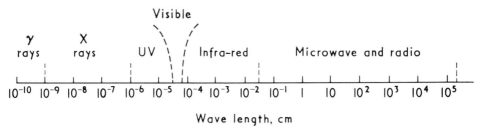

Wave length, cm

Figure 109. The spectrum of electromagnetic waves.

quanta." (In theoretical physics, "photon" and "light-quantum" have slightly different connotations, but we may use the words interchangeably.) There is an invariant relationship between the wave length, and the energy of each photon associated with any electromagnetic radiation.* With numerical values substituted for fundamental physical constants, this is written conveniently as

$$\lambda V = 12.4 \qquad\qquad \text{(A–1)}$$

* Obviously, something cannot simultaneously be a particle and a wave. The resolution of this difficulty, probably the most profound problem of modern physics, happily need not concern us here. Equation A-1 merely relates the wave length which must be attributed to the wave when the wave model is used, and the energy which must be attributed to each radiated particle when the particle model is used. The reconciliation of the two models in a deeper theory is irrelevant to our purposes.

266

Here the wave length λ is expressed in Ångstrom units, Å, and the photon energy V is expressed in kilo-electron volts, keV. ($1Å = 10^{-8}$ cm and 1 keV $= 1.6 \times 10^{-9}$ ergs.)

Radiation is characterized equally well by the specification of its wave length or its quantum energy, and we shall be switching freely between these terminologies.

A.2. THE ENERGY LEVELS OF ELECTRONS IN ATOMS

Atoms consist of positively charged nuclei surrounded by "clouds" of negatively charged electrons, which are bound to the nuclei by electrical attraction. In each atom, electrons are stably bound only in certain "orbits," and for each stable orbit there is a specific "energy level" or "ionization energy," the minimum energy which must be imparted to an electron in that orbit to knock it free of the atom. The energy levels of the orbits are characteristic of the chemical elements; all atoms of the same type have the same set of energy levels, and the levels differ from one chemical element to another.* As we shall see, the existence of discrete characteristic energy levels is fundamental to x-ray physics.

The electronic orbits are conventionally classified by means of the letters "K," "L," etc. In general, an atom has two electrons, called the K electrons, in orbits which are the closest to the nucleus. As the K electrons are exposed most strongly to the nuclear attraction, they have the greatest ionization energies. There is a group of eight orbits of similar ionization energies, much less than the energy of the K levels; these orbits are referred to as the L shell. The next group, of still smaller ionization energy, is named the M shell, and so on.

A.3. THE PRODUCTION OF X-RAYS

The x-rays which are significant in x-ray microscopy are generated by two processes: "bremsstrahlung," and the removal of electrons bound to inner atomic orbits ("ionization"). We shall describe these processes first as they occur within x-ray tubes.

In a conventional x-ray tube (Fig. 110), electrons are released from a hot filament, are accelerated by the voltage applied between filament and anode, and then impinge on the anode. If the electrical potential applied to

---

* This statement is quite accurate for gases, where the atoms are isolated and exert only negligible forces on each other. In liquids and solids, the energy levels of electrons bound to one nucleus are affected by the force fields of neighboring atoms. The effect is generally quite small for the inner electronic orbits, which are the most important in x-ray work. Consequently, for these orbits in liquids and solids, the electronic energy levels of an atom are almost, but not entirely independent of the environment, i.e. of the chemical state of the atom.

the tube is designated as E kilovolts, the energy of each impinging electron is E kev.* In one type of collision with an atom within the anode, an electron may lose any part or all of its energy, which then appears as an x-ray photon, only an insignificant amount of energy being transferred to the atom. Hence photons are produced with any energy V in the range $0 < V \leq E$ kev. The range of wave length of these x-rays, according to equation A – 1, is

$$\lambda \geqq \frac{12.4}{E} \qquad\qquad \text{(A–2)}$$

The radiation produced in this way is called "bremsstrahlung" †; it is also referred to as the "continuum" or "white radiation" since it may include any wavelength greater than 12.4/E.

Applied Potential

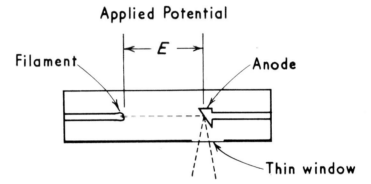

Figure 110. Components of a conventional x-ray tube.

In a different process of x-ray production, an impinging electron ionizes an atom by knocking away one of its inner electrons. An outer electron of the atom then "jumps" into the vacated orbit, thereby losing energy which may appear in the form of an x-ray photon.‡ The quantum energy and wavelength of this radiation are related to the energies of the initial and final electronic orbits, V_i and V_f, through the equation

$$V \approx \frac{12.4}{\lambda} = (V_f - V_i) \ keV \qquad\qquad \text{(A–3)}$$

Because the energies of the orbiting electrons are constants which are specific for each chemical element, it follows from equation A-3 that each ele-

---

* We shall use the symbol E to denote both electrical potential in kilovolts and energy in kev. The context will make the meaning clear.

† "Bremsstrahlung" is literally "braking-radiation": the radiation produced as the electron is decelerated.

‡ The energy lost in the transition between orbits does not always emerge as a photon. It is sometimes self-absorbed within the atom. The probability that the full energy escapes the atom in the form of a single photon is called the "fluorescence yield."

ment emits x-rays with certain characteristic quantum energies and wave lengths. The x-rays resulting from ionization are therefore referred to as "characteristic" radiation.

Characteristic photons are produced not only by the impact of electrons, but by any process which ionizes an inner atomic orbit. Notably, such ionization may be induced by the absorption of an incident x-ray ("photoelectric absorption"). Characteristic x-rays emitted under the impact of other x-rays are called "fluorescent."

TABLE A-I

DESIGNATIONS OF CHARACTERISTIC X-RAYS

Final Orbit of Electron	Initial Orbit of Electron	Name of X-ray
K	L_1	$K_{\alpha 1}$
K	L_2	$K_{\alpha 2}$
K	M_3	$K_{\beta 1}$
L_1	M_1	$L_{\alpha 1}$
L_2	M_2	$L_{\beta 1}$
L_1	N_3	$L_{\beta 2}$
L_2	N_4	$L_{\gamma 1}$

Characteristic x-rays are named according to the electronic "jump" or transition which created them. The initial and final electronic orbits are specified by the letters of their shells followed by subscripts which complete the identification. Several systems of notation are in use. Table A-I lists designations for the most important transitions in M. Siegbahn's notation, which we use in this book. Note that the name of the x-ray matches

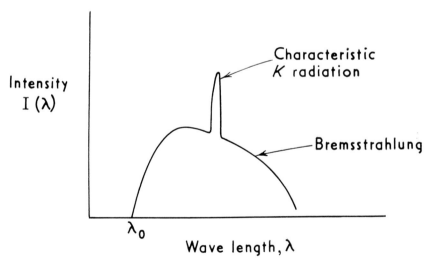

Figure 111. Spectrum from an x-ray tube. The energy radiated with wave length between λ and $\lambda + d\lambda$ is, by definition, $I(\lambda)\, d\lambda$. The shortest wave length λ_o is $\lambda_o = 12.4/V$ when V kilovolts are applied to the tube. The spectrum at long wave length is cut off by absorption in the window.

the final orbit of the electronic transition, which is the orbit where the ionization originally occurred.

The K_{a1} and K_{a2} radiations are by far the most important. Their wave lengths are so nearly equal that there is often no need to distinguish between them, and they may be referred to in common as the K_a radiation.

In sum, the output of an x-ray tube, generated in the anode, contains both continuum and characteristic radiation. The characteristic radiation results from ionizations produced directly by electron impact and also by photo-electric self-absorption of the anode's own radiation (this secondary process may account for a substantial fraction of the characteristic ouput). A typical spectrum of the radiation from an x-ray tube is shown in Figure 111. The anode usually consists of only one chemical element, and the output contains various characteristic "lines" corresponding to various electronic transitions within the atoms of that element.

X-rays may be generated in ways not yet mentioned, for example bremsstrahlung from charged particles such as protons and mesons, and in the annihilation reactions of electrons and positrons. But the processes described above, electron bremsstrahlung and electron transitions in ionized atoms, are the only ones relevant to our subject.

A.4. THE ABSORPTION OF X-RAYS

In x-ray microscopy, the absorption of x-rays is predominantly photo-electric. In this process, the full energy of the absorbed photon goes into the ionization of an atom.

Consider a beam of *parallel, monoenergetic* * photons impinging on a slab of material. As the photons progress through the slab, they are attenuated in random collisions with the absorbing atoms, but they do not change in direction or energy; the only change is a reduction in their number.†
Hence, at any depth within the slab, a layer of given thickness will always remove the same fraction of the beam incident upon it. The mathematical expression of this fact is the equation

$$N = N_o e^{-\mu x} \tag{A–4}$$

---

* Radiation which is all of one wave length is called "monochromatic" (by analogy with visible light). The photons of such radiation, in accord with Equation A-1, are monoenergetic, so the terms are interchangeable. We shall use the more common description "monochromatic" except when the physical processes are understood more readily in quantum terms.

† This statement is not strictly accurate because of two effects, scattering and x-ray fluorescence. Scattering is generally negligible in the conditions of x-ray microscopy; it is considered somewhat more extensively in Section A.5. Fluorescence is the process already described, the production of characteristic radiation as a consequence of the photoelectric absorption of the incident x-rays. This characteristic radiation is usually a negligible fracton of the total transmitted radiation; it will be considered more fully below in the special cases where it is important.

Here N_o is the number of photons in the beam striking the slab, N is the number reaching depth x, and μ is a quantity characteristic of the absorber.

We may visualize transmission and absorption in the slab as in Figure 112, with each atom in effect blocking off a certain area, σ_a. In a thin slab of

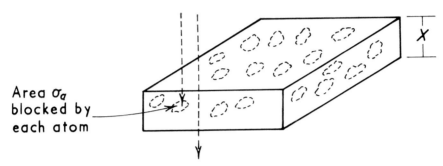

Area σ_a blocked by each atom

Figure 112. A simple model of transmission and absorption.

thickness x cm and area A cm², if there are n atoms per cm³, the total number of atoms is (nAx), the total blocked area is $(nAx\sigma_a)$, and the fraction absorbed from the incident beam is $nAx\sigma_a/A = nx\sigma_a$. This leads to an alternative form of Equation A-4:

$$N = N_o e^{-n\sigma_a x} \tag{A-5}$$

We may introduce the density of the absorber, ρ gm/cm³, and write

$$n\sigma_a x = \left(\frac{n\sigma_a}{\rho}\right)(\rho x)$$

The quantity ρx is the so-called "surface density," grams/cm² of surface facing the beam, which we shall denote as S. The quantity $n\sigma_a/\rho$, which we shall denote as σ, may readily be seen to be the area blocked off by each *gram* of absorber. The equation for absorption may now be put in a final form,

$$N = N_o e^{-\sigma S} \tag{A-6}$$

μ is called the linear absorption coefficient, σ_a the atomic absorption coefficient or atomic absorption cross section, and σ the mass absorption coefficient or mass absorption cross section. The relationship between them is summarized as

$$\mu = n\sigma_a = \rho\sigma \tag{A-7}$$

μ is a convenient quantity in empirical work, while σ_a is most closely associated with the underlying physical processes. However, the most useful form of the absorption equation is (A-6), which emphasizes that a given mass per unit area of an absorbing element has a fixed absorbing power regardless of how it is packed. This foreshadows the fact that the quantity directly measured by analytical absorption techniques is always a mass per unit area.

The equations (A-4) — (A-6) refer to absorbers containing only one element. For absorbers containing more than one element, the theory is readily extended to give

$$N = N_o e^{-(\sigma_1 S_1 + \sigma_2 S_2 + \ldots)} \tag{A-8}$$

where S_i is the mass of element i per unit area. Slightly modified forms of (A-8) are useful. If S is the total mass per unit area, and the weight-fractions of the constituent elements are f_i, then $S_i = f_i S$ and

$$\sigma_1 S_1 + \sigma_2 S_2 + \cdots = (f_1 \sigma_1 + f_2 \sigma_2 + \cdots) S \tag{A-9}$$

We can define an average mass absorption coefficient $\bar{\sigma}$ by

$$\bar{\sigma} = f_1 \sigma_1 + f_2 \sigma_2 + \cdots \tag{A-10}$$

and the exponent of (A-8) can then be written in the simple form $\bar{\sigma}S$. Also, it may be convenient to refer to total incident and transmitted energies, I_o and I, which are related to N_o and N simply by $I_o = V N_o$ and $I = V N$, V being the energy of each photon (recall that the discussion has been confined to monoenergetic radiation). Equation A-8 is then equivalent to these expressions for the transmission T of *monoenergetic* radiation:

$$T = I/I_o = N/N_o = e^{-(\sigma_1 S_1 + \sigma_2 S_2 + \ldots)} \tag{A-11}$$
$$= e^{-\bar{\sigma}S} \tag{A-12}$$
$$= e^{-(f_1 \sigma_1 + f_2 \sigma_2 + \ldots)S} \tag{A-13}$$

The absorption coefficients σ_i depend strongly on the quantum energy of the radiation and on the atomic number Z of the absorber. A typical dependence of σ on V for a given Z is shown in Figure 113. The sharp breaks in the curve, called absorption edges, occur at photon energies which barely suffice to expel an electron from one or another particular orbit. These are the ionization energies, V_K, V_L, etc., the subscript specifying the shell or orbit. On the graph, as the photon energy decreases below one of these critical values, absorption by energy transfer to the corresponding orbital electron is precluded and the absorption coefficient abruptly decreases. Between absorption edges, the coefficient varies strongly and continuously with photon energy and the dependence on photon energy and atomic number of the absorber may be expressed very approximately by the rule of thumb

$$\sigma = \text{constant} \times \frac{Z^3}{V^3} \tag{A-14}$$

The absorption-edge energies of any element are closely related to the energies of its characteristic photons. If the ionization energies of an inner and an outer orbit are respectively V_i and V_o, the absorption-edge energy for the inner orbit is simply V_i, while the jump of an electron from the one orbit to the other produces a characteristic photon of energy $(V_i - V_o)$. Thus the energy at an absorption edge is always somewhat greater than the energies of the corresponding characteristic photons. This has the important consequence that an element is relatively transparent to its own K radiation, as may be seen in Figure 113. The same is seen to be true of the least energetic L radiation, but the most energetic L lines are strongly absorbed in L orbits of lesser energy.

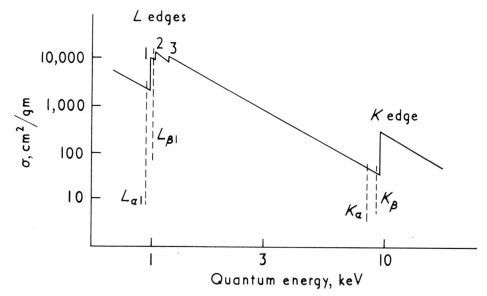

Figure 113. The dependence of the absorption coefficient of zinc (atomic number 30) on quantum energy. Both scales are logarithmic, so that Equation A-14 would appear as a straight line. The positions of some prominent emission lines, namely K_α, K_β, $L_{\alpha 1}$ and $L_{\beta 1}$, are also shown.

There are many inconsistencies and serious inaccuracies in the older tabulations of x-ray absorption coefficients, but there have been several sets of recent measurements to meet the need for reliable values for x-ray microanalysis and Heinrich (1966) has published an extensive, self-consistent table of coefficients based on weighted averages of almost all published data. Henke and his coworkers, most recently Henke and Elgin (1970), have presented excellent tabulations concentrating on x-ray energies below the range covered by Heinrich and most authors. For rough estimations relevant to this book, one may use Figure 114, which is based on the values of Heinrich and of Henke and Elgin.

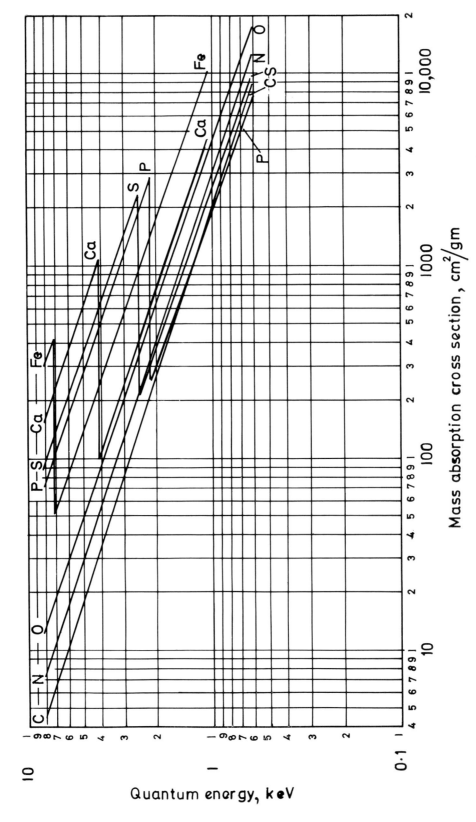

Figure 114. Dependence of absorption cross section on quantum energy, for several important elements.

A.5. THE SCATTERING OF X-RAYS

In addition to photoelectric absorption, the interactions of x-rays with matter include various types of scattering. For the wave lengths and specimens of interest in x-ray microscopy, the cross section for scattering is generally much less than that for absorption.* X-ray scattering within specimens will concern us only in fluorescence microanalysis, where the background of scattered rays sets a limit to the sensitivity of the method (Sec. 5.D.2). We must note here that in some scattering processes the scattered photon loses energy, and in others it does not. However, in fluorescence microanalysis, the loss of energy when it occurs is small and unimportant.

In substances which are crystalline rather than amorphous, the scattering observed in certain directions may be greatly intensified by the superposition, precisely in phase, of waves coming from the individual scattering centers. While we will not be concerned with the occurrence of this phenomenon within specimens, it is the basis of the selective detection of characteristic x-rays with diffracting crystals. This subject is discussed in Section A.7.

A.6. THE DETECTION OF X-RAYS

In microradiography, one needs a record of the spatial distribution of the X-rays which have passed through the specimen. For the contact method, the record is obtained on fine-grained photographic emulsion. In projection microradiography, the x-rays are again recorded most often directly on photographic emulsion, but good resolution does not require a very fine grain because the image is magnified.

For the operation of a projection x-ray microscope, it is quite desirable to have some means of forming an immediately visible image, which can guide the operator to the best setting of the controls. Such an image may be formed, just as in medical fluoroscopy, by letting the x-rays strike a fluorescent screen in place of the photographic emulsion. In practice the fluorescent images are often faint, and it may be necessary to couple the fluorescent screen to an image intensifier. Alternatively, the image may be formed and intensified by a television system, with the radiation striking

---

* Equation A-6 accurately describes the transmission of x-rays only when the cross section for scattering is much less than that for absorption. If this condition is not met, we can replace σ in (A-6) by the sum of the two cross sections; the resulting equation then describes *not* the total transmission, but the transmission of photons unchanged in direction. The total transmission no longer has even the form of equation A-6; the logic behind the derivation of that equation becomes inapplicable because the angular distribution of the beam changes as the beam penetrates the absorber. These considerations are important for the diagnostic applications of radiography and radioisotopes and for radiotherapy, but not for x-ray microscopy.

the face of a specially coated, x-ray-sensitive, television pickup tube. If image intensifiers or television systems are used, the final record may consist of photographs of the outputs of these devices.

In microanalysis, when there is no need for a record of the spatial distribution of the radiation but accurate measurement is desired over a wide range of intensities, one uses detectors which produce a discrete electrical pulse for each detected photon. The common detectors of this type are the gas counter, the scintillation counter and the semiconducting solid-state counter.

The most widely used type is the gas counter, exemplified in Figure 115. It consists of a gas-filled metal shell with a thin "window" to admit x-rays, and an insulated central wire. A voltage is applied between the shell and the wire. When a photon is absorbed in the gas inside the counter, the released photoelectron generally proceeds to ionize many more atoms, liber-

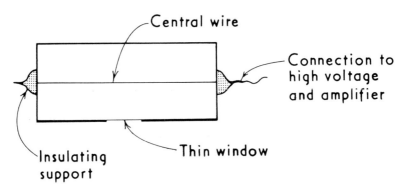

Figure 115. Components of a gas-filled x-ray counter.

ating "primary electrons." If the applied voltage is sufficient, these electrons produce additional ionizations as they approach the wire. Finally, all of the released electrons are collected at the wire. The process of collection induces an electrical pulse which may be amplified and counted.

When the applied voltage is high enough, this detector is called a Geiger counter and it produces one large electrical pulse, of invariant height, for each photon absorbed in the gas.

Over a range of lower voltages, the electric field is still strong enough for each primary electron to release many others, but the final pulse is smaller than a Geiger pulse. Photons of a given energy then produce pulses of proportional average height, and the detector is called a "proportional counter." The pulses are faster than Geiger pulses (approximately one microsecond as against approximately one millisecond), so that higher intensities may be counted without appreciable overlap. Another important advantage of proportional operation is that one can discriminate in favor of a particular characteristic radiation by adjusting the accessory circuits to count only pulses of the correct height.

The scintillation counter is shown schematically in Figure 116. A photon which is absorbed in the scintillator gives rises to visible or ultraviolet light, which may release one or more photoelectrons from the coating on the inside of the "photomultiplier" tube. Within the tube, the applied electrical potentials move the electrons along the paths shown, with collisions increasing their number at each electrode, so that a sizeable pulse is produced as they are finally collected. The scintillation counter is convenient and reliable, but is not widely used in x-ray microanalysis because of its inferior performance for x-rays of low quantum-energy.

Useful semiconducting solid-state detectors are a recent development. The principle of their operation is like the gas counter's: an x-ray photon is absorbed in the solid material; the photoelectron releases other electrons,

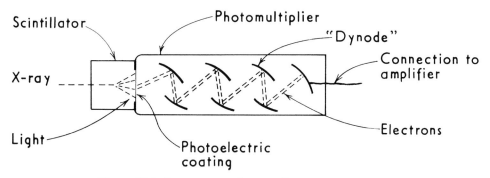

Figure 116. Components of a scintillation x-ray counter.

and an electrical pulse is produced by the movement of the liberated charges under an applied electrical potential.

A gas counter does not produce a pulse for every photon incident upon its window. A photon of high energy may traverse the counter without being absorbed in the gas, while a photon of low energy may be absorbed in the window and never reach the gas. Semiconductor and scintillation counters are more suitable for photons of quite high energy (for example, more than 100 keV). However, one can efficiently count even the most energetic characteristic photons generally encountered in x-ray microanalysis by filling a gas counter with a gas of high absorption cross section; and low-energy photons can be efficiently counted by means of special techniques which permit the use of extremely thin windows. Consequently, none of the three types of counter has a decisive advantage with respect to simple detection, and the choice of the best type for a given application depends largely on their capabilities to *discriminate* between photons of different energies. Discrimination is part of the subject of x-ray spectroscopy, which is discussed in the next section.

A.7. X-RAY SPECTROSCOPY

The x-ray transmission spectrum and the x-ray emission spectrum of a specimen may both be used for chemical analysis. We shall consider first the general features of the method, and then the nature of the instrumentation.

In principle x-ray spectroscopy is quite similar to the spectroscopy of ultraviolet and visible light, but important differences follow from the fact that the UV and visible radiations arise from transitions between outer electronic orbits, whereas the characteristic x-rays involve the inner orbits. Fortunately for x-ray analysts, the energy levels of the inner electrons are relatively few in number, and they are related to atomic number in a simple way.

X-ray spectroscopy is especially suitable for local chemical *elemental* analysis *in situ* for these reasons:

1. The energy levels of the inner electrons, and hence the x-ray spectra of the elements, are almost independent of the state of chemical binding. X-ray measurements *in situ* therefore indicate the total local concentration of an element regardless of its chemical state or states.*

2. In x-ray transmission spectroscopy, there is no need to use colorimetric complexing reagents with attendant destruction or, at the least, an indeterminable modification of the specimen.

3. X-ray emission spectra may be excited nondestructively, in contrast to the ashing procedures and the arc- or spark-excitation used in the emission spectroscopy of UV and visible light.

In x-ray emission spectroscopy, one studies the wave-length distribution and intensities of the x-rays which are generated within specimens. The emission spectra of the constituents of a specimen are excited by the incidence of fast electrons or other x-rays as already discussed (Sec. A.3). For the excitation of a particular characteristic line in any element, it is necessary only that the energy of the incident electron or photon should exceed the energy of the corresponding absorption edge. The characteristic quantum energies depend on the atomic number Z in a conveniently simple way, especially for the K radiations: a rough approximation for the quantum energies of the K_a emission lines is

---

* Chemical state does have a small but significant effect on the wave lengths of the K x-rays of elements of low atomic number (approx. $Z \leqq 15$), and on the L radiations from somewhat heavier elements. This effect, called "chemical shift," may complicate analysis for elements of low atomic number, especially when a spectroscope of high resolving power is used, but at the same time it opens up possibilities for the identification of chemical states *in situ*.

$$V_{K\alpha} \text{ (in } keV) \cong \left(\frac{Z - 0.6}{10}\right)^2 \qquad \text{(A-15)}$$

Some actual quantum energies and wave lengths of the $K_{\alpha 1}$ and $L_{\alpha 1}$ emission lines and the K and L_1 absorption edges are plotted in Figure 117 for a wide range of atomic numbers.

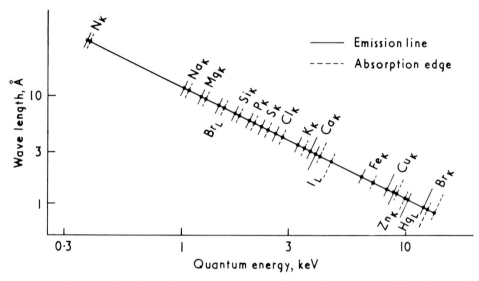

Figure 117. Wave lengths and quantum energies of some important emission lines and absorption edges. Both scales logarithmic. Only K and L_1 absorption edges, and K_α and $L_{\alpha 1}$ emission lines, are shown.

A specimen containing several elements may yield an x-ray emission spectrum similar in form to the curves obtained in other types of spectroscopy. Figure 118 is an example. The qualitative identification of the major constituents is obviously straightforward. Quantitative analysis by x-ray emission is widely practised, but for accurate work the method may be complicated (cf. Secs. 4.D.3 and 5.E.7).

In x-ray transmission spectroscopy, specimens are irradiated with x-rays and one studies the wavelength dependence of the transmission of the impinging radiation.* This dependence is not like the pattern observed with ultraviolet or visible light because x-ray absorption "lines" do not exist. In-

---

* As long as scattering and fluorescence are negligible, the intensities of the incident radiation I_o, the transmitted radiation I_t and the absorbed radiation I_a are related simply by $I_o = I_t + I_a$; the absorption, $A = I_a/I_o$ and the transmission, $T = I_t/I_o$, are related simply by $A = I - T$, and it is immaterial whether one talks of "absorption" or of "transmission" spectroscopy. Although "absorption" is the more common terminology, we have chosen to use "transmission spectroscopy" because the quantity actually measured is not absorption but transmission, and the published spectra are almost always graphs of the wave-length dependence of transmission, not of absorption.

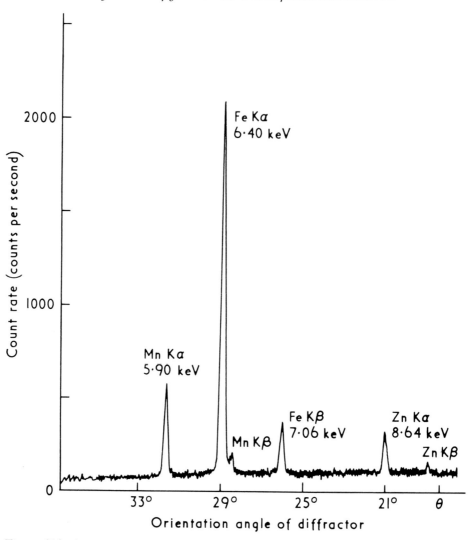

Figure 118. An x-ray emission spectrum from a manganese-zinc ferrite. Spectrum obtained by a diffractor, as explained below (cf. Fig. 123 and Equation A-16). Each orientation angle of the diffractor selects a particular wave length in accord with Equation A-16.

stead, the transmission spectrum of a specimen exhibits "edges" conforming to the curve of Figure 113. Figure 119 is an example. Whereas an emission spectrum like Figure 118 permits a relatively simple quantitative analysis because the signal at a "peak" wavelength may be attributed virtually entirely to one element, in a transmission spectrum like Figure 119 such an attribution is usually not possible. Nevertheless, x-ray transmission spectroscopy can provide both qualitative and quantitative analysis through the so-called "straddling method" which is discussed in Section 5.A.3. As

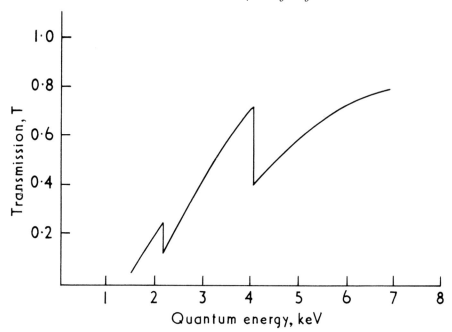

Figure 119. A transmission spectrum. Calculated for a hypothetical specimen, 3 mg/cm², 50% $Ca_3 (PO_4)_2$ and 50% protein.

shown there, in favorable cases this method may be practised without specifically spectroscopic apparatus, and it is not even necessary to record a full spectrum; two points of the transmission curve may suffice for the assay of a single element.

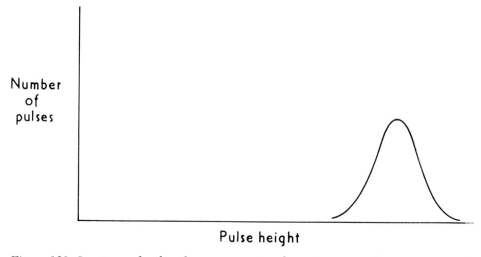

Figure 120. Spectrum of pulses from a proportional counter exposed to *monoenergetic* photons.

For refined x-ray transmission spectroscopy and for x-ray emission spectroscopy, one needs spectroscopic devices which can selectively respond to a particular characteristic radiation while discriminating against other characteristic lines and against the x-ray continuum. We have already noted that the gas proportional counter can provide some discrimination since photons of a given energy produce electrical pulses of proportional average height. However, the capacity of the proportional counter to distinguish between photons of different energies is quite limited. When a counter is exposed to *monoenergetic* photons and one records the numbers of pulses of various heights in a histogram, a "pulse-height distribution" like Figure 120 is obtained. The heights show a considerable spread about the average, largely because there is a random division of energy between ioniza-

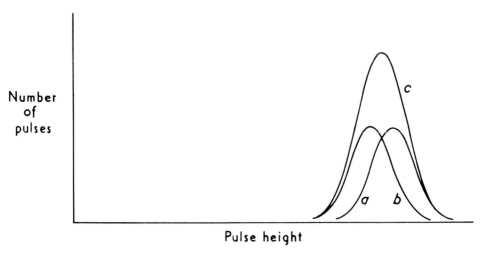

Figure 121. Overlapping pulse-height spectra. Curve (a): Produced by zinc K_a photons (Z = 30). Curve (b): Produced by copper K_a photons (Z = 29). Curve (c): Produced by an equal mixture of (a) and (b).

tion and competing processes in the counter gas, so that there is a statistical fluctuation in the total number of electrons released per photon. Figure 121 shows the pulse-height distributions for the K radiations from two elements of neighboring atomic numbers. The overlap obviously makes it impossible to accept the pulses from one element while rejecting completely the pulses from a neighboring element.[*] It is also clear that a proportional counter system inevitably responds to a fairly wide energy-band of the x-ray continuum.

Scintillation counters do not offer superior performance in x-ray microscopy. Intrinsically inferior to the proportional counter in spread of pulse

[*] Nevertheless, with suitable calibration it is possible to "unscramble" the pulses from a proportional counter so as to determine accurately the number of photons of each type in a mixture of characteristic radiations from two neighboring elements. See Sec. 5.D.2.

heights, they have been valuable as discriminators only at higher quantum
energies (approx. ≥ 50 keV), where statistical fluctuations in pulse height
are not so detrimental and it is difficult to provide efficient proportional
counters.

In semiconducting solid-state counters, the number of primary electrons
released per photon is greater than in gas proportional counters. The consequent reduction in the statistical spread of pulse heights implies a po-
tentially superior quantum-energy discrimination. But the noise level of
the associated amplifiers—a constant background independent of quantum
energy—sets a limit to the discrimination actually achieved. Over the past

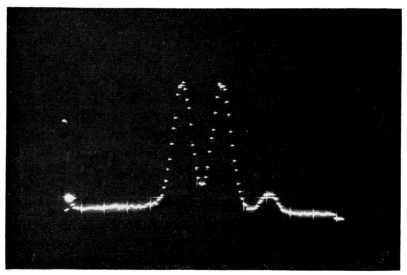

Figure 122. Pulse-height spectra obtained with a cooled silicon semiconducting detector.
The main peaks were produced by the K_α radiations of nickel ($Z = 28$) and copper
($Z = 29$); the small peak at the right was due to copper K_β radiation. There is little
overlap of the pulse-height spectra.

few years the noise levels have been remarkably reduced with a corresponding increase in the applicability of these detectors, and continuing
improvement is anticipated. The net result at present (July, 1971) is that
the semiconductors have superior discrimination at high quantum energies,
where they are displacing the scintillators, and their discrimination is already superior to the gas counters' through most of the range encountered
in x-ray microscopy, down to quantum energies of approximately one keV.
Figure 122 illustrates the performance already available a little while ago.
The K x-rays of adjacent elements are obviously much more "separable"
than with a gas counter (cf. Fig. 121). While a number of technical problems remain to be solved before the semiconductors can be fully exploited
in x-ray microscopy, it is clear that these new detectors have already con-

siderably extended the capability for x-ray spectroscopy based entirely on pulse counters.

When the concentration of a chemical element in a specimen is too low, a pulse counter is unable to pick out its characteristic photons against the background of more intense characteristic radiations and continuum. Superior discrimination may then be obtained with diffracting crystals.

Figure 123 depicts the action of a diffracting crystal. We suppose that the crystal contains planes of scattering centers separated by the distance *d*, and is exposed to a beam of x-rays of wave length λ and sharply defined direction θ. It can be shown, for any distant point of observation *P* located

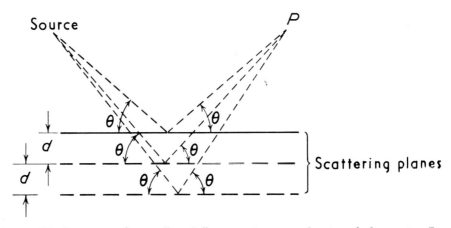

Figure 123. Geometry of crystalline diffraction. Source and point of observation *P* are sufficiently remote so that all angles "θ" can be virtually equal.

on the "reflection" of the direction of incidence, that the radiation scattered from the different centers will all arrive in phase if

$$2d \sin \theta = n\lambda \qquad\qquad \text{(A–16)}$$

where *n* is an integer. Equation A-16 is the famous Bragg condition.

If, now, the incident radiation of sharply defined direction contains a wide assortment of wave lengths, the radiation satisfying (A-16) will appear with a high intensity at *P* while radiation of other wave lengths will be very weak there. With a good crystal, nearly 100 percent of the radiation satisfying (A-16) is diffracted. Other radiation may reach *P* by ordinary scattering from one or another atom, but the background intensity from this process is very low.

A diffracting crystal is incorporated into a system suitable for x-ray microscopy as illustrated in Figure 6 (Sec. 1.B) and Fig. 97 (Sec. 5.E). Some of the x-rays from the specimen strike the crystal; radiation satisfying (A-16) is diffracted to the detector and counted. In order to select different wave lengths at will, there must be a mechanical arrangement which enables one to set the crystal to the angle required by (A-16) and which simultaneously moves the detector into the path of the diffracted rays.

In x-ray microanalysis, the x-rays usually issue from a small region, virtually a point source. The radiation from a point cannot strike all areas of a plane at the same angle. For a flat crystal, the angle could be sufficiently constant to satisfy (A-16) only if the crystal were small or far from the source. The sensitivity would then be poor. Therefore suitably curved crystals must be used, enabling the Bragg condition to be satisfied for a much larger fraction of the characteristic radiation coming from the specimen.

Equation A-16 sets a limit to the wave lengths which can be analyzed with a particular set of planes in a given crystal; since $n \geq 1$ and $sin\ \theta \leq 1$, the wave length must not exceed $2d$. Until recently, diffractors with adequate spacing were not available for the characteristic radiations of long wave length from elements of atomic number $Z < 11$. Now synthetic "crystals" (really multilayered films), most often barium or lead stearate, are standard and have extended the range to $Z \geq 5$. Special diffraction gratings also are being intensively developed for the K radiations of elements $Z \geq 4$. The gratings are just becoming standard at the time of writing and may prove superior for the low-Z radiations (Davidson *et al.*, 1970). There is no difficulty in the analysis of the radiations from elements of very *high* atomic number; while the utilization of K radiation of very short wave length is awkward for various reasons, one can always observe the L or M radiations instead. The net result is that diffractors are now available for the routine analysis of characteristic x-rays from all elements of atomic number $Z \geq 5$.

In order to decide whether or not to use a diffractor for a given analytical problem, one must weigh two dominant considerations against each other: wave-length selectivity, and intensity. Although diffractors are not perfectly selective,* they cleanly separate the characteristic radiations of neighboring elements and are vastly more disciminating than pulse counters. On the other hand, diffractors are impractical when the x-ray intensity is quite low. Because of the stringent requirement concerning angle of incidence, it is difficult to provide a diffractor which intercepts and passes on more than 0.1% of the monoenergetic radiation from a point source; the available percentage is much less in most diffracting systems. In contrast, it is usually simple to put the window of a proportional counter close enough to intercept 1% or more of the radiation. Figure 124, taken

---

* There are several reasons why a crystal is not perfectly "sharp": Basic theory shows that even a perfect crystal of finite size is not perfectly discriminating; real crystals have defects (atoms out of place, and entire regions slightly out of alignment with each other); and, most importantly, in all practical designs, radiation from a point source does not strike all parts of the crystal at exactly the same angle. Hence, if Equation A-16 is satisfied for one wave length at one point on the crystal face, it will also be satisfied for a slightly different wave length at some other point.

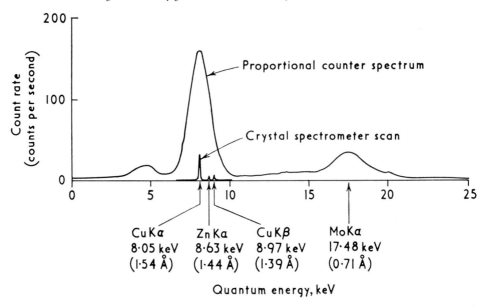

Figure 124. Comparison of x-ray fluorescence spectra taken with a gas proportional counter spectrometer (nondiffractive) and with a crystal spectrometer (diffractive). (From Zeitz and Lee, 1966. Courtesy of Academic Press.)

from a study by Zeitz and Lee (1966), gives a vivid comparison of the collecting power and the wavelength selectivity of a diffracting and a non-diffracting spectrometer in one particular x-ray fluorescence instrument, where the detector is a gas proportional counter.

The wave-length discrimination of a nondiffractive spectrometer can be greatly improved by the use of a solid-state detector rather than a gas counter. In standard electron microprobe analyzers (Sec. 5.E), it is usually difficult to fit the solid detector close to the specimen, so that counting rates are likely to be little improved over those from high-quality diffracting spectrometers (cf. Kimoto *et al.*, 1970), and the main advantage of the solid-state nondiffractive system is in producing all of the elemental peaks simultaneously. However, the solid detectors may be fitted close to the specimen in most commercial scanning electron microscopes, to give a big gain in intensity with a wave-length discrimination intermediate between the performance of diffracting crystals and gas counters. The potential of this instrumental arrangement has been analyzed recently by Sutfin and Ogilvie (1970), and reassessed for still more modern solid-state detectors by Russ (1971).

In the literature, x-ray spectroscopic systems which employ a diffractor are referred to as "dispersive," and systems in which the radiation goes directly from the specimen to a pulse counter are called "nondispersive" (or else "energy dispersive" since the discrimination is based on quantum energy). The term dispersive is appropriate in optical spectroscopy where the analyzing components, gratings or prisms, diffract or refract all of the incident radiation simultaneously, sending each constituent wave length off in its own direction. But the term is not appropriate when diffracting crystals are used to analyze x-rays, because a three-dimensional periodic structure does not simultaneously diffract a continuum of radiation into different directions; it selectively diffracts the very restricted wave lengths satisfying Equation A-16 while the remainder of the radiation is mainly absorbed or else scattered indiscriminately in all directions. We shall therefore not use the terms "dispersive" and "nondispersive," but shall refer to diffractive and nondiffractive x-ray spectroscopy.

A.8. BIBLIOGRAPHY

We hope that the appendix provides all of the x-ray physics needed for this book. However, for the reader who wants a fuller exposition of one or another topic, a bibliography is listed at this point.

General Texts

Clark, G. L. (1955): *Applied X-Rays*, 4th ed. New York, McGraw-Hill.

Clark, G. L. (Ed.) (1963): *Encyclopedia of X-Rays and Gamma Rays*. New York, Reinhold.

Compton, A. H., and Allison, S. K. (1935): *X-Rays in Theory and Experiment*, 2nd ed. New York, D. van Nostrand.

Cosslett, V. E., and Nixon, W. C. (1960): *X-Ray Microscopy*. Cambridge (England), Cambridge University Press.

Kaelble, E. F. (Ed.) (1967): *Handbook of X-Rays*. New York, McGraw-Hill.

Specialized References

Microradiography

Engström, A. (1966): X-Ray Microscopy and X-Ray Absorption. In Pollister, A. W. (Ed.): *Physical Techniques in Biological Research*. Cells and Tissues. New York, Academic Press, Vol. IIIA, pp. 87–171.

Nondiffractive X-ray Spectroscopy With Semiconductor Detectors

ASTM Special Technical Publication No. 485 (1971). Energy Dispersion X-Ray Analysis: X-Ray and Electron Probe Analysis (J. C. Russ, Ed.). Philadelphia, American Society for Testing and Materials.

Electron Microprobe Analysis

Hall, T. A. (1971): The microprobe assay of chemical elements, In Oster, G. (Ed.): *Physical Techniques in Biological Research*, 2nd ed. Optical Techniques. New York, Academic Press, Vol. 1A, pp. 157–275.

REFERENCES

Amprino, R., and Camanni, F. (1956): Historadiographic and autoradiographic researches on hard dental tissues. *Acta Anat, 28:* 217–258.

Amprino, R., and Engström, A. (1952): Studies on x-ray absorption and diffraction of bone tissue. *Acta Anat, 15:*1–22.

Amrehn, H. (1956): Energieverteilung im Spektrum der Röntgen-Bremsstrahlung dünner Antikathoden in Abhängigkeit von der Ordnungszahl. II. *Z Physik, 144:*529–537.

Amrehn, H., and Kulenkampff, H. (1955): Energieverteilung im Spektrum der Röntgen-Bremsstrahlung dünner Antikathoden im Abhängigkcit von Ordnungszahl und Spannung. *Z Physik, 140:*452–464.

Andersen, C. A. (1967a): An introduction to the electron probe microanalyzer and its application to biochemistry. *Methods of Biochemical Analysis.* New York, Interscience, vol. 15., pp. 147–270.

Andersen, C. A. (1967b): The quality of x-ray microanalysis in the ultra-soft x-ray region. *Br J Appl Phys, 18:*1033–1043.

Anderton, H. (1966a): An X-ray image intensification system for use with a point projection x-ray microscope. *Advances in Electronics and Electron Physics,* New York, Academic Press, vol. 22B, pp. 919–925.

Anderton, H. (1966b): Ph.D. Thesis, Cambridge, England, University of Cambridge.

Anderton, H. (1967): Point-projection x-ray microscopy. *Sci Progr (Oxford),* 55:337–356.

Angervall, L., Hansson, G., and Röckert, H. (1960): Pulmonary siderosis in electrical welder. *Acta Pathol Microbiol Scand, 49:*373–380.

Applebaum, E. (1932): Incipient dental caries. *J Dent Res, 12:*619–627.

Applebaum, E. (1938): Grenz ray studies of the calcification of enamel. *J Dent Res, 17:*181–190.

Applebaum, E. (1948a): Tissue changes in dental caries. *Ann Dent, 7:*1–14.

Applebaum, E. (1948b): Grenz ray and histologic studies of tissue changes in dental caries. *Ann Dent, 7:*67–90.

Applebaum, E., Hollander, F., and Bodecker, C. F. (1933): Normal and pathological variations in calcification of teeth as shown by the use of soft x-rays. *Dent Cosmos, 75:*1097–1105.

Appleton, T. C. (1970): Towards electron microscope autoradiography of diffusible substances. *Proc Roy Micr Soc, 5:*190.

Appleton, T. C. (1971): Dry ultrathin frozen sections for electron microscopy: The cryostat approach. *J Microsc* (in press).

Atkinson, H. F. (1952): Ground sections of enamel. *Br Dent Ann, 1:*144–149.

Bahr, G. F., Johnson, F. B., and Zeitler, E. (1965): The elementary composition of organic objects after electron irradiation. *Lab Invest, 14*:1115–1133.

Banfield, W. G., Tousimis, A. J., Hagerty, J. C., and Padden, T. R. (1969): Electron probe analysis of human lung tissues. In Earle, K. M., and Tousimis, A. J. (Eds.): *Progress in Analytical Chemistry.* New York, Plenum Press, vol. 3, pp. 9–34.

Barclay, A. E. (1947): Microarteriography. *Br J Radiol, 20*:394–404.

Barclay, A. E. (1948): Microarteriography. *Am J Roentgenol, 60*:1–12.

Barclay, A. E. (1949): The vascularisation of the human stomach. A preliminary note on the shunting effect of trauma. *Br J Radiol, 22*:62–67 (also in *Gastroenterology, 12*:177–183).

Barclay, A. E. (1951): *Microarteriography.* Oxford, England, Blackwell, and Springfield, Thomas, 102 pp.

Baud, C. A., and Lobjoie, D. P. (1966a): Biophysical investigations on the mineral phase in the superficial layers of human dental enamel. *Helv Odontol Acta, 10*:40–46.

Baud, C. A., and Lobjoie, D. P. (1966b): L'emploi du radioanalyseur a microsonde électronique pour l'étude de la distribution du fluor dans les coupes d'émail dentaire. *Ann Histochim, 11*:151–156.

Baud, C. A., Kimoto, S., and Hashimoto, H. (1963): Etude de la distribution du calcium dans l'os haversien avec le radioanalyseur à microsonde électronique. *Experientia, 19*:525.

Baud, C. A., Bang, S., Lee, H. S., and Baud, J.Ph. (1968): X-ray studies of strontium incorporation into bone mineral *in vivo. Calcif Tissue Res, 2* (*Suppl.*): 6.

Bax, D., and van der Linden, L. W. J. (1969): Electron probe microanalysis of filled human teeth. In Mollenstedt, G. and Gaukler, K. H. (Eds.): *X-Ray Optics and Microanalysis.* Berlin, Heidelberg and New York, Springer-Verlag, pp. 597–600.

Beaman, D. R., Nishiyama, R. H., and Penner, J. A. (1969): The analysis of blood disease with the electron microprobe. *Blood, 34*:401–413.

Belanger, L. F., Robichon, J., and Belanger, C. (1963): Mounting hard tissues for microradiography. *Experientia, 19*:163–164.

Bell, M. A. (1969): The technic of x-ray histochemistry applied to neurohistology. *Experientia, 25*:837–841.

Bellman, S. (1953): Microangiography. *Acta Radiol* (Suppl. 102), pp. 104.

Bellman, S., and Adams-Ray, J. (1956): Vascular reaction after experimental cold injury. A microangiographic study on rabbit ears. *Angiology, 7*:339–367.

Bellman, S., and Engfeldt, B. (1955a): Kidney lesions in experimental hypervitaminosis D; a microradiographic and microangiographic study. *Am J Roentgenol, 74*:288–294.

Bellman, S., and Engfeldt, B. (1955b): A microangiographic study on the effect of gronadotrophin upon the blood vessel system of the ovaries of rabbits. *Acta Radiol, 43*:459–468.

Bellman, S., and Frank, H. A. (1958): Intercoronary collaterals in normal hearts. *J Thorac Cardiovasc Surg, 36*:584–603.

Bellman, S., and Oden, B. (1957): Experimental micro-lymphangiography. *Acta Radiol* (Stockh.) *47*:289–307.

Bellman, S., and Oden, B. (1958–59): Regeneration of surgically divided lymph vessels: An experimental study on the rabbit's ear. *Acta Chir Scand, 116*:99–117.

Bellman, S., and Strombeck, J. O. (1960): Transformation of the vascular system in cold-injured tissue in the rabbit's ear. *Angiology, 11*:108–125.

Bellman, S., and Velander, E. (1959): Vascular transformation in experimental tubed pedicles. *Br J Plast Surg, 12*:1–21.

Bellman, S., Block, E., and Odeblad, E. (1953): A microangiographic study of the minute ovarian blood vessels in albino rats. *Br J Radiol*, 26:584–588.

Bellman, S., Frank, H. A., Lambert, P. B., and Roy, A. J. (1959): Studies of collateral
vascular responses. *Angiology*, 10:214–232.

Bergman, G., Hammarlund-Essler, E., and Lysell, L. (1958): Microradiographic study
of caries in deciduous teeth. *Acta Odontol Scand*, 16:113–126.

Bergman, G., Linden, L-A., and Röckert, H. (1966): An attempt to analyse the enamel
fluid. *Adv Fluorine Res Dental Caries Prevention*, 4:163.

Berkley, C., Langer, A. M., and Baden, V. (1967): Instrumental analysis of inspired
fibrous pulmonary particulates. *Trans N Y Acad Sci* Series II, 30 (No. 2):331–350.

Besic, F. C., Knowles, C. R., Wiemann, M. R., and Keller, O. (1969): Electron probe
microanalysis of noncarious enamel and dentin and calcified tissues in mottled teeth.
J Dental Res, 48:131–139.

Bessen, I. I. (1957): A high resolution PMR x-ray microscope with electron microscope
conversion. *Norelco Reporter*, 4:119–123.

Bethe, H. A. (1933): *Handbuch der Physik*, 24:491–523.

Bettencourt, J. M., and Mirabeau-Cruz, J. (1963): La circulación hepatica normal y
patologica. Madrid, Libraria Cientifica Medica Española, 329 pp.

Birks, L. S. (1963): *Electron Probe Microanalysis*. New York, Interscience.

Bishop, H. (1965): Ph.D. Thesis, Cambridge, England, University of Cambridge.

Björkerud, S. (1966): Enzymaktivitetsmätningar i human arteriell intima. *Nord Med*,
75:722.

Björkerud, S., and Rosengren, B. (1966): The determination of succinate oxidation and
mass of micro samples of human arterial intima. *J Atheroscler Res*, 6:195.

Blackwood, H. J. J. (1965): Cell differentiation in the mandibular condyle of the rat
and man. In *Proceedings of the Second European Symposium on Calcified Tissues*.
Collection des Colloques de l'Université di Liège, pp. 23–29.

Bohatirchuk, F. (1942): Die Fragen der Microröntgenographie. *Fortschr Röntgenstr*,
65:253–261.

Bohatirchuk, F. (1953): Historoentgenography (microradiography) of impregnated
cancer tissue. *Am J. Roentgenol*, 70:119–125.

Bohatirchuk, F. (1954): Some microradiographical data on bone aging. *Br J Radiol*,
27:177–182.

Bohatirchuk, F. (1961): Medico-biologic research by microradiography. In *The Encyclopedia of Microscopy*. New York, Reinhold. pp. 591–627.

Bohatirchuk, F. (1963): Microradiology of mammalian bone. *J Can Ass Radiol*, 14:29–
38.

Bohatirchuk, F. (1966): Calciolysis as the initial stage of bone resorption. A stain
historadiographic study. *Am J Med*, 41:836–846.

Bohr, H., and Dollerup, E. (1960): Microradiographic investigations on bones in
osteomalacia and rickets. Engström, A., Cosslett, V. E., and Pattee H., Jr.: In *X-Ray
Microscopy and X-Ray Microanalysis*. Amsterdam and New York, Elsevier, pp. 184–
190.

Borom, M. P., and Hanneman, R. E. (1967): Local compositional changes in alkali
silicate glasses during electron microprobe analysis. *J Appl Physics*, 38:2406–2407.

Boyde, A. (1967): A single-stage carbon replica method and some related techniques
for the analysis of the electron-microscope image. *J Roy Microscop Soc*, 86:359–370.

Boyde, A., and Switsur, V. R. (1963): Problems associated with the preparation of biological specimens for microanalysis. Pattee, H. H., Jr., Cosslett, V. E., and Engström,

A.: In *X-Ray Optics and X-Ray Microanalysis*. New York, Academic Press, pp. 499–506.

Boyde, A., and Wood, C. (1969): Preparation of animal tissues for surface-scanning electron microscopy. *J Microsc, 90* (Part 3): 221–249.

Boyde, A., Switsur, V. R., and Fearnhead, R. W. (1961): Application of the scanning electron-probe X-ray microanalyzer to dental tissues. *J Ultrastruct Res, 5:*201–207.

Boyde, A., Switsur, V. R., and Stewart, A. D. G. (1963): An assessment of two new physical methods applied to the study of dental tissues. *Proc. 9th ORCA Cong.* New York, Pergamon Press, pp. 185–193.

Brash, J. C. (1955): Neuro-vascular hila of limb muscles. Edinburgh, Livingstone, 100 pp.

Brattgård, S. O., and Hydén, H. (1954): The composition of the nerve cell studied with new methods. *Int Rev Cytol, 3:*455–476.

Brattgård, S. O., and Hydén, H. (1952): Mass, lipids, pentose, nucleoproteins and proteins determined in nerve cells by x-ray microradiography. *Acta Radiol (Stockh) (Suppl),* 94:44 pp.

Brattgård, S. O., Hallén, O., and Hydén, H. (1953): A critical evaluation of x-ray microradiography. *Biochim Biophys Acta, 10:*486–487.

Breine, U., Johansson, B., and Röckert, H. (1965): Human autogenous bone grafts. II. Studies of the fate of grafts to the hard palate in clefts by using tetracycline ultraviolet fluorescence technique and x-ray microscopy. *Acta Chir Scand, 129:*250.

Broek, S. L. van den (1957): A simple contact microradiography unit with sealed-off x-ray tube. In Cosslett, V. E., Engström, A., and Pattee, H., Jr.: *X-Ray Microscopy and Microradiography*. New York, Academic Press, pp. 64–71.

Brooks, E. J., Tousimis, A. J., and Birks, L. S. (1962): The distribution of calcium in the epiphyseal cartilage of the rat tibia measured with the electron probe x-ray microanalyzer. *J Ultrastruct Res, 7:*56–60.

Brown, J., Felber, F., Richards, J. and Saxon, D. (1948): On making thin Nylon films. *Rev Sci Instrum, 19:*818.

Brown, A. C., Gerdes, R. J., and Johnson, J. (1971): S.E.M. and microprobe analysis of congenital hair defects. In Johari, O. (Ed.): *Proc. 4th Annual SEM Symposium.* Part I, pp. 369–376. Chicago, Illinois Institute of Technology Research Institute.

Burch, G. J. (1896): Recent researches on Röntgen rays. Plant structure revealed by Röntgen rays. *Nature (Lond), 54:*109 and 111.

Burhop. E. H. S. (1952): *The Auger Effect and Other Radiationless Transitions* (Sec. 3.7). Cambridge, England, Cambridge University Press.

Burny, F., Wollast, R., Balimont, P., and de Marnette, R. (1968): Etude de la minéralisation osseuse à l'échelle de la microstructure au moyen de la microsonde électronique. *Calcif Tissue Res, (Suppl., August):*3.

C and N: See Cosslett, V. E., and Nixon, W. C. (1960).

Campbell, A. J. (1963): K x-ray yields from elements of low atomic number. *Proc R Soc A, 274:*319–342.

Campbell, J., and Pennefather, C. M. (1919): An investigation into the blood supply of muscle, with special reference to war surgery. *Lancet, I.:*294–296.

Carlisle, E. M. (1970): Silicon: A possible factor in bone calcification. *Science, 167:*279–280.

Carlson, L. (1957): Evaluation of microradiograms for dry weight determination. *Exp Cell Res (Suppl), 4:*193–196.

Carroll, K. G., and Tullis, J. I. (1968): Observations on the presence of titanium and zinc in human leukocytes. *Nature, 217:*1172–1173.

Carroll, K. G., Spinelli, F. R., and Goyer, R. A. (1970): Electron probe microanalyzer localization of lead in kidney tissue of poisoned rats. *Nature*, 227:1056.

Carroll, K. G., Mulhern, J. E., and O'Brien, V. L. (1971): Microprobe analysis of localized concentrations of metals in various human tissues. *Oncology*, 25:11–18.

Castaing, R., and Pichoir, F. (1966): Analyse non dispersive des éléments très légers. *In* Castaing, R., Deschamps, P., and Philibert, J. (Eds.): *X-Ray Optics and Microanalysis*. Paris, Hermann, pp. 454–461.

Charles, A. (1966): Preparation of films with holes for electron microscopy. *Nature*, 212:106.

Christensen, K. (1969): A way to prepare frozen thin sections of fresh tissue for electron microscopy. In Roth, L. J., and Stumpf, W. E. (Eds.): Autoradiography of Diffusible Substances. New York, Academic Press, pp. 349–362.

Churchill-Davidson, I. (1964): The use and effects of high-pressure oxygen in radiotherapy. Clinical application of hyperbaric oxygen. Boerema, Elsevier, 140 pp.

Clark, G. L. (1947): Medical, biological and industrial applications of monochromatic radiography and microradiography. *Radiology*, 49:483–495.

Clark, G. L. (1955): Microradiography. In: *Applied X Rays*, 4th ed. London and New York, McGraw-Hill, pp. 238–262.

Clarke, J. A., Salsbury, A. J., and Wolloughby, D. A. (1970): Application of electron probe microanalysis and electron microscopy to the transfer of antigenic material. *Nature*, 227:69–71.

Collette, J. M. (1955): Microradiography. Foundations and principles of the technique. A new application—microlymphangiography. *J Belge Radiol*, 36:293–312.

Combée, B. (1955): An x-ray tube for microradiography. *Philips Tech Rev*, 17:45–46.

Combée, B., and Engström, A. (1954): A new device for microradiography and a simplified technique for the determination of the mass of cytological structures. *Biochim Biophys Acta*, 14:432–434.

Combée, B., and Recourt, A. (1957): A simple apparatus for contact microradiography between 1.5 and 5 kV. *Philips Tech Rev*, 19:221–233.

Compton, A. H., and Allison, S. K. (1935): *X Rays in Theory and Experiment*. New York, Van Nostrand, 828 pp.

Cordier, A., Durou, C., and Verdier, P. (1965): Microradiographie par projection. *J Microsc 4*:573–586.

Cosslett, V. E. (1959): The comparative merits of different methods of microradiography. *Appl Sci Res*, 7:338–344.

Cosslett, V. E. (1965): X-ray microscopy. *Rep Progr Physics*, 28:381–410.

Cosslett, V. E., and Nixon, W. C. (1952): An experimental x-ray shadow microscope. *Proc R Soc B, 140*:422–431.

Cosslett, V. E., and Nixon, W. C. (1960): *Cambridge Monographs on Physics: X-Ray Microscopy*. Cambridge, England, Cambridge University Press, 406 pp.

Cosslett, V. E., and Switsur, V. R. (1963): Some biological applications of the scanning microanalyzer. In Pattee, H. H., Jr., Cosslett, V. E., and Engström, A. *X-Ray Optics and X-Ray Microanalysis*. New York, Academic Press, pp. 507–512.

Crewe, A. V., Eggenberger, D. N., Wall, J., and Welter, L. M. (1968): Electron gun using a field emission source. *Rev Sci Instrum*, 39:576–583.

Culling, C. F. A. (1963): *Handbook of Histopathological Technic*, 2nd ed. London, Butterworth, 553 pp.

Cunningham, G. J. (1960): Microradiography. In *Tools of Biological Research*, 2nd Series. Oxford, Blackwell, pp. 155–175.

Daniel, P. M., and Pritchard, M. M. L. (1951): Variations in the circulation of the portal venous blood within the liver. *J Physiol, 114:*521–537.

Dauvillier, A. (1927): An x-ray tube that produces x-rays with an effective wavelength of 8 Å. *C R Acad Sci (Paris), 185:*1460–1462.

Dauvillier, A. (1930): The production of high definition microradiographs. *C R Acad Sci (Paris), 190:*1287.

Davidson, E., Hartwick, A. J., and Taylor, J. M. (1970): Expanded wave-length coverage with digitally controlled x-ray spectrometers. *Proceedings of the National Conference of Electron Microprobe Analysis,* 5th. New York City, 1970. Paper 42.

Dietrich, J. (1960): Application de la microradiographie par contact a l'étude de la cellule végétale. In Engström, A., Cosslett, V. E., and Pattee, H. H., Jr.: X-Ray Microscopy and Microanalysis." Amsterdam and New York, Elsevier, pp. 306–310.

Dietrich, J. (1966): La microradiographie par contact, méthode d'analyse cytologique. In Castaing, R., Deschamps, P., Philibert, J.: *X-Ray Optics and Microanalysis.* Paris, Hermann, pp. 658–663.

Dolby, R. M. (1961): Ph.D. Thesis. Cambridge, England, University of Cambridge.

Dolby, R. M. (1963): X-ray microanalysis of the light elements. *J Sci Instrum, 40:*345–351.

Doran, F. S. A. (1951): Aetiology of chronic gastric ulcer. Observations on the blood supply of the human gastric mucosa; with a note on the arterio-venous shunt. *Lancet, 260:*199–202 and 635–636.

Dreyfus, F., and Frank, R. M. (1964): Microradiographie et microscopie électronique du cement humain. *Bull Group Int Rech Sci Stomatol, 7:*167–181.

Dreyfus, F., Frank, R. M., and Gutmann, B. (1964): La sclérose dentinaire. *Bull Group Int Rech Sci Stomatol, 7:*207–229.

Duncumb, P. (1959): The x-ray scanning microanalyzer. *Br J Appl Physics, 10:*420–427.

Duncumb, P. (1966): Precipitation studies with Emma—A combined electron microscope and x-ray microanalyzer. In McKinley, T. D., Heinrich, K. F. J., and Wittry, D. B. (Eds.) *The Electron Microprobe.* New York, Wiley, pp. 490–499.

Dyson, N. A. (1956): Ph.D. Thesis. Cambridge, England, University of Cambridge. The absorption tables in this work have been reproduced in Cosslett and Nixon (1960), p. 372.

Eastman Kodak Co. (1955): *Bibliography of Microradiography and Soft X-ray Radiography.* Prepared by the X-ray Division of Kodak Research Laboratories. Eastman Kodak Co., Rochester, New York.

Echlin, P. (1971): The examination of biological material at low temperatures. In *Scanning Electron Microscopy/1971.* Part I, pp. 225–232. Proceedings 4th Annual Scanning Electron Microscope Symposium, IIT Research Institute, Chicago, Ill.

Ehrenberg, W. (1947): X-ray optics. *Nature (London), 160:*330.

Ehrenberg, W., and Spear, W. E. (1951): An electrostatic focusing system and its application to a fine focus x-ray tube. *Proc Phys Soc, 64B:*67–75.

Ely, R. V. (1963): *Micro X-Radiography and Analysis (1913–1962).* Guildford, Surrey, England, Micro X-Ray Consultants (G.B.), 122 pp.

Engel, W. K., Resnick, J. S., and Martin, E. (1968): The electron probe in enzyme histochemistry. *J Histochem Cytochem, 16:*273–275.

Engfeldt, B. (1958): Recent observations on bone structure. *J Bone Joint Surg, 40A:*698–706.

Engfeldt, B., and Engström, A. (1954): Biophysical studies on bone tissue. 12. Experimentally produced ectopic bone tissue. *Acta Orthop Scand, 24:*85–100.

Engfeldt, B., Engström, A., Helander, C. G., Wilton, A., and Zetterström, R. (1952): Biophysical studies on bone tissue. I. Paget's disease. *Acta Pathol Microbiol Scand,* 31:256–261.

Engfeldt, B., Bergman, G., and Hammarlund-Essler, E. (1954): Studies on mineralized dental tissues. I. A microradiographic and autoradiographic investigation of teeth and tooth germs of normal dogs. *Exp Cell Res,* 7:381–392.

Engfeldt, B., Bjornerstedt, R., Clemedson, R., and Engström, A. (1954a): A preliminary study of the *in vivo* and *in vitro* uptake of Sr 90 in bone tissue and the osseous localization of radioactive fission products in atomic explosions. *Acta Orthop Scand,* 24:100–114.

Engfeldt, B., Engström, A., and Zetterström, R. (1954b): Biophysical studies of the bone tissue in osteogenesis imperfecta. *J Bone Joint Surg (Br),* 36:654–661.

Engfeldt, B., Engström, A., and Datta, N. (1955): The distribution of dry mass in the walls of normal and pathological blood vessels. *Exp Cell Res (Suppl),* 3:110–116.

Engström, A. (1946): Quantitative micro and histochemical elementary analysis by Roentgen absorption spectrography. *Acta Radiol Stockh (Suppl),* 63:106 pp.

Engström, A. (1950): Use of soft x rays in the assay of biological material. In *Progress in Biophysics.* London, Butterworth, pp. 164–196.

Engström, A. (1951): Note on the cytochemical analysis of elements by Roentgen rays. *Acta Radiol,* 36:393–396.

Engström, A. (1957): Contact microradiography—a general survey. In Cosslett, V. E., Engström, A., and Pattee, H., Jr.: *X-Ray Microscopy and Microradiography.* New York, Academic Press, pp. 24–33.

Engström, A., and Glick, D. (1950): The mass of gastric mucosa cells measured by x-ray absorption. *Science, 111:*379–380.

Engström, A., and Lindström, B. (1950): A method for the determination of the mass of extremely small biological objects. *Biochim Biophys Acta,* 4:351.

Engström, A., and Lundberg, B. (1957): A simple midget x-ray tube for high resolution microradiography. *Exp Cell Res,* 12:198–200.

Engström, A., and Weissbluth, M. (1951): Absorption of x-rays in inhomogeneous histo- and cytological samples. *Exp Cell Res,* 2:711–714.

Engström, A., Greulich, R. C., Henke, B., and Lundberg, B. (1957): High resolution contact microradiography with ultra-soft polychromatic x-rays. In Cosslett, V. E., Engström, A., and Pattee, H., Jr.: *X-Ray Microscopy and Microradiography.* New York, Academic Press, pp. 218–233.

Engström, A., Lundberg, B., and Bergendahl, G. (1957): High resolution microradiography with ultra-soft x-rays. *J Ultrastruct Res,* 1:147–157.

Ericsson, S. G. (1965): Quantitative microradiography of cementum and dentine. *Acta Radiol (Suppl.):*246.

Feindel, W. H., and Perot, P. (1965): Red veins in the brain. *J Neurosurg,* 22:315–325.

Feindel, W. H., Yamamoto, Y. L., and Hodge, C. P. (1967): Intracarotid fluorescein angiography: A new method for examination of the epicerebral circulation in man. *Can Med Assoc J,* 96:1–7.

Fitzgerald, P. J. (1956). Applications of soft x-ray methods to a study of normal and neoplastic cells. *Ann N Y Acad Sci,* 63:1141.

Fitzgerald, P. J. (1957): Changes in mass concentration of the pancreatic acinar cell during degeneration, death and regeneration following ethionine. In Cosslett, V. E., Engström, A., and Pattee, H., Jr.: *X-Ray Microscopy and Microradiography.* New York, Academic Press, pp. 507–519.

Francois, J., Collette, J. M., and Neetens, A. (1955): Microradiographic study of the internal wall of Schlemm's canal. *J Belge Radiol*, 38:1–15.

Frank, R. M., Capitant, M., and Goni, J. (1966): Electron probe studies of human enamel. *J Dent Res*, 45:672–682.

Frazier, P. D. (1966): Electron probe analysis of human teeth: Some problems in sample preparation. *Norelco Reporter*, 19:25–27.

Frazier, P. D. (1967): Electron probe analysis of human teeth. Ca/P ratios in incipient carious lesions. *Arch Oral Biol*, 12:25–33.

Fyfe, F. W. (1960a): Histological effects of intermittent pressure on the rabbit's upper tibial epiphysis. *Anat Rec*, 136:336.

Fyfe, F. W. (1960b): Artificial torsion in growing rabbit tibiae. *Proc Can Fed Biol Soc*, 3:24.

Gabbiani, G., Badonnell, M. C., and Baud, C. A. (1969): The role of iron in experimental cutaneous calcinosis. *Fed Proc*, 28:552.

Galle, P. (1966): Cytochimie sur coupes ultra fines par spectrographie des rayons X. Sixth International Congress for Electron Microscopy, Kyoto. pp. 79–80.

Galle, P. (1967a): Microanalyse des inclusions minérales du rein. Proceedings 3rd International Congress Nephrology, Washington, 1966. Basel/New York, Karger, Vol. 2, pp. 306–319.

Galle, P. (1967b): Les néphrocalcinoses: Nouvelles données d'ultrastructure et de microanalyse. *Actualités néphrologiques de l'hôpital Necker*. 1967, pp. 303–315.

Galle, P., and Morel-Maroger, L. (1965): Les lésions rénales du saturnisme humain et expérimental. *Nephron*, 2:273–286.

Galle, P., Berry, J. P., and Stuve, J. (1968): New applications of electron microprobe analysis in renal pathology. *Proceedings National Conference Electron Microprobe Analysis*, 3rd. Chicago, 1968. Paper 41.

Gardner, D. L., and Hall, T. A. (1969): Electron-microprobe analysis of sites of silver deposition in avian bone stained by the v. Kossa technique. *J Pathol*, 98:105–109.

Glimcher, M. J., and Krane, S. M. (1962): Studies on the interactions of collagen and phosphate: I. The nature of inorganic orthophosphate binding in radioisotopes and bone. Oxford, Blackwell.

Glimcher, M. J., and Krane, S. M. (1964): The incorporation of radioactive inorganic orthophosphate as organic phosphate by collagen fibrils in vitro. *Biochemistry*, 31:195–202.

Goby, P. (1913): A new application of x Rays—Microradiography. *C R Acad Sci (Paris)*, 156:686–688.

Goldfischer, S., and Moskal, J. (1966): Electron probe microanalysis of liver in Wilson's disease. *Am J Pathol*, 48:305–315.

Goldsztaub, S., and Schmitt, J. (1960). Utilisation des rayonnements de fluorescence en microradiographie de contact. Ses applications en minéralogie et pétrographie. In Engstrom, A., Cosslett, V. E., and Pattee, H., Jr.: *X-Ray Microscopy and X-Ray Microanalysis*. Amsterdam and New York, Elsevier, pp. 149–152.

Goldsztaub, S., and Schmitt, J. (1963): Some mineralogical and technical applications of contact microradiography with fluorescent x rays. In Pattee, H., Jr., Cosslett, V. E., and Engström, A. (Eds.): *X-Ray Optics and X-Ray Microanalysis*. New York, Academic Press, pp. 127–132.

Göthmann, L. (1960a): The normal arterial pattern of the rabbit's tibia. A microradiographic study. *Acta Chir Scand*, 120:201–210.

Göthmann, L. (1960b): The arterial pattern of the rabbit's tibia after the application of an intramedullary nail. *Acta Chir Scand*, 120:211–219.

Green, M., and Cosslett, V. E. (1961): The efficiency of production of characteristic X radiation in thick targets of a pure element. *Proc Physical Soc*, 78:1206–1214.

Greulich, R. C. (1960): Application of high resolution microradiography to qualitative experimental morphology. In Engström, A., Cosslett, V. E., and Pattee, H., Jr., (Eds.): *X-Ray Microscopy and Microanalysis*. Amsterdam and New York, Elsevier pp. 273–287.

Greulich, R. C., and Engström, A. (1956): A new approach to high resolution micro-radiography using extremely soft x rays. *Exp Cell Res*, 10:251–254.

Gros, Ch. M., and Girardie, J. (1964): Etude radiologique du tissu mammaire. Histo-radiographie. Microradiographie par contact en rayonnement de fluorescence. *J Radiol Electrol*, 45:563–570.

Gueft, B., Kikkawa, Y., and Moskal, J. (1964): Electron probe studies of tissues in Wilson's disease, lead poisoning and bunamiodyl nephropathy. *J Appl Physics*, 35:3077.

Hagberg, S., Haljamäe, H., and Röckert, H. (1967): Shock reactions in skeletal muscle. II. Intracellular potassium of skeletal muscle before and after induced haemorrhagic shock. *Acta Chir Scand*, 133:265.

Hale, A. J. (1962): Identification of cytochemical reaction products by scanning x-ray emission microanalysis. *J Cell Biol*, 15:427–435.

Hale, A. J., Hall, T., and Curran, R. C. (1967): Electron microprobe analysis of calcium phosphorus and sulphur in human arteries. *J Pathol Bacteriol*, 93:1–17.

Haljamäe, H. (1967): Shock reactions in skeletal muscle cells. I. A method for the de-termination of potassium content in skeletal muscle cell. *Acta Chir Scand*, 133:259–263.

Haljamäe, H., and Röckert, H. (1969): Comparison between x-ray fluorescence and ultra-micro flame photometric analyses of the electrolyte content in single cells. In Möllenstedt, G., and Gaukler, K. H.: *X-Ray Optics and Microanalysis*. Heidelberg and New York, Springer-Verlag, pp. 561–568.

Hall, T. A. (1958): A non-dispersive x-ray fluorescence unit for the analysis of biological tissue sections. *Advances in X-Ray Analysis*. New York, Plenum Press, vol. 1, p. 297.

Hall, T. A. (1961): X-ray fluorescence analysis in biology. *Science*, 134:449–455.

Hall, T. A. (1963): Nondispersive x-ray fluorescence spectrometry. In Clark, G. L. (Ed.): *Encyclopedia of X Rays and Gamma Rays*. New York, Reinhold, pp. 653–655.

Hall, T. A. (1968): Some aspects of the microprobe analysis of biological specimens. In Heinrich, K. F. J. (Ed.): *Quantitative Electron Probe Microanalysis*. Washington, D. C., National Bureau of Standards, Special Publication 298, pp. 269–299.

Hall, T. A. (1971): The microprobe assay of chemical elements. In Oster, G. (Ed.): *Physical Techniques in Biological Research*, 2nd ed. Optical Techniques New York, Academic Press, Vol. IA, pp. 157–275.

Hall, T. A., and Höhling, H. J. (1969): The application of microprobe analysis to biol-ogy. In Mollenstedt, G., and Gaukler, K. H. (Eds.): *X-Ray Optics and Microanalysis*. Heidelberg and New York, Springer-Verlag, pp. 582–591.

Hall, T. A., and Werba, P. R. (1971): Quantitative microprobe analysis of thin speci-mens; continuum method. *Proceedings of the 25th Anniversary Meeting of the Electron Microscopy and Analysis Group of the Institute of Physics, at Cambridge*. London and Bristol, The Institute of Physics, pp. 146–149.

Hall, T. A., Hale, A. J., and Switsur, V. R. (1966): Some applications of microprobe analysis in biology and medicine. In McKinley, T. D., Heinrich, K. F. J. and Wittry, D. B. (Eds.): *The Electron Microprobe*. New York, Wiley, pp. 805–833.

Hall, T. A., Höhling, H. J., and Bonucci, E. (1971): Electron probe x-ray analysis of osmophilic globules as possible sites of early mineralization in cartilage. *Nature,* 231:535–536.

Hallén, O. (1956): On the cutting and thickness determination of microtome sections. *Acta Anat, 26 (Suppl)*: 25.

Hallén, O., and Röckert, H. (1960): The preparation of plane-parallel sections of desired thickness of mineralized tissues. In Engström, A., Cosslett, V. E., and Pattee, H., Jr. (Eds.): *X-Ray Microscopy and Microanalysis.* Amsterdam and New York, Elsevier, pp. 169–176.

Hamberger, A., and Röckert, H. (1964): Intracellular potassium in isolated nerve cells and glial cells. *J Neurochemistry, 11:757–760.*

Hancox, N. M. (1957): Experiments on the fundamental effects of freeze substitution. *Exp Cell Res, 13:263–275.*

Harrison, R. G. (1951): Selection and injection of contrast media. In Barclay, A. E. (Ed.): *Microarteriography.* Oxford, Blackwell, chap. 5, pp. 52–60.

Heath, J. C. (1960): The histogenesis of malignant tumours induced by cobalt in the rat. *Br J Cancer, 14:478–482.*

Heinrich, K. F. J. (1964): Interrelationships of sample composition, backscatter co-efficient, and target current measurement. *Advances in X-Ray Analysis.* New York, Plenum Press. Vol. 7, pp. 325–339.

Heinrich, K. F. J. (1966): X-ray absorption uncertainty. In McKinley, T. D., Heinrich, K. F. J., and Wittry, D. B. (Eds.): *The Electron Microprobe.* Appendix. New York, Wiley, pp. 350–377. Note: The Heinrich absorption tables are reproduced in Salter, W. J. M. (1970): *A Manual of Quantitative Electron Probe Microanalysis.* London, Structural Publications, pp. 91–146.

Henke, B. L. (1959): Microstructure, mass and chemical analysis with 8 to 44 Ångstrom X-radiation. *Advances in X-Ray Analysis.* New York, Plenum Press, vol. 2, pp. 117–155.

Henke, B. L. (1965): Some notes on ultrasoft x-ray fluorescence analysis—10 to 100 Å region. *Advances in X-Ray Analysis.* New York, Plenum Press, vol. 8, pp. 269–284.

Henke, B. L. (1967): Techniques of low energy x-ray and electron physics. 50 to 1000 eV region. *Norelco Reporter, 14:75–83 and 98.*

Henke, B. L. (1970): An introduction to low energy x-ray and electron analysis. *Advances in X-Ray Analysis.* New York, Plenum Press, vol. 13.

Henke, B. L., and Elgin, R. L. (1970): X-ray absorption tables for the 2-to-200 Å region. *Advances in X-Ray Analysis.* New York, Plenum Press, vol. 13.

Henke, B. L., White, R., and Lundberg, B. (1957): *J Appl Physics, 28:98.* (The absorption tables in this work have been reproduced in part in Cosslett and Nixon, 1960, p. 375.)

Henke, B. L., Elgin, R. L., Lent, R. E., and Ledingham, R. B. (1967): X-ray absorption in the 2-to-200 Å region. *Norelco Reporter, 14:112–116.*

Hewes, C. G., Nixon, W. C., Baez, A. V., and Kampmeier, O. F. (1956): X-ray microscopy of veins of the skull. *Science, 124:129.*

Hiroaka, T., and Glick, D. (1963): Histochemical x-ray absorption measurement of lipid in the adrenal gland of the rat. In Pattee, H., Jr., Cosslett, V. E., and Engström, A. (Eds.): *X-Ray Optics and X-Ray Microanalysis.* New York, Academic Press, p. 107.

Hodson, S., and Marshall, J. (1970): Ultracryotomy: A technique for cutting ultrathin sections of unfixed frozen biological tissues for electron microscopy. *J Microsc, 91:105–117.*

Hodson, S., and Marshall, J. (1971): Migration of potassium out of electron microscope specimens. *J Microsc, 93:49–53.*

Höhling, H. J., Hall, T. A., Boothroyd, B., Cooke, C. J., Duncumb, P., and Fitton-Jackson, S. (1967a): Untersuchungen der Vorstadien der Knochenbildung mit Hilfe der normalen und electronenmikroskopischen electron probe x-ray microanalysis. *Naturwissenschaften, 54*:142–143.

Höhling, H. J., Hall, T. A., and Fearnhead, R. W. (1967b): "Electron Probe X-Ray Microanalysis" als quantitative histologische Methode zur Analyse von Vorstadien der altersbedingten Aorten-Mineralisierung. *Naturwissenschaften, 54*:93–94.

Höhling, H. J., Hall, T. A., Boyde, A., and von Rosenstiel, A. P. (1968): Combined electron probe and electron diffraction analysis of prestages and early stages of dentine formation in rat incisors. *Calcif Tissue Res, 2 (Suppl)*, paper No. 5.

Höhling, H. J., Schöpfer, H., Höhling, R. A., Hall, T. A., and Gieseking, R. (1970): The organic matrix of developing tibia and femur, and macromolecular deliberations. *Naturwissenschaften, 7*:357.

Höhling, H. J., Kriz, W., Schnermann, J., and von Rosenstiel, A. P. (1971): Messungen von Elektrolyten in Nierenschnitten mit der Mikrosonde: Methode. Anat. Anz. Suppl. ad *128*, 217–220.

Höhling, H. J., Hall, T. A., Kriz, W., von Rosenstiel, A. P., Schnermann, J., and Zessack, U. (1972): Loss of mass in biological (kidney) specimens during electron probe x-ray microanalysis. *Proceedings International Symposium on Modern Technology in Physiological Sciences*. Munich, July, 1971 (in press).

Holmstrand, K. (1957). Biophysical investigations of bone transplants and bone implants. *Acta Orthop Scand (Suppl)*, 26.

Hydén, H. (1959): Quantitative assay of compounds in isolated fresh nerve cells and glial cells from control and stimulated animals. *Nature (Lond), 184*:433.

Hydén, H. (1960): The neuron. In *The Cell*. New York, Academic Press, vol. 4, chap. 5, pp. 216–323.

Ichinokawa, T., Kobayashi, H., and Nakajima, M. (1969): Density effect of x-ray emission from porous specimens in quantitative electron probe microanalysis. *Jap J Appl Physics, 8*:1563–1568.

Ingram, M. J., and Hogben, A. M. (1968): Procedures for the study of biological soft tissue with the electron microprobe. In *Developments in Applied Spectroscopy*. New York, Plenum Press, vol. 6, pp. 43–54.

Istock, J. T., Miller, C. W., Chambers, F. W., and Lyon, H. W. (1960): Historadiography of hard and soft tissues. *Armed Forces Med J, 11*:497–506.

Jackson, C. K. (1957): Projection microradiographs of developing rat bone. Cosslett, V. E., Engström, A., and Pattee, H., Jr. (Eds.): In *X-Ray Microscopy and Microradiography*. New York, Academic Press, pp. 459–461.

Johnson, A. R. (1969): The distribution of strontium in the rat femur as determined by electron microprobe analysis. *Proceedings National Conference on Electron Microprobe Analysis*, 4th. Pasadena, 1969. Paper No. 38.

Jongebloed, W. L. (1966): Application possibilities of projection microradiography. In Castaing, R., Deschamps, P., and Philibert, J. (Eds.): X-Ray Optics and Microanalysis. Paris, Hermann, pp. 636–641.

Jowsey, J. (1955): The use of the milling machine for preparing bone sections for microradiography and microautoradiography. *J Sci Instrum, 32*:159–163.

Jowsey, J., Kelly, P. J., Riggs, B. L., Bianco, A. J., Scholz, D. A., and Gershon-Cohen, J. (1965): Quantitative microradiographic studies of normal and osteoporotic bone. *J Bone Joint Surg, 47A*:785.

Juster, M., Fishgold, H., and Laval-Jeantet, M. (1963): Premiers éléments de microradiographie clinique (crane et os longs). Paris, Masson et Cie, pp. 186.

Key, J. A. (1950): Blood vessels of a gastric ulcer. *Brit Med J*, 2:1464–1465.

Kimoto, S., Hashimoto, H., and Tagata, T. (1970): Dispersive x-ray spectrometer for scanning electron microscope. *Proceedings National Conference Electron Microprobe Analysis*, 5th. New York City, 1970, Paper 51.

Kinley, C. E., and Saunders, R. L. de C. H. (1968): An x-ray microscope study of myocardial revascularisation. *Can J Surg*, 11:27–35.

Kirkpatrick, P., and Baez, A. V. (1948): Formation of optical images by x-rays. *J Opt Soc Am* 38:766–774.

Kormano, M. (1967): Technical aspects in microangiography using barium sulphate (micropaque) suspension. *Invest Radiol*, 2:14–16.

Kramer, I. R. H. (1951): A technique for the injection of blood vessels in the dental pulp using extracted teeth. *Anat Rec, 111*:91–100.

Kriz, W. (1971): Messungen intrazellulärer Elektrolytgehalte in der Rattenniere mit der Mikrosonde. Habil. Schrift, Münster.

Kriz, W., Höhling, H. J., Schnermann, J., and Rosenstiel, A. P. von (1971): Messungen von Elektrolyten in Nierenschnitten mit der Mikrosonde: Erste Ergebnisse. *Anat Anz (Suppl)* 128:221–225.

Kulenkampff, H. (1922): Über das kontinuierliche Röntgenspektrum. *Ann Physik*, 69:548–596.

Lamarque, P. (1936): Histology—historadiography. *CR Acad Sci*, 202:684–687.

Lamarque, P. (1936): Historadiography; new application of x-rays. *Radiology*, 27:563–568.

Lamarque, P. (1936): Historadiographic technique. *CR Assoc Anat*, 31:197–207.

Lamarque, P. (1936): Historadiography, new application of Roentgen rays. *Montpellier Medical, 11*:47–68.

Lamarque, P. (1936): L'historadographie. Nouvelle technique d'examen des coupes microscopiques. *Presse Méd, 44*:478–481.

Lamarque, P. (1938): Historadiography. *Br J Radiol, 11*:425–435.

Lamarque, P., and Turchini, J. (1936): First results obtained by the technique of historadiography. Degree of transmission of tissue components for soft x rays of wavelength about 5 Å *CR Soc Biol, 122*:294–295.

Lamarque, P., and Turchini, J. (1937): Une nouvelle méthode d'utilisation des rayons X, l'historadiographie. De quelques applications de l'historadiographie intéressant plus spécialement les sciences médicales. *Extrait de Montpellier Médical* (No. 2 Fev.): 35 pp.

Laüchli, A., Spurr, A. R., and Wittkopp, R. W. (1970): Electron probe analysis of freeze-substituted epoxy resin embedded tissue for ion transport studies in plants. *Planta (Berl)*, 95:341–350.

Leborgne, F., Leborgne, R., and Leborgne, F. (Jn.) (1957): Microradiography in pathology of the breast. In Cosslett, V. E., Engström, A., and Pattee, H., Jr. (Eds.): *X-Ray Microscopy and Microradiography*. New York, Academic Press, pp. 531–533.

Lechene, C. (1970): The use of the electron microprobe to analyse very minute amounts of liquid samples. *Proceedings National Conference on Electron Microprobe Analysis*, 5th. New York City, 1970. Paper 32.

LePoole, J. B., and Ong Sing Poen (1956): *Appl Sci Res, B5*:454.

LePoole, J. B., and Ong Sing Poen (1957). Description of the Delft x-ray microscope. In Cosslett, V. E., Engström, A., and Pattee, H., Jr. (Eds.): *X-Ray Microscopy and Microradiography* New York, Academic Press, pp. 91–95.

Libanati, C. M., and Tandler, C. J. (1969): The distribution of the water-soluble inorganic orthophosphate ions within the cell: Accumulation in the nucleus. Electron probe microanalysis. *J Cell Biol*, 42:754–765.

Lindström, B. (1955): Roentgen absorption spectrophotometry in quantitative cytochemistry. *Acta Radiol (Suppl)*, 125:206 pp.

Lineweaver, J. L. (1963): Oxygen outgassing caused by electron bombardment of glass. *J Appl Physics*, 34:1786.

Ljungqvist, A. (1963): Fetal and postnatal development of the intrarenal arterial pattern in man. A micro-angiographic and histologic study. *Acta Paediatr*, 52:443–464.

Ljungqvist, A., and Lagergren, C. (1962): Normal intrarenal arterial pattern in adult and ageing human kidney. A micro-angiographical and histological study. *J Anat (Lond)*, 96:285–300.

Long, J. V. P. (1958): X-ray absorption microanalysis with fine-focus tubes. *J Sci Instrum*, 35:323–329.

Long, J. V. P., and Röckert, H. (1963): X-ray fluorescence microanalysis and the determination of potassium in nerve cells. In Pattee, H., Jr., Cosslett, V. E., and Engstrom, A. (Eds.): *X-Ray Optics and X-Ray Microanalysis*. New York, Academic Press, p. 513.

Lundberg, B., and Engström, A. (1957): A simple midget x-ray tube for high resolution microradiography. *Exp Cell Res*, 12:198–200.

Lundberg, B., and Henke, B. L. (1956): Slide rule for radiographic analysis. *Rev Sci Instrum*, 27:1043.

Machado, F. A. S. da S. (1965): Micro-linfangiografia hepatica. Dissertacao de Licenciatura, Faculdade de Medecina de Lisboa, Portugal, 92 pp., 23 figs.

Mallett, G. R., Fay, M. J., and Mueller, W. M. (Eds.) (1966): *Advances in X-Ray Analysis*. New York, Plenum Press, vol. 9.

Mallett, G. R., and Newkirk, J. B. (Eds.) (1970): *Advances in X-Ray Analysis*. New York, Plenum Press, vol. 13.

Marshall, D. J. (1967), Ph.D. Thesis. Cambridge, England, University of Cambridge.

Marshall, D. J., and Hall, T. A. (1968): Electron probe x-ray microanalysis of thin films. *Br J Appl Physics (J Phys D)*, Ser. 2, 1:1651–1656.

Meek, A. (1896): A biological application of Röntgen photography. *Nature (Lond)*, 54:8.

Mellors, R. C. (1964): Electron probe microanalysis. I. Calcium and phosphorus in normal human cortical bone. *Lab Invest*, 13:183–195.

Mellors, R. C. (1966): Electron microprobe analysis of human trabecular bone. *Clin Orthop*, 45:157–167.

Mellors, R. C., and Carroll, K. G. (1961): A new method for local chemical analysis of human tissue. *Nature*, 192:1090–1092.

Mitchell, G. A. G. (1950): Microradiographic demonstration of tissues treated by metallic impregnation. *Nature*, 165:429.

Mitchell, G. A. G. (1951): Microradiography in biology. *Br J Radiol*, 24:110–117.

Mitchell, J. R. A., and Schwartz, C. J. (1965): *Arterial Disease*. Oxford, Blackwell, 90 pp.

Mitra, S. K., and Hall, T. A. (1971): A new network scheme for pulse-height analysis applied to non-dispersive x-ray emission microanalysis. *J Physics D: Appl Phys*, 4:748–752.

Molenaar, I. (1960): A new apparatus for the preparation of very thin ground sections for high resolution mircoradiography. In Engström, A., Cosslett, V. E., and Pattee, H., Jr.: *X-Ray Microscopy and X-Ray Microanalysis*. Amsterdam and New York, Elsevier, pp. 177–183.

Moniz, E. (1931): *Diagnostic des tumeurs cérébrales et épreuve de l'encéphalographie artérielle.* Paris, Masson, 512 pp.

Moniz, E. (1934): *L'angiographie cérébrale. Ses applications et résultats en anatomie, physiologie et clinique.* Paris, Masson, 327 pp.

Morel, F., Roinel, N., and Le Grimellec, C. (1969): Electron probe analysis of tubular fluid composition. *Nephron, 6:*350–364.

Mosley, V. M., Scott, D. B., and Wyckoff, R. W. G. (1957): X-ray microradiography of tissue sections with magnesium radiation. *Biochim Biophys Acta, 24:*235–237.

Nelms, A. T. (1956): National Bureau of Standards (U. S.) Circular No. 577.

Nicholson, W. A. P. (1971): Ph.D. Thesis. Cambridge, England, University of Cambridge.

Nikiforuk, G., and Sreebny, L. (1953): Demineralization of hard tissues by organic chelating agents at neutral pH. *J Dent Res, 32:*859–867.

Nilsonne, U. (1959): Biophysical investigations of the mineral phase in healing fractures. *Acta Orthop Scand (Suppl):* 37.

Nixon, W. C. (1960): Projection x-ray microscopy with forward scattered electron focusing. In Engström, A., Cosslett, V. E., and Pattee, H., Jr. (Eds.): *X-Ray Microscopy and Microanalysis.* Amsterdam and New York, Elsevier, pp. 105–109.

Nixon, W. C. (1961): X-ray microscopy. *Contemp Physics, 2:*183–197.

Oden, B. (1960): A micro-lymphangiographic study of experimental wounds healing by second intention. *Acta Chir Scand, 120:*100–114.

Oden, B. (1961): Micro-lymphangiographic studies of experimental skin homografts. *Acta Chir Scand, 121:*233–241.

Oden, B., Bellman, S., and Fries, B. (1958): Stereo-microlymphangiography. *Br J Radiol, 31:*70–80.

Oderr, C. (1964): Architecture of the lung parenchyma. *Am Rev Resp Dis, 90:*401–410.

Oderr, C., and Dauzat, M. (1964): An x-ray microscope designed for study of lung parenchyma. *Am Rev Resp Dis, 90:*448–452.

Okawa, J., and Trombka, J. L. (1956): The technic of making microangiograms of rabbit bone marrow. *Am J Clin Pathol, 26:*758–764.

Oliver, M. F., and Boyd, G. S. (1959): Effect of bilateral ovariectomy on coronary artery disease and serum-lipid levels. *Lancet, II:*690.

Omnell, K. Å. (1957): Quantitative roentgenologic studies on changes in mineral content of bone in vivo. *Acta Radiol (Suppl):148.*

Ong Sing Poen (1959): *Microprojection With X Rays.* Delft, Hoogland en Waltman, 132 pp.

Pattee, H. H., Garron, L. K., McEwen, W. K., and Feeney, M. L. (1957): Stereomicroradiography of the limbal region of human eye. In Cosslett, V. E. Engström, A., and Pattee, H., Jr. (Eds.): *X-Ray Microscopy and Microradiography.* New York, Academic Press, pp. 534–538.

Pearse, A. G. E. (1968): *Histochemistry,* 3rd ed. London, Churchill, vol. I.

Pfeifer, R. (1928): *Die Angioarchitektonik der Grosshirnrinde.* Berlin, Springer, 157 pp.

Pfeifer, R. (1930): *Grundlegende Untersuchungen für Angioarchitektonik des menschlichen Gehirns.* Berlin, Springer, 220 pp.

Philibert, J. (1969): Etat actuel des méthodes quantitatives d'analyse par sonde électronique. In Mollenstedt, G., and Gaukler, K. H. (Eds.): *X-Ray Optics and Microanalysis.* Berlin, Heidelberg and New York, Springer-Verlag, pp. 114–131.

Philips, F. S. (1961): Relations between zinc content and selective cytotoxicity of diphenylthiocarbazone. *Fed Proc (Suppl), 10:*129–131.

Randaccio, M. (1964): Indagini istologiche mediante microradiografia. *Minerva Radiol,* 9:512–517.

Ranwez, F. (1896): Application de la photographie par les rayons Röntgen aux recherches analytiques des matières végétales. *C.R. Acad Sci (Paris),* 122:841–842.

Ranzetta, G. V. T., and Scott, V. D. (1966): Specimen contamination in electron-probe microanalysis and its prevention using a cold trap. *J Sci Instrum,* 43:816–819.

Recourt, A. (1957a): Osmium tetroxide fixation observed by means of microradiography. In Cosslett, V. E., Engström, A., Pattee, H., Jr.: *X-Ray Microscopy and Microradiography.* New York, Academic Press, pp. 475–483.

Recourt, A. (1957b): Contact microradiography below 1 kV. In Cosslett, V. E., Engström, A., and Pattee, H., Jr. (Eds.): *Microscopy and Microradiography.* New York, Academic Press, pp. 234–239.

Reed, S. J. B. (1968): The relative intensities of K and L x-ray lines. *Br J Appl Physics* (Ser. 2), 1:1090.

Remagen, W., Caesar, R., and Heuck, F. (1968): Elektronenmikroskopische und mikroradiographische Befunde am Knochen der mit Dihydrotachysterin behandelten Ratte. *Virchows Arch,* 345:245–254.

Remagen, W., Höhling, H. J., Hall, T. A., and Caesar, R. (1969): Electron microscopical and microprobe observations on the cell sheath of stimulated osteocytes. *Calcif Tissue Res,* 4:60–68.

Rhodes, J. R. (1968): Private communication (from Texas Nuclear Corp., Allandale Station, Austin, Tex. 78756).

Robertson, J. K. (1956): *Radiology Physics.* London, MacMillan, chap. 10, pp. 150–176.

Robertson, A. J., Rivers, D., Nagelschmidt, G., and Duncumb, P. (1961): Stannosis—benign pneumoconiosis due to tin dioxide. *Lancet,* May 20, 1961, pp. 1089–1093.

Robison, W. L., and Davis, D. (1969): Determination of iodine concentration and distribution in rat thyroid follicles by electron probe microanalysis. *J Cell Biol,* 43:115–122.

Röckert, H. (1955): Some observations correlated to obliterated dentinal tubules and performed with microradiographic technique. *Acta Odontol Scand,* 13:271–275.

Röckert, H. (1955): Microradiographic studies of teeth. *Experientia,* 9:143–146.

Röckert, H. (1956): Variations in the calcification of cementum and dentin as seen by the use of microradiographic technique. *Experientia,* 12:16–22.

Röckert, H. (1958): A quantitative x-ray microscopical study of calcium in the cementum of teeth. *Acta Odont Scand (Suppl 25),* 16. 68 pp., 20 figs.

Röckert, H. (1963): X-ray absorption and x-ray fluorescence micro-analyses of mineralized tissue of rats which have ingested fluoridated water. *Acta Pathol Microbiol Scand,* 59:32–38.

Röckert, H. (1965): Quantitative measurements of microscopic changes in the vascular bed. *Experientia,* 21:169.

Röckert, H. (1969): Biological work using microfluorescence analysis. In Mollenstedt, G., and Gaukler, K. H. (Eds.): *X-Ray Optics and Microanalysis.* Heidelberg and New York, Springer-Verlag, pp. 546–549.

Röckert, H., and Rosengren, B. (1966): X-ray fluorescence microanalysis of intracellular potassium in nervous tissue after γ-irradiation at increased oxygen partial pressure. In Castaing, R., Deschamps, P. and Philibert, J. (Eds.): *X-Ray Optics and Microanalysis.* Paris, Hermann, pp. 650–652.

Röckert, H., and Saunders, R. L. de C. H. (1958): A microradiographic study of ovarian dermoid teeth. *Experientia,* 14:59–61.

Röckert, H., and Zettergren, L. (1963): Tissue reaction to barium sulphate contrast medium. *Acta Pathol Microbiol Scand*, 58:445–450.

Röckert, H., Engström, A., Hallén, O., Herberts, G., Lidén, G., Norlund, B., and Shea, J. J. (1965): Otosclerosis—studied with x-ray microscopy and fluorescence microscopy after administration of tetracyclines. *J Laryngol Otol*, 79:305.

Rogers, G. L., and Palmer, J. (1969): The possibilities of x-ray holographic microscopy. *J Microsc*, 89:125–135.

Rogers, T. H. (1952): Production of monochromatic X-radiation for microradiography by excitation of fluorescent characteristic radiation. *J Appl Physics*, 23:881–887.

Röntgen, W. C. (1895). *Über eine neue Art von Strahlen*. Sitzungsber. Phys.-Med. Ges. Wurzburg, p. 132.

Rosengren, B. H. O. (1959): Determination of cell mass by direct x-ray absorption. *Acta Radiol (Suppl)*, 178.

Rosenstiel, A. P. von, and Zeedijk, H. B. (1968): An electron probe and electron microscope investigation of asbestos bodies in lung sputum. In *Proceedings National Conference Electron Microprobe Analysis*, 3rd. Chicago, 1968. Paper 43.

Rosenstiel, A. P. von, Vahl, J., and Kyselova, J. (1968): Röntgenmikroanalytische Untersuchungen von Ag-Sn-Cu-Leigierungen. Abstracts of the Vth Internat. Congress on X-Ray Optics and Microanalysis, Tübingen, September 1968. (Presented at the Congress but not included in the published proceedings.)

Rosenstiel, A. P. von, Höhling, H. J., Schermann, J., and Kriz, W. (1970): Electron probe microanalysis of electrolytes in kidney slices. *Proceedings National Conference Electron Microprobe Analysis*, 5th, New York City, 1970. Paper 33.

Rosenstiel, A. P. von, Kriz, W., Höhling, H. J., and Schnermann, J. (1971): Messungen von Elektrolyten in Nierenschnitten mit der Mikrosonde. *Mikrochim Acta* (Wien), 1971:697–716.

Ross, P. A. (1928): *J Opt Soc Am* and *Rev Sci Instrum*, 16:433.

Rosser, H., Boyde, A., and Stewart, A. D. G. (1967): Preliminary observations of the calcium concentration in developing enamel assessed by scanning electron-probe x-ray emission microanalysis. *Arch Oral Biol*, 12:431–440.

Rubin, B., and Saffir, A. (1970): Calcification and the ground substance: Precipitation of calcium phosphate crystals from a nutrient gel. *Nature*, 225:78–79.

Rubin, P., Casarett, G. W., Kurohara, S. S., and Fujii, M. (1964): Microangiography as a technique. Radiation effect versus artifact. *Am J Roentgenol*, 92:378–387.

Russ, J. C. (1971): Progress in the design and application of energy dispersion x-ray analyzers for the SEM. In Johari, O. (Ed.): *Proc of the Fourth Annual SEM Symposium*. Chicago, *Illinois Institute of Technology Research Institute*, Part I, pp. 65–72.

Russ, J. C., and McNatt, E. (1969): Copper localization in cirrhotic rat liver by scanning electron microscopy. In Arceneaux, C. J. (Ed): *Proccedings Electron Microscopy Society Amer* Baton Rouge, Claitor's, 27:38.

Russ, J. C. (Ed.) *et al.* (1971): *Energy Dispersion X-Ray Analysis: X-Ray and Electron Probe Analysis*. Special Technical Publication 485, Philadelphia, American Society for Testing and Materials.

Ruthmann, A. (1970): *Methods in Cell Research*. London, Bell and Son.

Saffir, A. J., and Ogilvie, R. E. (1967): The effects of diet on the microstructure of teeth. *Transactions National Conference Electron Microprobe Analysis*, 2nd. Boston, 1967. Paper No. 38.

Saffir, A. J., and Rubin, B. (1970): The growth of calcium phosphate crystals by controlled diffusion in bio-simulating silica gels. In Scanning Electron Microscopy/1970.

Proceedings of Third Annual Scanning Electron Microscope Symposium. Chicago, I.I.T. Research Institute, pp. 201–208.

Saffir, A. J., Ogilvie, R. E., and Harris, R. S. (1969): Simultaneous multi-element analysis with a single x-ray spectrometer. *Proceedings National Conference Electron Microprobe Analysis*, 4th, Pasadena, Cal. 1969. Paper 53.

Saffir, A. J., Ogilvie, R. E., and Harris, R. S. (1970a): Electron microprobe analysis of fluorine in teeth. International Association for Dental Research, 47th General Meeting, March 1970.

Saffir, A. J., Ogilvie, R. E., Alvarez, C. J., and Harris, R. S. (1970b): Electron microprobe study of Ca/P ratio and fluorine distribution in rat teeth following feeding of Na fluoride and Na trimetaphosphate. International Association for Dental Research, 48th General Meeting, March 1970, New York.

Salmon, J. (1957): Microradiographic techniques and plant histology. In Cosslett, V. E., Engström, A., and Pattee, H., Jr. (Eds.): *X-Ray Microscopy and Microradiography*. New York, Academic Press, pp. 465–472.

Salmon, J. (1957): Contribution to microradiography in plant biology: Cancerization and tissue growth. In Cosslett, V. E., Engstrom, A., and Pattee, H., Jr. (Eds.): *X-Ray Microscopy and Microradiography*. New York, Academic Press, pp. 484–486.

Salmon, J. (1961): Perspectives et bilan des techniques microscopiques par rayons X, dérivées de la microradiographie, en particular dans le domaine végétal. In *Bulletin de microscopie appliquée*. Paris, Editions Revue d'Optique, pp. 17–67.

Saunders, R. L. de C. H. (1957a): Microradiographic studies of human adult and fetal dental pulp vessels. Cosslett, V. E., Engström, A., and Pattee, H., Jr. (Eds.): *X-Ray Microscopy and Microradiography*. New York, Academic Press, pp. 561–571.

Saunders, R. L. de C. H. (1957b): X-ray microscopy of human dental pulp vessels. *Nature (London)*, 180:1353–1354.

Saunders, R. L. de C. H. (1957c): Unpublished observations on tin miner's lung.

Saunders, R. L. de C. H. (1959): Vascular studies with the x-ray projection microscope. *J Anat Soc India*, 8:1–6.

Saunders, R. L. de C. H. (1960a): Histology by the projection microscope. In *Encyclopedia of Microscopy*. New York, Reinhold; also London, Chapman and Hall, pp. 582–586.

Saunders, R. L. de C. H. (1960b): Microangiography with the projection microscope. In *Encyclopedia of Microscopy*. New York, Reinhold; also London, Chapman and Hall, pp. 627–635.

Saunders, R. L. de C. H. (1960c): Microangiography of the brain and spinal cord. In Engström, A., Cosslett, V. E., and Pattee, H., Jr. (Eds.): *X-Ray Microscopy and X-Ray Microanalysis*. Elsevier, Amsterdam and New York, pp. 244–256.

Saunders, R. L. de C. H. (1961): X-ray projection microscopy of the skin. In *Advances in Biology of Skin*, Vol. 2, *Blood Vessels and Circulation*, pp. 38–56. Oxford, London and New York, Pergamon Press.

Saunders, R. L. de C. H. (1962): Medical applications of x-ray microscopy. *Can J Surg*, 5:299–310.

Saunders, R. L. de C. H. (1964): The contribution of microscopes to the study of living circulation: Radiographic techniques. *J R Microscop Soc*, 83:55–62.

Saunders, R. L. de C. H. (1966): X-ray microscopy of the periodontal and dental pulp vessels in the monkey and man. *Oral Surg*, 22:503–518.

Saunders, R. L. de C. H. (1967a): Studies with a new high voltage x-ray microscope (XMPJ) coupled to an x-ray sensitive Vidicon. *Can J Spectroscopy*, 12:3–8.

Saunders, R. L. de C. H. (1967b): Microangiographic studies of periodontic and dental pulp vessels in monkey and man. *J Can Dent Assoc, 33*:245–252.

Saunders, R. L. de C. H. (1969): Biological applications of projection x-ray microscopy. In Mollenstedt, G., and Gaukler, K. H. (Eds.): *X-Ray Optics and Microanalysis.* Heidelberg and New York, Springer-Verlag, pp. 550–560.

Saunders, R. L. de C. H., and Carvalho, V. R. C. (1963): X-ray microscopy of the microvascular system of the human lung. In Pattee, H., Jr., Cosslett, V. E., and Engström, A. (Eds.): *X-Ray Optics and X-Ray Microanalysis.* New York, Academic Press, pp. 109–122.

Saunders, R. L. de C. H., and Ely, R. V. E. (1966): Cerebral microangiography and "in vivo" studies with an x-ray microscope cum x-ray sensitive Vidicon. In Castaing, R., Deschamps, P., and Philibert, J. (Eds.): *X-Ray Optics and Microanalysis.* Paris, Hermann, pp. 642–649.

Saunders, R. L. de C. H., and James, R. (1960): X-ray microscopy of arterio-venous anastomoses. *Acta Anat, 43*:272–276.

Saunders, R. L. de C. H., and Montagna, W. (1964): X-ray microscopy and microangiography: Their potential value. *Arch Dermatol, 89*:451–454.

Saunders, R. L. de C. H., and Röckert, H. (1967): Vascular supply of dental tissues, including lymphatics. In *Structural and Chemical Organization of Teeth.* New York, Academic Press, Vol. 1, pp. 199–245.

Saunders, R. L. de C. H., and Van der Zwan, L. (1960): Exploratory studies of tissue by x-ray projection microscope. In Engström, A., Cosslett, V. E., and Pattee, H., Jr.: *X-Ray Microscopy and X-Ray Microanalysis.* Amsterdam and New York, Elsevier, pp. 293–305.

Saunders, R. L. de C. H., Lawrence, J., and MacIver, D. A. (1957a): Microradiographic studies of the vascular patterns in muscle and skin. In Cosslett, V. E., Engström, A., and Pattee, H., Jr. (Eds.): *X-Ray Microscopy and Microradiography.* New York, Academic Press, pp. 539–550.

Saunders, R. L. de C. H., Lawrence, J., MacIver, D., and Nemethy, N. (1957b): The anatomic basis of the peripheral circulation in man. In Redisch, W., and Tangio, F. F. (Eds.): *Peripheral Circulation in Health and Disease.* New York, Grune and Stratton, pp. 113–145.

Saunders, R. L. de C. H., Feindel, W. H., and Carvalho, V. R. (1965): X-ray microscopy of the blood vessels of the human brain. Parts I and II. *Med Biol Illus, 15*:108–122 and 234–246.

Saunders, R. L. de C. H., Bell, M. A., and Carvalho, V. R. C. (1969): X-ray histochemistry used for simultaneous demonstration of neurones and capillaries in the human brain. In Möllenstedt, G., and Gaukler, K. H. (Eds.): *X-Ray Optics and Microanalysis.* Heidelberg and New York, Springer-Verlag, pp. 569–578.

Scheuer, P. J., Thorpe, M. E. C., and Marriott, P. (1967): A method for the demonstration of copper under the electron microscope. *J Histochem Cytochem, 15*:300–301.

Schnermann, J., Rosenstiel, A. P. von, Kriz, W., and Höhling, H. J. (1970): Electron probe microanalysis of intracellular electrolytes in kidney sections. *Pflügers Arch, 319*:80.

Schrodt, G. R., Hall, T. A., and Whitmore, W. F. (1964): The concentration of zinc in diseased human prostate glands. *Cancer, 17*:1555–1566.

Scott, D. B., Nylen, M. U., and Pugh, M. H. (1960): Contact microradiography as an adjunct to electron microscopy. *Norelco Reporter, 7*:32–35 and 43.

Shackleford, J. M. (1965): Microradiographic interpretations of histologic structure. *Ala J. Med Sci, 2*:127–136.

Sheble, A. M., and Ocumpaugh, D. (1971): Electron microprobe measurement of fluorine pickup in human teeth. *Proceedings National Conference Electron Microprobe Analysis,* 6th. Pittsburgh, 1971.

Sikorski, J., Moss, J. S., Newman, P. H., and Buckley, T. (1968): A new preparation technique for examination of polymers in the scanning electron microscope. *J Sci Instrum* (Series 2), *1*:29–31.

Sims, R. T., and Hall, T. (1968a): X-ray emission microanalysis of the density of hair proteins in kwashiorkor. *Br J Dermatol 80*:35–38.

Sims, R. T., and Hall, T. (1968b): X-ray emission microanalysis of proteins and sulphur in rat plantar epidermis. *J Cell Sci, 3*:563–572.

Sims, R. T., and Marshall, D. J. (1966): Location of nucleic acids by electron probe x-ray microanalysis. *Nature, 212*:1359.

Sissons, H. A., Jowsey, J., and Stewart, L. (1960a): Quantitative microradiography of bone tissue. In Engström, A., Cosslett, V. E., and Pattee, H., Jr.: *X-Ray Microscopy and X-Ray Microanalysis.* Amsterdam and New York, Elsevier, pp. 199–205.

Sissons, H. A., Jowsey, J., and Stewart, L. (1960b): The microradiographic appearance of normal bone tissue at various ages. In Engström, A., Cosslett, V. E., and Pattee, H., Jr. (Eds.): *X-Ray Microscopy and X-Ray Microanalysis.* Amsterdam and New York, Elsevier pp. 206–215.

Soila, P. (1963): Microangiography in living human subjects. In Pattee, H., Jr., Cosslett, V. E., and Engström, A. (Eds.) *X-Ray Optics and X-Ray Microanalysis.* New York, Academic Press, pp. 123–126.

Söremark, R., and Grøn, P. (1966): Chlorine distribution in human dental enamel as determined by electron probe microanalysis. *Arch Oral Biol, 11*:861–866.

Sousa, A. de, and Mirabeau-Cruz, J. (1957): Aspectos morfo-functionais das estruturas finas hepaticas. Lisboa, Bertrand (Irmaos) Lda., 64 pp.

Sousa, A. de, Mirabeau-Cruz, J., and de Morais, J. B. (1958): Microangiopneumographie. *Ann Radiol, 1*:9–24.

Spalteholz, W. (1914): Über das Durchsichtigmachen von menschlichen und tierischen Präparaten und seine Theoretischen Bedingungen, 2nd ed. Leipzig, S. Hirzel, 93 pp.

Spector, W. S. (1956): Handbook of Biological Data. Philadelphia, Saunders, 584 pp.

Splettstosser, H. R., and Seemann, H. E. (1952): Application of fluorescence x rays to metallurgical microradiography. *J Appl Physics,* 23:1217–1222.

Stenn, K., and Bahr, G. F. (1970): Specimen damage caused by the beam of the transmission electron microscope. *J Ultrastruc Res, 31*:526–550.

Stroke, G. W. (1966): Attainment of high resolutions in image-forming x-ray microscopy with "lensless" Fourier-transform holograms and correlative source-effect compensation. In Castaing, R., Deschamps, P., and Philibert, J. (Eds.): *X-Ray Optics and Microanalysis.* Paris, Hermann, pp. 30–46.

Suga, S. (1967): Application of x-ray microanalysis in the study of biological tissues, especially hard tissues. *Odontology, 55*:217–224 (in Japanese).

Suga, S. (1969): Electron microprobe studies on the mineralization process of tooth and bone. In Mollenstedt, G., and Gaukler, K. H. (Eds.): *X-Ray Optics and Microanalysis.* Heidelberg and New York, Springer-Verlag, pp. 592–596.

Sundström, B. (1966): The technique of preparing thin ground sections of hard tissues; tooth and bone. *Acta Odontol Scand, 24*:159.

Sutfin, L. V., and Ogilvie, R. E. (1970): A comparison of x-ray analysis techniques available for scanning electron microscopes. In Scanning Electron Microscopy/1970. *Proceedings Third Annual Scanning Electron Microscope Symposium.* Chicago, IIT Research Institute, pp. 17–24.

Swift, J. A., and Brown, A. C. (1970a): An environmental cell for the examination of wet biological specimens at atmospheric pressure by transmission scanning electron microscopy. *J Physics E (J Sci Instrum)*, 3:924–926.

Swift, J. A., and Brown, A. C. (1970b): Transmission scanning electron microscopy of sectioned biological materials. In Scanning Electron Microscopy/1970. *Proceedings Third Annual Scanning Electron Microscope Symposium.* Chicago, IIT Research Institute, pp. 113–120.

Swift, J. A., Brown, A. C., and Saxton, C. A. (1969): Scanning transmission electron microscopy with the Cambridge Stereoscan Mk. II. *J Phys E: Sci Instrum*, 2:744–746.

Tandler, C. J., Libanati, C. M., and Sanchis, C. A. (1970): The intracellular localization of inorganic cations with potassium pyroantimonate. Electron microscope and electron microprobe analysis. *J Cell Biol*, 45:355–366.

Thewlis, J. (1940): *Medical Research Council Special Report 238.* 82 pp.

Tirman, W. S., Caylor, C. E., Banker, H. W., and Caylor, T. E. (1951): Microradiography. Its application to the study of the vascular anatomy of certain organs of the rabbit. *Radiology*, 57:70–80.

Tousimis, A. J. (1963): Electron-probe microanalysis of biological specimens. In Pattee, H., Jr., Cosslett, V. E., and Engström, A. (Eds.): *X-Ray Optics and X-Ray Microanalysis.* New York, Academic Press, pp. 539–557.

Tousimis, A. J. (1966): Applications of the electron probe x-ray microanalyzer in biology and medicine. *Am J Med Electronics*, 5:15–23.

Tousimis, A. J. (1971): Sulfur and copper in tissues from patients with hepatolenticular degeneration: Wilson's disease revisited. *Proceedings National Conference Electron Microprobe Analysis*, 6th. Pittsburgh, 1971.

Tousimis, A. J., and Adler, I. (1963): Electron probe x-ray microanalyzer study of copper within Descemet's membrane in Wilson's disease. *J Histochem Cytochem*, 11:40.

Turchini, J. (1936): Historadiography. Principal histological applications. *CR Assoc Anat*, 31:314–322.

Turchini, J. (1937): De quelques applications de l'historadiographie intéressant plus spécialement les sciences médicales. *Montpellier Méd*, 11:69–77.

Turchini, J. (1937): Applications of Historadiography. *Bull Histol Appl*, 14:17–28.

Turchini, J. (1937): Contribution to the study of metallic impregnation using the technique of historadiography. *CR Assoc Anat*, 32:429–431.

Turchini, J. (1938): New historadiographic observations; opacity to soft x rays and density of tissue structure. *CR Assoc Anat*, 33:460–462.

Turchini, J. (1942): Radiography of biological material. *Schweiz Z Path Bakt*, 5:137–149.

Urist, M. R. (1966): Origins of current ideas about calcification. *Clin Orthop*, 44:13–39.

Van Huysen, (1960). The microstructure of young dentine. *Arch Oral Biol.* 3–4, 157–160

Vassamillet, L. F., and Caldwell, V. E. (1968): Electron probe microanalysis of alkali metals in glasses. *Proceedings National Conference Electron Microprobe Analysis*, 3rd. Chicago, 1968. Paper No. 40.

Victoreen, J. H. (1950). *Medical Physics.* Chicago, Year Book Publishers, vol. 2, pp. 887–894.

Wallgren, G. (1957a): Microradiographic studies of developing bone in human embryos. In Cosslett, V. E., Engström, A., and Pattee, H., Jr.: *X-Ray Microscopy and Microradiography.* New York, Academic Press, pp. 448–461.

Wallgren, G. (1957b): Biophysical analyses of the formation and structure of human fetal bone. *Acta Paediatr (Suppl), 113.*

Wallgren, G., and Holmstrand, K. (1957): Technical considerations on quantitative microradiography on bone. *Exp Cell Res, 12:*188–191.

Wei, S. H. Y., and Ingram, M. J. (1968): Electron microprobe analysis of the silver amalgam-tooth interface. *Proceedings National Conference Electron Microprobe Analysis,* 3rd. Chicago, 1968. Paper No. 44.

Wörner, H., and Wizgall, H. (1969): Zerstörungsfreie Analyse von Zahnhartsubstanzen mit der Electronenstrahlmikrosonde. In Mollenstedt, G., and Gaukler, K. H. (Eds.): *X-Ray Optics and Microanalysis.* Heidelberg and New York, Springer-Verlag, pp. 601–607.

Wyckoff, R. W. G., and Crossant, O. (1963): Microradiography of dentine using characteristic x rays. *Biochim Biophys Acta, 66:*137–143.

X-Ray Microscopy and Microradiography (1957). Proceedings of a Symposium held at the Cavendish Laboratory, Cambridge, 1956. Ed. by Cosslett, V. E., Engström, A. and Pattee, H., Jr. New York, Academic Press, 645 pp.

X-Ray Microscopy and X-Ray Microanalysis (1960). Proceedings of the Second International Symposium held at Stockholm, 1959. Ed. by Engström, A., Cosslett, V. E., and Pattee, H., Jr. Amsterdam, London, New York and Princeton, Elsevier, 540 pp.

X-Ray Optics and X-Ray Microanalysis (1963). Proceedings of the Third International Symposium held at Stanford University, Stanford, California, 1962. Ed. by Pattee, H. H., Jr., Cosslett, V. E., and Engström, Arne. New York and London, Academic Press, 622 pp.

X-Ray Optics and Microanalysis: Optique des Rayons X et Microanalyse (1966). Proceedings of the Fourth International Symposium held at Orsay, France, 1965. Ed. by Castaing, R., Deschamps, P., and Philibert, J. Paris, Hermann, 707 pp.

X-Ray Optics and Microanalysis (1969). Proceedings of the Fifth International Congress on X-Ray Optics and Microanalysis held at Tübingen, Germany, 1968. Ed. by Mollenstedt, G., and Gaukler, K. H., Berlin, Heidelberg and New York, Springer-Verlag, 612 pp.

Zajicek, J. and Zeuthen, E. (1956). Quantitative determination of cholinesterase activity in individual cells. *Experimental Cell Res. 11,* 568.

Zeigler, D. C., Zeigler, W. H., Harclerode, J. E., and White, E. W. (1969): Electron microprobe analysis of thyroidal iodine in the white-throated sparrow. *Proceedings National Conference Electron Microprobe Analysis,* 4th. Pasadena, Cal., 1969, Paper No. 37.

Zeitz, L. (1962): The design of an x-ray microprobe analyzer for biological specimens. Biophysics Laboratory, Palo Alto, California, Stanford University, Report No. 67.

Zeitz, L. (1969): X-ray emission analysis in biological specimens. *Progress in Analytical Chemistry. X-Ray and Electron Probe Analysis in Biomedical Research.* New York, Plenum Press, vol. 3, pp. 35–73.

Zeitz, L., and Andersen, C. A. (1966): X-ray microprobe analysis for zinc in the rat prostate. In Castaing, R., Deschamps, P., and Philibert, J. (Eds.): *X-Ray Optics and Microanalysis.* Paris, Hermann, pp. 691–698.

Zeitz, L., and Lee, R. (1966): Zinc analysis in biological specimens by x-ray fluorescence. *Anal Biochem, 14:*191–204.

Zuppinger, A. (1935): *Die theoretischen Grundlagen und Möglichkeiten der Röntgendiagnostischen Weichteiluntersuchung.* Leipzig, Georg Thieme Verlag, 99 pp.

Zweifach, B. W. (1959): The microcirculation of the blood. *Sci Am, 200:*54–60.

INDEX

Note: Entries in *italics* are personal names. Page numbers in **boldface** identify definitions or major sections for the entries in question.